Byron Howell
Tyler Junior College

**Student Solutions Manual
And
Study Guide**
To Accompany

General, Organic, and Biological Chemistry:
An Integrated Approach

Kenneth W. Raymond
Eastern Washington University

WILEY
JOHN WILEY & SONS, INC.

Cover Photo by Cris Benton, Kite Aerial Photography

To order books or for customer service please, call 1-800-CALL WILEY (225-5945).

ISBN-13   978- 0-471-73771-1
ISBN-10   0-471-73771-2

Printed in the United States of America

10 9 8 7 6 5 4 3 2 1

Printed and bound by Bind Rite Graphics

Table of Contents

Acknowledgement

I want my family to know how much I cherish their love and support, so thanks to my wife Jacque and my two daughters Mindy and Michelle.

I certainly would not have had the opportunity to write this study guide if Ken Raymond had not written the textbook; thank you Ken for allowing me to be a part of this project. Thank you Richard Treptow and Joan Kalkut for your assistance in making sure this book is as error free as possible and for your presentation suggestions.

Last, but certainly not least, thank you Jennifer Yee, both for selecting me to me a part of this project and for your assistance and patience.

Introduction

As you embark on what very well could be your first chemistry course, you may be asking yourself if you will actually be able to pull it off and pass the course. Students often arrive in chemistry class thinking it is one of those courses that is "really hard." We are often afraid of the unknown and so if you have not had any (or very little) experience with the subject it is only natural for you to be concerned. Be confident that you can learn chemistry. You actually live and breathe chemistry every day. Everyday chemistry such as doing laundry, cooking, digestion, and living come so naturally to you that you never stop to realize that you are constantly practicing chemistry.

The material covered in this course will at times be challenging but it should be no harder for you to absorb that any other subject you've encountered for the first time. Just be sure to read your assignments, take good notes in class, complete the assigned homework, and study for your test as you would for any course. This study guide has been prepared to assist you in studying and preparing for success.

Each Study Guide chapter is divided into three major components:

1. Knowledge and Comprehension. In this section, you will be provided an opportunity to review fundamental concepts that primarily require memorization to achieve the chapter objectives.
2. Application. Here, sample problems will assist you in reviewing concepts that require a higher level of mastery through practice.
3. Practice test. These tests are designed with a variety of questioning styles chosen at random addressing all of the chapter's objectives to simulate an in-class test.

When sample problems require that you recall multiple concepts, those concepts are reviewed as either italicized paragraphs or in table format. These problems have been specifically chosen to help you better understand the concepts presented in the textbook. They also provide a concise review of the chapter objectives.

The answers to the sample problems offer you detailed solutions. Whenever a math concept or chemical application is required to achieve the answer, each step of the solution process is worked out for you.

Now it is up to you. Your teacher, your textbook, the videos, the interactive learningware problems, and this book are all here to help you. Take advantage of all that is offered and you should have a very successful learning experience.

Byron Howell, Ed.D.
Coordinator, SCRMCC/Chemistry Instructor
Coordinator A.C.E. Science Program
email bhow@tjc.edu

Chapter 1
Science and Measurements

Chapter Overview

If you were building a house, you wouldn't even attempt to start if your only tools and supplies were one board, a hammer, and a few nails. To that same end, it is important that you begin your study of chemistry with the proper tools and supplies. Chapter 1 is designed to provide you with the proper math tools and guidance to insure you have a proper foundation for building future concepts. The first part of the chapter helps you to understand the thought process, sometimes called "Scientific Method," that scientists use to help resolve a problem. The second part of the chapter details how scientists describe or classify the materials they work with each day. You would need to know the difference between a window and a door in order to build that house. Finally, it is impossible to use the tools provided without a sense of the materials available. To that end, the final part of the chapter explains many of the math tools that scientists use to help them measure or calculate what they have and what they need in order to come to a reasonable answer or conclusion. Learning these concepts will give you a firm foundation for you to understand the concepts presented in the remaining chapters.

Knowledge/Comprehension

Typically, there are new terms or phrases that you must memorize in order for the lesson to make sense to you. Look up this information and memorize their explanations and definitions:

scientific method	physical properties
law	physical changes
theory	accuracy
hypothesis	precision
experiment	potential energy
matter	kinetic energy
energy	unit
temperature	significant figures
conversion factors	exact numbers

The following activity is designed to help you master the objectives outlined in the beginning of Chapter 1 and then summarized at the end of the chapter. To make the best use of your study time, *read* Chapter 1 before continuing. Next, work the matching activity provided as though you are taking a test *without first* referring to the answer page. Put your answers on a separate sheet of paper so that you can make more than one attempt without seeing your previous answers. Once you have completed the activity, check your answers. Make *flash cards* for those you missed. Have a study buddy drill you on them if you missed more than half of the terms. Once you have reviewed the ones

you missed for at least 15 minutes try the activity again. If you miss fewer than four, write in the answers so that you will have them to review just before taking the test.

Match the terms or phrases to the correct meaning

_____1. Scientific method

_____2. Physical properties

_____3. Law

_____4. Physical changes

_____5. Theory

_____6.Accuracy

_____7. Hypothesis

_____8. Precision

_____9.Experiment

_____10. Potential energy

_____11. Matter

_____12. Kinetic energy

_____13. Energy

_____14. Unit

_____15. Temperature

_____16. Significant figures

_____17. Conversion factors

_____18. Exact numbers

a. anything that has mass and occupies space

b. a tentative explanation that is based on present facts

c. symbol that stands for the quantity being measured

d. those digits in a measurement that are reproducible when the measurement is repeated, plus the first doubtful digit

e. have an unlimited number of significant figures

f. average kinetic energy measured in °F, °C, or K

g. two equivalent measurements used in a ratio to transform one unit into another

h. statements describing things that are consistently and reproducibly observed

i. an experimentally tested explanation of an observed behavior

j. the ability to do work and transfer heat

k. stored energy

l. the energy of motion

m. controlled tests to collect data relating to a question

n. characteristics that can be determined without changing the composition of the matter.

o. the process used to gather and interpret information

p. related to how close a measured value is to a true value

q. changing matter without altering its chemical composition

r. a measure of reproducibility of a measurement

2

In addition to knowing the terms or phrases, you should be able to compare, contrast, or list examples that show you understand the following concepts:

- Describe the three states of matter.
- Describe the two forms of energy.
- Describe and give examples of physical properties and physical changes.
- Explain the difference between the terms "accurate" and "precise".

Once you know the appropriate tools for Chapter 1, you are ready to evaluate your understanding of those tools and their relationships. In this type of concept building, it is necessary to compare or contrast terms or concepts. In some cases you may be required to give examples or recognize a term or concept when it is depicted in a picture or sample scenario.

Sample Problems

1. Which of the pictures below have the same states of matter?

a.   b.   c.   d.

Notice that this type of problem requires you to have knowledge of the states of matter. You must know that the three states are solid, liquid, and gas. You must know what each of these states look like. Solids have a definite shape and volume, liquids have a definite volume but its shape can vary depending on its container, and gases have neither definite shape or form.

With this in mind, look at problem 1 and recognize that there is a definite shape to choice a, the microscope; it is a solid. Choice b is a cloud and therefore is a gas. Choice c. has both a cloud and rain so it has both a gas and a liquid. Finally, choice d. has a man and a car, both with a definite shape and volume, making them solid. This makes the correct answers a and d; they are the same state of matter since they are both solids.

2. The spacecraft shown is ready for take-off. Which form of energy does it currently contain?

To answer this question you must be familiar with the two forms of energy; potential energy is energy that is stored; kinetic energy is energy due to the motion of the object. In this case the rocket is not moving; therefore it cannot have kinetic energy. The rockets have chemicals in them that can react to provide energy. **Therefore, the spacecraft has stored energy called *potential energy*.**

3. Describe the energy changes that the ball shown below is undergoing.

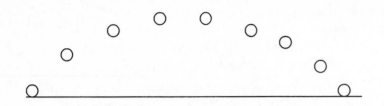

Since the ball is starting out on the floor, it does not have potential or kinetic energy. As it is kicked into the air, the energy imparted to put it in motion is in the form of kinetic energy. As the ball rises, it increases in its potential to fall and the kinetic energy is converted into potential energy. When the ball has slowed to a stop, it is as high as it can rise; it has maximum potential energy and zero kinetic energy. As gravity starts the ball back toward the floor, the potential energy is being converted back into kinetic energy. As the ball falls, its energy becomes more and more kinetic and less and less potential. When the ball comes to rest on the floor, its kinetic energy is transferred to the floor and both its kinetic and potential energies are zero again.

4. Sort the following terms or phrases as being either a physical property or an example of a physical property.

        a. color   b. square   c. size   d. odor   e. at 25 $^\circ$C   f. green

        g. density h. large   i. stinks j. temperature k. 1.40 g/cm^3 l. shape

To begin the sorting process, you must first remember that a physical property is a broad category that is used to group similar characteristics such as shape. There are many different shapes of matter. An example of a physical property is much more specific to a particular form of matter such as the ball is round. Round would be an example of a physical property. It is these more specific examples of an object's physical properties that are generally used to describe it.

Physical Properties	Example of Physical Property
a. color	f. green
c. size	h. large
d. odor	i. stinks
g. density	k. 1.40 g/cm^3
j. temperature	e. at 25 $^\circ$C
l. shape	b. square

5. Which of the following statements do *not* describe a physical change?

       a. The water in the tea kettle begins to boil and steam rises from the spout.
       b. After two hours the temperature still read 35.0 °C.
       c. As the water began to freeze, its density went from 1.00 g/cm^3 to 0.985 g/cm^3.
       d. The compactor smashed the car into the shape of a cube.
       e. The newspaper was torn into shreds.

To answer this question you must first know that a physical change is a change in one or more of the physical properties of a material without changing the nature of the material.

In choice a, the water changed from liquid to steam. In choice c two physical changes occurred, the liquid water was changing to ice (solid water) and the density was decreasing. In choice d the car changed shape from that of a car to that of a cube. In choice e the newspaper went from large sheets of paper to smaller sheets of paper. **The answer is b since the temperature did not change.**

6. Three students are conducting the same experiment but using three different balances to measure the mass of their final product. Examine the data tables shown below. If the correct mass (true mass) should have been 16.345 grams, which student is most likely using a balance that was not zeroed or calibrated properly?

Student #1	Student #2	Student #3
16.340	16.700	16.000
16.347	16.710	16.740
16.344	16.705	15.789
16.346	16.707	17.456

In order to explain the difference between accuracy and precision you must first know their definitions. Accuracy is defined as how close the measured value is to a true or accepted value. Precision is defined as how well the measured value can be reproduced. The precision observed in an experiment usually depends on how well the measuring device is able to reproduce a measurement. This is often determined by the graduations on the device or its calibration.

The answer is Student #2. Note that student #3 has data that is not consistent nor is it close to the accepted value. Student # 1 had fairly reproducible results (good precision) but also had numbers which were all close to the accepted value (good accuracy). Since student #2 had very reproducible results (good precision) but was not close to the acceptable answer (poor accuracy) the balance probably had a consistent error indicating it was not zeroed properly or was not properly calibrated.

Application

To apply what you've already learned and master the concepts, you must utilize all of the knowledge, understanding, facts, and/or techniques that you have gathered up to this point. Simply knowing an equation or process is not enough to master the concept; you develop these by *practice*. Solve as many different types of problems with as many different variations as you *strive to master the skill*. Your textbook provides a very good assortment of problems at the end of the chapter. Work as many as needed for skill mastery, not just those assigned. An online PowerPoint review of scientific notation, metric system, exponents, and conversion units can be accessed at http://chemistry.tyler.cc.tx.us/measurements.htm .

- Express values using scientific notation and metric prefixes.
- Report calculation results involving measured quantities using the correct number of significant figures.
- Convert from one unit of measurement into another.

In order to truly master any skill you must *practice*. In this section example problems related to the application objectives are discussed. Remember: the more you practice, the better your chances of being able to successfully use these objectives to solve problems. Additional practice problems are at the end of the chapter in your textbook and in the online interactive lessons.

Sample Problems

Scientific notation, also called exponential notation, is a way of expressing either very large or very small approximations by using a standard form that places the decimal just to the right of the first non-zero digit (for example: 3.5) that is multiplied by a power of ten (for example: 10^4). The number would then appear as 3.5×10^4. If there is a negative in the power of ten, this means the number represents a fraction (for example: 1.5×10^{-3} is the same as the fraction 0.0015). The power of ten indicates how many places the decimal was moved to make the number be in standard form. A positive power indicates the decimal of the original number had to be moved to the left. A negative power means the decimal of the original number was moved to the right.

1. Convert 0.0000567 into scientific notation.

To solve, first recognize that the first non-zero digit is to the *right* of the decimal. Since the decimal will move to the right you know the exponent will be *negative*. Count the number of places to determine the exponent number. 00.000567 one, 000.00567 two. 0000.0567 three, 00000.567 four, and then to standard form 000005.67 makes five. The number written in standard scientific notation is **5.67×10^{-5}**.

2. Convert the number 13,000 into scientific notation.

To solve, first recognize that the first non-zero digit is to the *left* of the decimal. Since the decimal will move to the left you know the exponent will be *positive*. Count the number of places to determine the exponent number. 1300.<u>0</u> one, 130.<u>00</u> two. 13.<u>000</u> three, and then to standard form 1.<u>3000</u> makes four. The number written in standard scientific notation is **1.3 x 10^4**.

In order to take a number from scientific notation to regular number (sometimes called a floating decimal number on your calculator) you simply reverse the process. The power of ten tells you the number of places the decimal will move and the sign of the exponents tells you the direction. If the sign is positive, the decimal moves right; if the sign is negative, the decimal moves left. If the decimal moves past the non-zero digits then use zeroes as placeholders to indicate you moved past that place.

3. Convert 4.56 x 10^3 to a regular number.

To solve, first note that the exponent is a positive 3. This means that the decimal should be moved three places to the right. 45.<u>6</u> x 10^2 one, 4<u>56</u>. x 10^1 two, and then 4<u>560</u> makes three. **So the answer is 4560** (Note that when zero placeholders are needed the decimal is no longer written.)

4. Convert 2.5 x 10^{-4} to a regular number.

To solve, first note that the exponent is a negative four. This means that the decimal should be moved four places to the left. .<u>25</u> x 10^{-3} one, .<u>025</u> x 10^{-2} two, .<u>0025</u> x 10^{-1} three and .<u>00025</u> makes four. (Note that when zero placeholders are needed, an extra zero is added to the left of the decimal so that the decimal can be seen clearly. The zero to the left of the decimal does not change the number's value.)**Answer is 0.00025**

There are two skills to master when reporting calculation results involving measured quantities using the correct number of significant figures:
   1. Correctly using significant figures in addition and subtraction calculations
   2. Correctly using significant figures in multiplication and division calculations
Following the rules for using significant figures guides you to calculate the answer at the same level of precision that the original measures allow. They also give you a process for knowing how to round numbers when your calculator includes a multitude of decimal places (for example: 12.34576589) and you need to know how many of the digits to record in your answer).

First make sure you can recognize a significant figure when you see one.

5. How many significant figures are in each of the following?

        a. 0.00023   b. 123.43   c. 789000   d. 2001

In a, there are two significant figures. In b, there are five significant figures. In c, there are three significant figures because ending zeroes are not significant unless there is a decimal point in the number. In d, there are four significant figures because zeroes between two non-zero digits are always significant.

When performing addition or subtraction calculations, the number with the fewest significant places determines the number of places allowed in the answer.

6. Perform the following calculations representing measured values:

        a. 7.85 + 10.123
        b. 12.5 – 0.457

a. The calculated answer is 17.973 but 7.85 has only two decimal places while 10.123 has three decimal places. Therefore the answer can only have two decimal places. Since the digit, 3, after the hundredths place value (**17.973**) is less than five, it is dropped. **Answer is 17.97**

b. The calculated answer is 12.043 but 12.5 has only one decimal place while 0.457 has three decimal places. Therefore the answer can only have one decimal place. Since the digit, 4, after the tenths place (**12.043**) is less than five, it and the number after it are dropped. **Answer is 12.0**

When performing multiplication or division calculations, the number of significant digits kept in the answer should be the same as the number with the fewest significant figures used in the calculation. (Note: Here is where you apply the previous skill of counting significant figures. Watch those zeroes!)

7. Perform the following calculations representing measured values:

        a. 546.785 ÷ 310

        b. 23.567 x 0.11

a. The calculation yields the number 1.763822581. Note that the number 546.785 has six significant figures but the number 310 only has two. Therefore, the answer must be rounded to two significant figures. Since the number following the second significant figure is larger than five, the tenths place is rounded up by one. **1.763822581** becomes 1.8. **Answer is 1.8**

b. The calculation yields the number 2.59237. Note that the number 23.567 has five significant figures but the number 0.11 only has two. Therefore, the answer must be rounded to two significant figures. Since the number following the second significant figure is larger than five, the tenths place is rounded up by one. **2.5**9237 becomes 2.6. **The correct answer is 2.6**

Often after a measurement has been made in science it is necessary to compare it to a standard or accepted value. However, the measuring device you're using may not use the same units as the reference measurement. A measured quantity called for in an experiment may not match the units on the available measuring device. When this occurs it is necessary to be able to use the factor-label method and conversion factors to convert from one unit to another.

To accomplish this, equivalent measurement of each unit is used to convert the number and unit given to the number and unit desired. Although there are some measured conversion factors that your instructor requires you to memorize, for the purpose of this study review skill we'll assume that the instructor gives you the needed conversion factors for each conversion.

8. Determine the weight of 135.0 kg woman in pounds. (1 kg = 2.20 lb)

When setting up a conversion problem, the factor-label method instructs you to place the number to be converted first in the equation. This is followed by a ratio of units with the unit you are converting from on bottom and the unit you wish to convert to on top. The numbers that go in front of the units in the ratio are given in the equivalency statements (like 1 Kg = 2.20 lb).

$$135.0 \text{ kg} \times \frac{2.20 \text{ lb}}{1 \text{ kg}} = \textbf{297 lb}$$

9. An experiment calls for 0.175 L of solution. How many milliliters would that require? (1 L = 1 x 10³ mL)

When obtaining this answer be sure to use proper exponent rules as studied above. The final answer should be in standard form.

$$0.175 \text{ L} \times \frac{1 \times 10^3 \text{ mL}}{1 \text{ L}} = \textbf{1.75 x 10}^2\textbf{mL}$$

10. A student's car requires 15.0 gallons of gasoline to fill the tank. How many liters of gasoline would be required? (1 gal = 4 qt) and (1qt = 0.946 L)

If the conversion factor that directly converts from one unit to another is not available, the factor-label method can be used in steps. First, use the conversion factor you have (1 gal = 4 qt) to change from gal to qt. Then use a second conversion factor (1qt = 0.946 L) to convert from qt to L. In these cases, think of using the factor-label method as trying to cross a stream. If you cannot get over it in one step, move from stone to stone until you get to the other side. The calculated number is 56.76 but is rounded to 56.8 because the original measurement had only three significant figures.

$$15.0 \text{ gal} \times \frac{4 \text{ qt}}{1 \text{ gal}} \times \frac{0.946 \text{ L}}{1 \text{ qt}} = \textbf{56.8 L}$$

Temperature conversions cannot be accomplished by using conversion factors. These conversions use equations that give the relationship between the two measurements. A summary of those relationships is shown below. Most instructors will expect you to know these and they are not typically given to you during test periods.

To convert.........	.....use the equation
°F → °C	°F = (1.8 x °C) + 32
°C → °F	°C = $\dfrac{°F - 32}{1.8}$
°C → K	K = °C + 273.15
K → °C	°C = K − 273.15

11. Convert 245.0 °F to °C

Start with the conversion equation °F = (1.8) °C + 32 and replace the variable °F with the number given. Then solve for °C.

245.0 = (1.8) °C + 32
245.0 − 32 = (1.8) °C
213 ÷ 1.8 = °C
118.3 = °C   Note that the answer has the same number of decimal places as the original measurement when converting temperatures.

12. Convert 135 °C to K

Start with the conversion equation K = °C + 273.15 and replace the variable °C with the number given. Then solve for K.

K = 135 + 273.15 = 408.15 which rounds to 408.

K = 408

13. Convert 358 K to °F

Start with the conversion equation K = °C + 273.15 and replace the variable K with the number given. Then solve for °C.

358 = °C + 273.15

$358 - 273.15 = {}^{\circ}C = 84.85$ which rounds to 85
$85 = {}^{\circ}C$

Next, use the conversion equation ${}^{\circ}F = (1.8){}^{\circ}C + 32$. Replace the variable ${}^{\circ}C$ with the number given, then solve for ${}^{\circ}F$.

${}^{\circ}F = (1.8)(85) + 32$
${}^{\circ}\mathbf{F} = \mathbf{185}$

Chapter 1 Practice Test

Now that you've had the chance to study and review Chapter 1's practice problems provided in your textbook, you are ready to practice problems as you might see them in a testing situation. The questions and problems below are randomly selected in much the same way your instructor will select them; that is, not necessarily ordered in the same way as they appear in the textbook. To make best use of this practice, have your paper, pencil and calculator ready as you would for a class exam. Use a clock and give yourself 45 minutes to complete this activity. Note that taking this test under similar conditions as a class exam provides two benefits:

1. Testing your knowledge of the chapter's contents and your mastery of the required skills.
2. Practicing working under the stress of a timed exam.

Most instructors have a type of testing format that they prefer and will most likely tell you about it before the test. For practice, this self-test uses a variety of test question formats such as matching, fill-in-the-blank, listing, multiple-choice, and essay questions. *Be sure to have a periodic table!*

Note the time and begin.

1. $324.5 \text{ K} = $ _____ ^0C ?

    a. 423.5   b. 51.4   c. 597.5   d. 162.5   e. 25

2. How many significant figures are in the number 0.000780?

    a. 7   b. 6   c. 5   d. 4   e. 3

3. If a piece of metal had a mass of 20.0 g and a density of 2.35 g/ml, what volume would it occupy?

    a. 8.51 ml   b. 20.0 ml   c. 47.0 ml   d. .117 ml   e. 235 ml

4. How many significant figures are in the measured quantity 0.002004?

    a. 7   b. 6   c. 5   d. 4   e. 3

5. The prefix kilo- stands for:

    a. 10^3 b. 10^{-3} c. 10^6 d. 10^{-2} e. 10^0

6. What is the answer to the following operation including correct significant figures?

$$324.55-(6104.5/22.3)$$

     a. 51   b. 50   c. 50.8   d. 50.81   e. -259

7. 30 °C = _____ °F

     a. 303   b. 62   c. 86   d. 54   e. - 1.1

8. 32 km converted to mm is

     a. 3.2×10^3 b. 3.2×10^7 c. 0.0032   d. 32000   e. 3.2×10^{-9}

9. The sum of 2.5×10^{-2} and 5.22×10^{-1} is:

     a. 7.72×10^{-3} b. 7.72×10^{-1} c. 5.47×10^{-1} d. 5.47×10^{-2} e. NCR

10. How many significant figures are in the number 235,000?

     a. 6   b. 5   c. 4   d. 3   e. 2

Give two reasons why the study of chemistry is important.

11.

12.

Express the values in scientific notation.

13. 12,000,000 _____

14. 634,000 _____

Express the following numbers in standard values.

15. 4.53×10^4 _____

16. 2.50×10^{-3} _____

17. Define the nature of chemistry in your own words.

18. You are an astronaut aboard the space station Mir. The solar panel that supplies oxygen to your station has just stopped working. Describe how you could "scientifically" search for a solution.

Match the following terms or phrases to the correct meaning

19._____Scientific method

20._____ Energy

21._____Potential energy

22._____Law

23._____Significant figures

24._____Theory

25._____ Kinetic energy

a. statements that describe things that are consistently and reproducibly observed

b. an experimentally tested explanation of an observed behavior

c. the ability to do work and transfer heat

d. stored energy

e the energy of motion

f. the process used to gather and interpret information

g. those digits in a measurement that are reproducible when the measurement is repeated, plus the first doubtful digit

PRACTICE TEST Answers

1. b
2. e
3. a
4. d
5. a
6. a
7. c
8. b
9. c
10. d
11. Everything around us is made of chemicals (sample answer)
12. Sample answer: Chemistry helps us to better understand what chemicals are and how to use them.
13. 1.2×10^7
14. 6.34×10^5
15. 453×10^4
16. 0.00250
17. Chemistry is the study of matter and how it changes from one form to another. (sample answer)
    You may also illustrate your knowledge of this concept by giving specific examples of chemistry in use, such as how gasoline is burned and used as fuel.
18. I would first find as much information as possible about the oxygen supply from the ship's manuals or Control. Then based on that information and the supplies on hand, I would propose a hypothesis about the cause of the problem. I would then design a way to fix it. If it works, then great! If not, I would revise my hypothesis and try again.
19. f
20. c
21. d
22. a
23. g
24. b
25. e

ANSWER PAGE FOR BUILDING KNOWLEDGE

Match the terms or phrases to the correct meaning

___o___ 1. Scientific method

___n___ 2. Physical properties

___h___ 3. Law

___q___ 4. Physical changes

___i___ 5. Theory

___p___ 6. Accuracy

___b___ 7. Hypothesis

___r___ 8. Precision

___m___ 9. Experiment

___k___ 10. Potential energy

___a___ 11. Matter

___l___ 12. Kinetic energy

___j___ 13. Energy

___c___ 14. Unit

___f___ 15. Temperature

___d___ 16. Significant figures

___g___ 17. Conversion factors

___e___ 18. Exact numbers

a. anything that has mass and occupies space

b. a tentative explanation that is based on present facts

c. symbol that stands for the quantity being measured

d. those digits in a measurement that are reproducible when the measurement is repeated, plus the first doubtful digit

e. have an unlimited number of significant figures

f. average kinetic energy measured in °F, °C, or K

g. two equivalent measurements used in a ratio to transform one unit into another

h. statements that describe things that are consistently and reproducibly observed

i. an experimentally tested explanation of an observed behavior

j. the ability to do work and transfer heat

k. stored energy

l. the energy of motion

m. controlled tests to collect data relating to a question

n. characteristics that can be determined without changing what the matter is made of

o. the process used to gather and interpret information

p. related to how close a measured value is to a true number

q. changing matter without altering its chemical composition

r. a measure of reproducibility of a measurement

Chapter 2
Atoms and Elements

Chapter Overview

If you want to learn more about how your car works, it wouldn't make sense to go to a car dealership and sort the cars by color, size or utility. It would also do very little good to sit by the roadside and watch how other cars perform on the road, or find out where they were going. Instead, you would pop the hood and take a look at the parts that make up the car.

Similarly, in order to better understand how and why matter does what it does, it is necessary to take a closer look at the individual components of matter. Chapter 2 has been designed to introduce you to the fundamental component of matter, the atom. The first part of the Chapter 2 describes the basic structure of the atom and how chemists identify and classify the atom in nature. The next part of the chapter describes the relationship between the atomic parts and how they help chemists tell one atom from another. Then in order to work with the atom you will also learn about the relationship between mass and a given number of atoms. Finally, you will learn how atomic parts can come and go in a process known as radiation and some of the medical uses of radiation.

Knowledge/Comprehension

Typically there are new terms or phrases that you must memorize in order for the lesson to make sense to you. Look up this information and memorize their explanations and definitions:

atoms	protons	neutrons
electrons	amu	atomic symbol
periodic table	atomic weight	nonmetals
semimetals	group	representative element
mass number	period	transition metals
lanthanide elements	actinide elements	mole
atomic number	radioisotopes	nuclear radiation
alpha	beta	positron
gamma rays	half-life	element
metals		

The following activity is designed to help you master the objectives outlined in the beginning of Chapter 2 and then summarized at the end of the chapter. To make the best use of your study time, read Chapter 2 before continuing. Next, work the matching activity provided as though you are taking a test *without* first referring to the answer page. Put your answers on a separate sheet of paper so that you can make more than one attempt without seeing your previous answers. Once you have completed the activity, check your answers. Make *flash cards* for those you've missed. Have a study buddy drill you on them if you missed more than half of the terms. Once you have reviewed the ones you missed, try the activity again for at least another 15 minutes. If you miss fewer than four, write in the answers so that you will have them to review just before taking the test.

Match the terms or phrases to the correct meaning

_____1. Atoms

_____2. Protons

_____3.Neutrons

_____4. Electrons

_____5. Element

_____6. amu

_____7. Atomic symbol

_____8. Periodic table

_____9. Atomic weight

_____10. Metals

_____11. Nonmetals

_____12. Semimetals

_____13. Group

_____14. Representative element

_____15. Mass number

_____16. Period

_____17. Transition metals

_____18.Lanthanide elements

_____19. Actinide elements

_____20. Mole

_____21. Atomic number

_____22. Radioisotopes

_____23. Nuclear radiation

a. time required for one-half of the atoms in a sample to decay

b. number of protons in an atom's nucleus

c. particle made up of 2 protons and 2 neutrons

d. elements with atomic numbers 90 through 103

e. horizontal row of the periodic table

f. the basic unit from which matter is constructed

g. have properties that are between the metals and nonmetals

h. part of atom with charge of (-1) and relatively zero mass

i. provides: atomic symbol, atomic number, and atomic weight

j. part of atom with charge of (+1) and relative mass of 1

k. atomic mass units

l. unstable atomic nuclei that emit nuclear radiation

m. counting unit that equals 6.02×10^{23} items

n. elements with atomic numbers 58 through 71

o. elements of the same vertical column of a periodic table

p. Groups 1A, 2A, 3A, 4A, 5A, 6A, 7A, and 8A

q. particle found in atom nucleus with mass =1, charge = 0

r. describes matter that consists of just one type of atom

s. usually one or two letter abbreviation of the element name

t. on the right of the heavy zigzag line on the periodic table

u. on the left of the heavy zigzag line on the periodic table

v. average mass in amu of the atoms of an element

w. a type of energy that travels as waves

_____24. Alpha   x. the Group B elements on the periodic table

_____25. Beta   y. total number of protons and neutrons in an atom's nucleus

_____26. Positron   z. electron that is ejected from the nucleus of a radioisotope

_____27. Gamma rays   aa. has the same mass as a beta particle, carries a +1 charge

_____28. Half-life   bb. Particles and energy released during a nuclear change

In addition to knowing the terms or phrases, you should be able to compare, contrast, or list examples that show you comprehend the following concepts:

- Describe the subatomic structure of an atom.
- Describe how isotopes of an element differ from one another.
- Describe the relationship between moles and atomic weight.
- Describe the four common types of radiation emitted by radioisotopes.

Once you know the appropriate tools for Chapter2, you are ready to evaluate your understanding of those tools and their relationships. In this type of concept building, it is necessary to compare or contrast terms or concepts. In some cases you may be required to give examples or recognize a term or concept when it is depicted in a picture or sample scenario.

Sample Problems

1. Which of the following is not found as a subatomic part of an atom?

a. nucleus   b. electron   c. proton   d. neutron   e. boron

Several definitions go into this solution. First you must recognize that subatomic means inside the structure of the atom. Then you must know the meanings of the choices given. In this case, the proton and neutron are subatomic particles that make up the nucleus of the atom and the electron is the subatomic particle that moves around the nucleus. Boron on the other hand is an element and therefore would not be a subatomic part of an atom. **Answer is e.**

2. Which of the following make up most of the mass of the atom?

a. electron   b. proton and neutron   c. neutron   d. proton   e. proton and electron

The electron has a relative mass of 0 amu while the proton and neutron each have relative mass of 1 amu. A combination of two of the particles would have a significant mass, therefore choices a, c, and d are eliminated. Choice e is not a good choice since the electron does not have a significant mass. Since protons and neutrons are of the same relative mass, the mass of the atom is equal to the sum of all the protons and neutrons in the nucleus. **Answer is b.**

3. Which of the following is not found in the center of the atom?

a. nucleus   b. electron   c. proton   d. neutron

By definition the nucleus is the center of the atom, which eliminates choice a. Protons and neutrons (choices c and d) are located in the nucleus and therefore in the center of the atom. Electrons (choice b) are found in a cloud type formation around the nucleus and are not in the center of the atom. **Answer is b.**

4. Isotopes of an element are identical except for the fact that they have a different number of:

a. atomic numbers   b. electrons   c. protons   d. neutrons

Choices a and c are similar answers since the atomic number is the number of protons in the atom. These are not correct because all atoms of the same element have the same number of protons. In order for an atom to be electrically neutral, the number of electrons must be equal to the number of protons, therefore answer b is eliminated. **Answer is d.**

5. In the isotope symbol $_1^2 H$, the number 2 stands for the:

a. atomic number   b. electrons   c. protons   d. neutrons   e. mass number

In the isotope symbol, the lower number is the number of protons for that element and is always the same for all of the isotopes of that element. The top number is equal to the number of protons and neutrons in the isotope or its mass number. **Answer is e.**

6. Which of the following are isotopes? (X stands for any element symbol)

a. $_7^{16} X$ and $_8^{16} X$ b. $_8^{16} X$ and $_7^{16} X$ c. $_8^{16} X$ and $_8^{15} X$

d. $_7^{15} X$ and $_8^{15} X$ e. $_8^{16} X$ and $_7^{16} X$

Choice a is eliminated because both isotopes have the same atomic number and atomic mass and would therefore be the same atom. Choice b is not correct because the two isotopes have different atomic numbers, 8 and 7, which means that they are not even the same element. In choice d they have the same atomic masses and different atomic numbers, which make them atoms of different elements with the same mass. In choice e the atoms have the same atomic numbers which mean they are the same element but have different mass numbers and are therefore isotopes. **Answer is c.**

7. How many atoms of magnesium are in 2.5 moles of magnesium?

a. 9.73 atoms b. 1.5×10^{24} atoms c. 24.32 atoms
d. 6.02×10^{23} atoms e. 2.5 atoms

First you must recall that 1 mole of any element contains 6.02×10^{23} atoms. This can be used as a conversion factor as follows:

$$\frac{2.5 \text{ mol}}{1} \times \frac{6.02 \times 10^{23} \text{ atoms}}{1 \text{ mol}} = 1.5 \times 10^{24} \text{ atoms}$$ **Answer is b.**

8. How many moles of atoms would be found in 135 grams of aluminum?

a. 4.98 mol b. 135 mol c. 27.10 mol
d. 6.02×10^{23} e. 3658.5 mol

In addition to equaling 6.02×10^{23}, 1 mole of atoms also equals the mass of the atomic weight expressed in grams. So, 1 mole of Al is equal to 27.10 g of Al. This can be used as a conversion factor as follows:

$$\frac{135 \text{ g}}{1} \times \frac{1 \text{ mol}}{27.10 \text{ g}} = 4.98 \text{ mol}$$ **Answer is a.**

Also note that the final answer has three significant figures just like the original measurement.

9. What would be the mass of 22.5×10^{23} atoms of oxygen be in grams?

a. 360. g b. 3.74 g c. 59.8 g d. 3.74×10^{23} g e. 16.00 g

This time both equivalences used in problem 8 are needed. 1 mol = 6.02×10^{23} and 1 mol of O is equal to 16.00 g of O. Two conversions are made as follows:

$$\frac{22.5 \times 10^{23} \text{ atoms of O}}{1} \times \frac{1 \text{ mol}}{6.02 \times 10^{23} \text{ atoms}} \times \frac{16.00 \text{ g of O}}{1 \text{ mol}} = 59.8 \text{ g of O}$$

Answer is c.

10. Which of the following is not a type of radiation emitted by radioisotopes?

a. alpha b. beta c. positron d. gamma ray e. radon

The alpha particle is a helium nucleus that is ejected from the nucleus of a radioisotope; the beta particle is an electron that is ejected from the nucleus of a radioisotope; a positron is a particle with the same mass as a beta particle but has a positive charge and is ejected from the nucleus of a radioisotope; gamma rays are high energy emissions that may be released from the nucleus of a radioisotope; radon is an element and is not ejected from a radioisotope. **Answer is e.**

11. The symbol $_{-1}^{0}\beta$ is used to designate the radioisotope emission called

a. alpha   b. beta   c. positron   d. gamma ray   e. beacon

The Greek symbol β is pronounced beta and the symbol $_{-1}^{0}\beta$ is used for the beta particle. **The answer is b.**

12. Which of the following radioisotope emissions does not have a charge or mass?

a. alpha   b. beta   c. positron   d. gamma ray   e. beacon

The symbol for the alpha particle is $_{2}^{4}\alpha$ which indicates it has a positive charge of 2 and a mass of 4; the symbol for a beta particle is $_{-1}^{0}\beta$ which means it has a charge of -1 and a mass of 0; the positron has the symbol $_{-1}^{0}\beta^{+}$ meaning it has a charge of +1 and a mass of 0; the symbol for the gamma ray is written $_{0}^{0}\gamma$ which means it has no charge or mass. Gamma radiation is energy not matter. **The answer is d.**

Application

To apply what you've already learned and master the concepts, you must utilize all of the knowledge, understanding, facts, and/or techniques that you have gathered up to this point. Simply knowing an equation or process is not enough to master the concept; you develop these by *practice*. Solve as many different types of problems with as many different variations as you *strive to master the skill*. Your textbook provides a very good assortment of problems at the end of the chapter. Work as many as needed for skill mastery, not just those assigned.

- Explain how elements are arranged on the periodic table.
- Explain how atomic number and mass number are used to indicate the makeup of an atom's nucleus.
- Explain how exposure to radiation can be controlled.

In order to truly master any skill you must *practice*. In this section example problems related to the application objectives are discussed. Remember the more you practice, the better your chances of being able to successfully use these objectives to solve problems. Additional practice problems are at the end of the chapter in your textbook and in the online interactive lessons.

Sample Problems

1. Which of the following elements has the symbol K?

a. oxygen   b. sodium   c. krypton   d. potassium   e. Kelvin

You were most likely assigned a certain number of elemental symbols that you need to memorize. Remember that while the symbols are generally made up of the first letter or the first letter and one other letter from the IUPAC name, some of the symbols are based on the Latin name of the element. In this case the symbol for potassium, K, stands for its Latin name kalium. **Answer is d.**

Note in order to apply your skill in reading information from the periodic table you must first be able to find the element. Most periodic tables use symbols only.

2. Which of the following is a metal?

a. silicon b. magnesium c. helium d. sulfur e. carbon

First make sure you know the symbols of each—silicon: Si, magnesium: Mg, helium: He, sulfur: S, carbon: C. Next, find the location of the symbols on the periodic table.

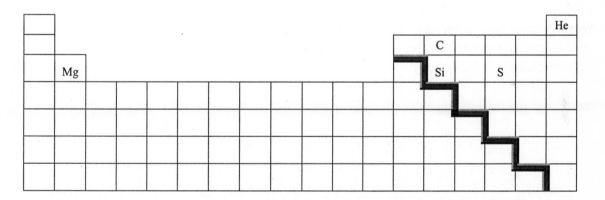

Silicon is found along the dark zigzagged line; it has properties of both metals and non-metals and is classified as a semimetal. Carbon, sulfur and helium are to the left of the zigzagged line and would have properties of non-metals. Only magnesium is found to the left of the zigzagged line and would be classified as a metal. **Answer is b.**

3. Which of the following sets of atoms is arranged from smallest to largest?

a. Al, Si, P   b. Cs, Rb, Na   c. B, C, Mg   d. F, Cl, Br   e. C, F, B

First, recall the rules regarding predicting size using the periodic table. Size generally increases as you go down a group (column) and from right to left in a period (row). Next locate the elements on the periodic table.

																	He
												B	C			F	
Na	Mg											Al	Si	P		Cl	
																Br	
Rb																	
CS																	

In choice d, (F, Cl, Br) the elements follow each other down a group and therefore their atomic size increases. In choice c, (B, C, Mg) there is no particular pattern since it changes rows (Mg is one row below B and C, meaning it is one complete energy level larger than the other two elements). In choice b, (Cs, Rb, and Na) the elements follow each other up a column, meaning that atomic size is decreasing. In choice a, (Al, Si, P) the elements are ordered from right to left on a period, which means that they are getting smaller in size. In choice e, (C, F, B) the elements are on the same row but not in the correct order. **Answer is d.**

4. How many protons and neutrons are located in the nucleus of $_{14}^{33}S$?

a. 16 protons and 16 neutrons
c. 16 protons and 17 neutrons
e. 33 protons and 16 neutrons

b. 16 protons and 33 neutrons
d. 17 protons and 16 neutrons

First you must recall that in an isotopic symbol, $_{14}^{33}S$, the lower number is the atomic number and the upper number is the atomic mass. Next, remember that the atomic number is the number of protons; this means you can eliminate choices d and e since they do not have 16 protons. Then, since atomic mass = protons + neutrons, you must subtract the atomic mass from the atomic number to get the number of neutrons. 33-16 = 17 which means there are 17 neutrons. **The answer is c.**

5. Identify the missing product in the nuclear equation below.

$$_{92}^{238}U \rightarrow \ ? \ + \ _{2}^{4}\alpha$$

Start by identifying the element symbol. To do this, subtract the atomic number of the alpha particle, $_{2}^{4}\alpha$, from the atomic number of the radioisotope given, $_{92}^{238}U$. (Remember, atomic number is the lower number) 92-2 = 90. This means that a new element is being produced. Th, which is element 90 on the periodic table is written as $_{90}Th$. Finally, determine the mass number of Th by subtracting the mass number of the alpha particle from the atomic mass value of the radioisotope given, 238- 24= 234. This is the atomic mass number for Th. **Answer is $_{90}^{234}Th$.**

6. Identify the missing product in the nuclear equation below.

$$^{15}_{8}O \rightarrow {}^{14}_{7}N + ?$$

In this variation of the problem, you are attempting to identify the nuclear particle that has been emitted. First, compare the product $^{14}_{7}N$ to the radioisotope given, $^{15}_{8}O$. Note that the mass has not changed as it is 15 for both. The atomic number has changed as it has gone down from 8 to 7. This means that you are looking for a nuclear emission which results from a proton becoming a neutron. This also means that the particle must have a mass of zero since the original mass does not change; however it must have a charge that accounts for the proton going from a positively charged particle to a neutral particle.

Examine the choices of particles: Losing the particle $^{4}_{2}\alpha$ would change both the number of protons and the number of neutrons thus changing the mass. Losing the particle $^{0}_{-1}\beta$ would mean that a negative charge was emitted, which would convert a neutron into a proton. It does not change the mass but would make the atomic number increase. The particle $^{0}_{1}\beta^{+}$ would result in a positive charge being lost, which would change a proton into a neutron while not changing the mass. The particle $^{0}_{0}\gamma$ is energy and does not change atomic mass or the atomic number.

Answer is the particle $^{0}_{1}\beta^{+}$ $^{15}_{8}O \rightarrow {}^{14}_{7}N + {}^{0}_{1}\beta^{+}$

Improving your ability to analyze problems involves connecting or explaining relationships between previously learned concepts. If you are not familiar with the terminology being used in the question or if components of the scenario seem disconnected to you, be sure to go back and review them. It is impossible to connect, compare, or explain concepts without first being familiar with them.

7. Which of the following radioisotope decay particles would not be used for diagnosis?

        a. $^{0}_{1}\beta^{+}$ b. $^{0}_{-1}\beta$ c. $^{0}_{0}\gamma$ d. $^{4}_{2}\alpha$

This type of question requires that you understand two concepts. Firstly, you must know how radioisotopes are used for diagnosis. When using radioisotopes for diagnosis, the radiation passes through the organs or tissues to some extent and captured on some type of photographic film. Secondly, you must know something about how exposure to radiation is controlled by using materials that can "block" the particles from moving through material. Because of their comparatively large mass, alpha particles lose energy after short distances and have very little penetrating power. The smaller β-type particles penetrate deeper and the gamma rays (high energy) easily penetrate the body. Therefore in this question you are looking for a radioisotope that doesn't work well, that is, you are looking for the one with the least penetrating ability. **Answer is d.**

8. Controlling exposure also involves knowing how much radiation is being produced. You swallowed 10.0 mg of the radioisotope $^{131}_{53}$I with a half life of 8.1 days, and 10.0 mg of $^{32}_{15}$P with a half life of 14.3 days. Which radioisotope would have produced the most radiation in 24 days?

You must first understand the concept of a half-life of a radioisotope. Remember that half-life is the time it takes for half of the radioactive material to decay. Since the half-life of $^{131}_{53}$I is 8.1 days and that divides into 24 days approximately three times, this radioisotope would have decayed through three half-life time periods. Therefore, after the first 8.1 days, 5.0 mg would have decayed, another 2.5 mg decays by the second half-life, followed by another 1.25 mg by the end of the third half-life period. This makes a total of 8.25 mg of $^{131}_{53}$I decayed in a 24-day period. For the $^{32}_{15}$P with a half-life of 14.3 days, 5.00 mg would have decayed by the first 14.3 days. Since the half-life of $^{32}_{15}$P is 14.3 days, you have not reached the second half life by day 24, and less than 2.5 mg will have decayed for a total of less than 7.5 mg. **This means that the $^{131}_{53}$I will produce the most radiation because it goes through more half-life transitions in the 24 days specified.**

9. Although radiation from $^{214}_{84}$Po might be useful for the treatment of a skin cancer, explain why it is not very useful as a diagnostic radioisotope.

To answer this question, you first need to know what type of decay particle comes from the radioisotope $^{214}_{84}$Po. Having found that it is an alpha particle ($^{4}_{2}\alpha$) emitter you must recall that alpha particles are relatively massive and do not have the ability to deeply penetrate the skin. **The fact that $^{214}_{84}$Po alpha particle emissions would mainly affect the surface makes it a good choice for treating a skin cancer since it would not penetrate tissue below the cancer to any great extent. For the same reason, however $^{214}_{84}$Po is a poor choice for diagnosis since for that use you need the particles to go through the body in order to photograph the organs.**

Chapter 2 Practice Test

Now that you've had the chance to study and review Chapter 2's practice problems provided in your textbook, you are ready to practice problems as you might see them in a testing situation. The questions and problems below are randomly selected in much the same way your instructor will select them; that is, not necessarily ordered in the same way as they appear in the textbook. To make best use of this practice, have your paper, pencil and calculator ready as you would for a class exam. Use a clock and give yourself 45 minutes to complete this activity. Note that taking this test under similar conditions as a class exam provides two benefits:

1. Testing your knowledge of the chapter's contents and your mastery of the required skills.
2. Practicing working under the stress of a timed exam.

Most instructors have a type of testing format that they prefer and will most likely tell you about it before the test. For practice, this self-test uses a variety of test question formats such as matching, fill-in-the-blank, listing, multiple-choice, and essay questions. **_Be sure to have a periodic table!_**

Note the time and begin.

1. Which symbol represents the element phosphorus?

A. P
B. Pd
C. Pb
D. Pt

2. Which element does the symbol K represent?

A. Potassium
B. Sodium
C. Chlorine
D. Mercury

3. Francium's symbol is:

A. Fe
B. Fr
C. F
D. Cf

4. Helium's symbol is:

A. H
B. He
C. Hg
D. Hs

5. What does the symbol Kr represent?

A. Potassium
B. Karatic
C. Krypton
D. Argon

6. What is the symbol for tin?

A. Tl
B. Ta
C. Ti
D. Sn

7. To what period and group does sulfur belong?

A. Period 3 and group 6A
B. Period 3 and group 8A
C. Period 6 and group 3A
D. Period 2 and group 8A.

8. Which group contains the Earth metals?

A. 1A
B. 8A
C. 2A
D. 5A

9. What is the name of the elements in group 8A?

A. Noble Gases
B. Alkali metals
C. Alkaline earth metals
D. Halogens

10. What group includes Boron?

A. 8A
B. 3A
C. 6A
D. 7A

11. How would you classify iodine?

A. Alkali metal
B. Alkaline earth metal
C. Halogen
D. Transition elements

12. How would you classify potassium?

A. Alkali metal
B. Alkaline earth metal
C. Halogen
D. Transition elements

13. How many electrons are in a gold atom?

A. 79
B. 118
C. 0
D. 197

14. How many protons are in argon?

A. 18
B. 0
C. 36
D. 34

15. Which of the following elements is a non-metal?

A. Mercury
B. Calcium
C. Sodium
D. Sulfur

16. All of the following are good conductors of electricity except:

A. Zn
B. Br
C. Hg
D. Al

17. Which of the following is electrically neutral?

A. Proton
B. Neutron
C. Electron

18. Which particle has the smallest mass?

A. Proton
B. Neutron
C. Electron

19. Which of the following has a positive charge?

A. Proton
B. Neutron
C. Electron

20. _____ spin(s) around the nucleus.

A. Protons
B. Neutrons
C. Electrons
D. Static Electricity

21. What is the atomic number of nitrogen?

A. 1
B. 5
C. 7
D. 79

22. What is the name of the element with the atomic number of 82?

A. Lead
B. Krypton
C. Aluminum
D. Tin

23. What is the symbol for the element that has an atomic number of 50?

A. Sn
B. Sb
C. Ag
D. O

24. How many protons are in the isotope $_{26}^{56}Fe$?

A. 34
B. 16
C. 26
D. 56

25. How many neutrons are in the isotope $_{17}^{54}Cl$?

A. 19
B. 37
C. 22
D. 54

26. How many electrons are in the isotope $_{20}^{49}Ca$?

A. 49
B. 20
C. 29
D. 19

27. Mercury is classified as a _____.

28. List the name and symbol of the four main types of radioisotope emissions.
    a.
    b.
    c.
    d.

29. Identify the missing particle in the following nuclear equation.

$$? \rightarrow {}_{18}^{40}Ar + {}_{1}^{0}\beta^{+}$$

30. Identify the missing particle in the following nuclear equation.

$${}_{8}^{14}O \rightarrow {}_{7}^{14}N + ?$$

31. List three common uses of radioisotopes in medicine.

     a.

     b.

     c.

32. Which type of radiation has no mass or charge? Why?

33. The exposure to radiation can cause damage to healthy tissue. Under what conditions might this be acceptable and when might the doctor want to consider an alternative treatment?

34. Rank the elements Fe, Ca, and Ni according to increasing atomic size.

35. List some physical properties of semi-metals.

PRACTICE TEST Answers

1. A	2. A	3. B	4. B	5. C
6. D	7. A	8. C	9. A	10. B
11. C	12. A	13. A	14. A	15. D
16. B	17. B	18. C	19. A	20. C
21. C	22. A	23. A	24. C	25. B
26. B				

27. Transition metal

28. a. $_2^4\alpha$ alpha particle   b. $_{-1}^0\beta$ beta particle   c. $_1^0\beta^+$ positron   d. $_0^0\gamma$ gamma rays

29. $_{19}^{40}K$

30. $_1^0\beta^+$

31. Examples of possible answers:
   a. diagnosing thyroid problems
   b. killing cancerous cells
   c. producing images of various organs or tissues

32. $_0^0\gamma$ gamma rays; gamma radiation is pure energy and therefore has no mass or charge.

33. Sample answer: Damage to healthy tissue might be acceptable if it prevents the need for surgical removal and eliminates the cancerous cells. Also, radiation can be acceptable if the area in question is not easily accessible by surgery. The doctor might want to consider an alternative treatment in the case of potentially damaging vital organs such as the eyes, ears, heart or reproductive organs. Sometimes the cancerous cells are so spread out that trying to radiate them would damage so much healthy tissue it would put the patient at risk of dying from the radiation.

   34. Ni, Fe, Ca

35. Semimetals have properties of both metals and non-metals so listing a mixture of the physical properties of each group would be acceptable. For example: Conducts electricity, brittle when solid, shiny or dull, moderate melting points.

ANSWER PAGE FOR BUILDING KNOWLEDGE

Match the terms or phrases to the correct meaning

___f__1. Atoms   a. time required for one-half of the atoms in a sample to decay

___j__2. Protons   b. number of protons in an atom's nucleus

___q__3. Neutrons   c. particle made up of 2 protons and 2 neutrons

___h__4. Electrons   d. elements with atomic numbers 90 through 103

___r__5. Element   e. horizontal row of the periodic table

___k__6. amu   f. the basic unit from which matter is constructed

___s__7. Atomic symbol   g. have properties that are between the metals and nonmetals

___t__8. Periodic table   h. part of atom with charge of (-1) and relatively zero mass

___v__9. Atomic weight   i. provides: atomic symbol, atomic number, and atomic weight

___u__10. Metals   j. part of atom with charge of (+1) and relative mass of 1

___t__11. Nonmetals   k. atomic mass units

___g__12. Semimetals   l. unstable atomic nuclei that emit nuclear radiation

___o__13. Group   m. counting unit that equals 6.02×10^{23} items

___p__14. Representative element   n. elements with atomic numbers 58 through 71

___y__15. .Mass number   o. elements of the same vertical column of a periodic table

___e__16. Period   p. Groups 1A, 2A, 3A, 4A, 5A, 6A, 7A, and 8A

___x__17. Transition metals   q. particle found in atom nucleus with mass =1, charge = 0

___n__18. Lanthanide elements   r. describes matter that consists of just one type of atom

___d__19. Actinide elements   s. usually one or two letter abbreviation of the element name

___m__20. Mole   t. on the right of the heavy zigzag line on the periodic table

___b__21. Atomic number   u. on the left of the heavy zigzag line on the periodic table

___l__22. Radioisotopes   v. average mass in amu of the atoms of an element

___bb__23. Nuclear radiation   w. a type of energy that travels as waves

___c__24. Alpha   x. the Group B elements on the periodic table

___z__25. Beta   y. total number of protons and neutrons in an atom's nucleus

___aa__26. Positron   z. electron that is ejected from the nucleus of a radioisotope

___w__27. Gamma rays   aa. has the same mass as a beta particle, carries a 1+ charge

___a__28. Half-life   bb. Particles and energy released during a nuclear change

Chapter 3
Compounds

Chapter Overview

After taking a closer look at the basic components of matter and discussing how their properties are classified and organized within the periodic table, you are now ready to manipulate elements and discover what it takes to predict a product of your own choosing. To do this you must first learn more about the basic building block, the atom. Similar to making a tinker toy creation, in which wooden spools have different numbers and placements of holes, atoms have different numbers and placement of electrons. You'll first learn how to determine the number of electrons and how those electrons are arranged around the nucleus of the atom. Then just as you would begin to build a house by linking together the tinker toy spools, you will discover how atoms link together by exchanging or sharing electrons. Once you have made your structure (called a compound) you'll learn how chemists manage to give every possible chemical combination its very own name. Finally, to actually make some of your newly modeled compounds, there must be a way to count the number of atoms needed (mole concept learned in Chapter 2) and convert that to a measurable mass using formula weight or molecular weight; atoms are far too small to count them in the laboratory.

Knowledge/Comprehension

Typically, there are new terms or phrases that you must memorize in order for the lesson to make sense to you. Look up this information and memorize their explanations and definitions:

Bohr model	orbitals	valence electrons
electron dot structure	ions	quantum mechanical model
monatomic	octet	cations
anions	ionic bond	covalent bond
ionic compound	molecules	binary
formula weight	molecular weight	

The following activity is designed to help you master the knowledge objectives outlined in the beginning of Chapter 3 and summarized at the end of the chapter. To make the best use of your study time, read Chapter 3 before continuing. Next, work the fill in the blank activity provided as though you are taking a test *without* first referring to the answer page first. Put your answers on a separate sheet of paper so that you can make more than one attempt without seeing your previous answers. Once you have completed the activity, check your answers. Make *flash cards* for those you've missed. Have a study buddy drill you on them if you missed more than half of the terms. Once you have reviewed the ones you missed, try the activity again for at least another 15 minutes. If you miss fewer than four, write in the answers so that you will have them to review just before taking the test.

Fill-in the term or terms needed to complete the statement

1. _____is the stable arrangement of eight valence electrons.

2. _____ _____describes electrons as circling the nucleus in fixed orbits.

3. _____ _____ _____describes electrons as in orbitals that surround the nucleus in energy levels with a maximum of electrons allowed per energy level.

4. _____are energy zones where electrons are believed to reside; each holds two electrons.

5. _____ _____are in the highest numbered, occupied energy level of an atom.

6. _____ are ions with a positive charge that.are formed when an atom loses electrons.

7. _____ _____ _____uses dots placed around the symbol of an element to represent the valence electrons of that element.

8. _____ are negatively charged ions that are formed when an atom gains electrons.

9. _____ are atoms or groups of atoms with an unequal number of protons and electrons caused by the atom or atoms either gaining or losing electrons.

10. _____ means formed from a single atom

11. _____ _____is the sum of the weights of the elements in the formula of a molecule.

12. _____ are an uncharged group of atoms joined by covalent bonds.

13. _____ _____is the sum of the weights of the elements in the formula of an ionic compound.

14. _____ _____ is the attraction between ions of opposite charges.

15. _____ _____is the sharing of electrons as a way of achieving eight outer electrons; typically occurs between nonmetals.

16. _____ _____are compounds formed with ionic bonds and named by combining the name of the cation with the name of the anion.

17. _____ are compounds that contain just two different elements.

In addition to knowing the terms or phrases, you should be able to compare, contrast, or list examples that show you understand the following concepts:

- Describe how the Bohr model and the quantum mechanical model of the atom differ in their view of how electrons are being arranged about an atom's nucleus.
- Describe electron dot structures.
- Explain the differences between an ionic bond and a covalent bond.

Once you know the appropriate tools for Chapter3, you are ready to evaluate your understanding of those tools and their relationships. In this type of concept building, it is necessary to compare or contrast terms or concepts. In some cases you may be required to give examples or recognize a term or concept when it is depicted in a picture or sample scenario.

Sample Problems

1. Which of the models below best depicts the Bohr model?

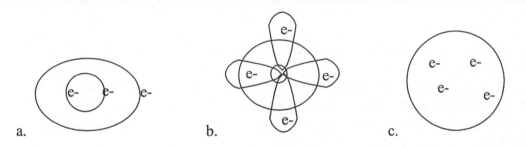

This objective is essentially contrasting two different theories about how the electrons are arranged in an atom. The first theory proposed by Niels Bohr was based on rather primitive photographic techniques. Bohr's model depicted the electron as moving in circles about the nucleus of the atom at distinct distances that corresponded to energy changes observed when light was emitted by energized hydrogen. This model is the one that many instructors still use today when trying to convey the general idea that electrons are found at specific energy levels. **Answer is a.**

2. Which of the models below best depicts the quantum mechanical model?

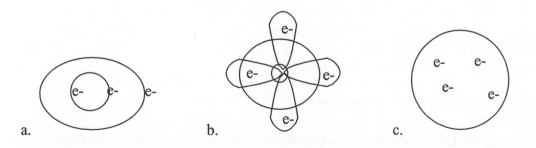

The quantum mechanical model is based on the ability to measure energy exchanges to a greater degree of accuracy. This model still uses the idea of discrete energy levels but takes that even further to say that within each energy level the electron resides in separate energy zones called orbitals. Each orbital represents an electron of particular energy. This allows for the maximum number of electrons to be calculated for each energy level using the equation: Maximum electrons $= 2n^2$, where n represents the energy level number. **Answer is b.**

3. According to the quantum mechanical model what is the maximum number of electrons that could reside in the third energy level?

        a. 2   b. 8   c. 18   d. 32   e. 50

Remember that according to the quantum mechanical theory the maximum number of electrons that can reside in an energy level is calculated using the equation: Maximum number of electrons $= 2n^2$, where n = the energy level number. Therefore, for energy level $= 3$, $2(3)^2 = 2 \times 9 = 18$. **Answer is c.**

The key to learning how to draw a dot structure is to remember that the dots represent the valence electrons for the element and that the number of valence electrons is given by the column number of the element on the periodic table. For example, oxygen is in group

VI; it has six valence electrons and has six dots drawn around the symbol O, $\cdot \overset{\cdot\cdot}{\underset{\cdot}{O}} :$
Since there are only eight groups, there are only eight types of electron dot structures. Remember that the quantum mechanical theory proposes the existence of orbitals: one s and three p orbitals per energy level. As electrons tend to repel each other, you place one dot per orbital until you have one on each side of the symbol and then start around again pairing them up.

4. Draw the dot structure for silicon.

Start by writing the correct symbol for silicon, Si. Look this symbol up to see which group it is located in on the periodic table. It is found in group IV, which tells you it has 4 valence electrons. Place your dots following this sequence:

$$\text{Si} \cdot \rightarrow \overset{\cdot}{\text{Si}} \cdot \rightarrow \cdot \overset{\cdot}{\text{Si}} \cdot \rightarrow \cdot \underset{\cdot}{\overset{\cdot}{\text{Si}}} \cdot$$

Answer is $\cdot \underset{\cdot}{\overset{\cdot}{\text{Si}}} \cdot$ although when not reacting (in ground state) some texts may show

two of the electrons paired to represent the filled s orbital or $\cdot \overset{\cdot}{\text{Si}} \colon$. Be sure to note how your instructor would like it to be drawn.

5. Draw the electron dot structure for bromine.

Start by writing the correct symbol for bromine, Br. Look this symbol up to see which group it is located in on the periodic table. It is found in group VII, which tells you it has 7 valence electrons. Place your dots following this sequence:

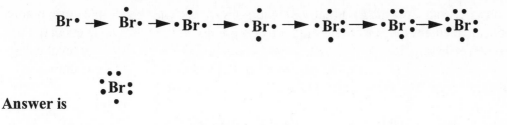

Answer is

6. Draw the dot notation for strontium.

Start by writing the correct symbol for strontium, Sr; then locate this symbol on the periodic table. It is found in group II which tells you it has 2 valence electrons. Place your dots following this sequence:

$$Sr\bullet \longrightarrow \bullet Sr\bullet$$

Answer is $\bullet Sr\bullet$ remember it may be written $Sr\overset{\bullet}{\bullet}$ as a stand-alone atom.

Recall that an ion is created when an atom either gains or loses an electron. The term covalent is derived from two parts: co meaning to share, and valence which means charge. Remember the outer electrons are called the valence electrons. Therefore, covalent means atoms share outer electrons. In a covalent bond, the outer energy levels of two atoms blend in the middle and the electrons move into shared orbitals between the two atoms. An ionic bond is the attraction force created when ions of opposite charge are next to each other. The attraction is somewhat like having opposite ends of magnets touching. The following questions illustrate ways your instructor might test your understanding of the differences between ionic bonds and covalent bonds.

7. Which of the following compounds would be most likely to have covalent bonds?

     a. Na_2S b. CaO c. $AlCl_3$ d. H_2O

In this case you are being asked to recognize that metals (recall from chapter 2 how to tell metals from nonmetals) have a tendency to lose electrons and nonmetals tend to gain them. Therefore, when a metal and nonmetal come in contact, the metal will give its outer electrons to the nonmetal, creating an ionic bond. When two nonmetals come in contact they both try to remove each other's outer electrons. This results in the electrons being held between them, creating a covalent bond. Choices a, b, and c, all have metal-nonmetal combinations and would make ionic bonds. Choice d has two nonmetals. **Answer is d.**

8. Which placement of the electrons below would most likely represent an ionic bond?

     a. X $\overset{e^-}{_{e^-}}X$ b. Xe^- e^-X c. X $\overset{e^-}{_{e^-}}X$

Here it is important that you know that a covalent bond occurs when the electrons are shared between the elements as in choice c. It is also important to recognize that sometimes electrons do not exchange or change at all as in choice b, meaning there is no bond (as in two metals in contact). Finally, it is important that you know that ionic bonds occur when one element gives its outer electron completely to the other element as in choice a. **Answer is a.**

9. Which of the following would you NOT see in a covalent bond?

  a. two atoms   b. electrons   c. nonmetal   d. metal

Metals tend to give off electrons when they react with a nonmetal, and the metals become ions. Metals do not form ionic or covalent bonds with other metals. Therefore you would not expect to find metals in a covalent bond. **Answer is d.**

Application

To apply what you've already learned and master the concepts, you must utilize all of the knowledge, understanding, facts, and/or techniques that you have gathered up to this point. Simply knowing an equation or process is not enough to master the concept. These you develop by *practice*. Solve as many different types of problems with as many different variations as you *strive to master the skill*. Your textbook provides a very good assortment of problems at the end of the chapter. Work as many as needed for skill mastery, not just those assigned.

- Explain how the electron dot structure of the representative element atom groups IA – VIIIA can be used to predict the charge of its monatomic ion.
- Name and write formulas of simple ionic compounds and binary molecules.
- Use formula weights and molecular weights in unit conversions involving moles and mass.

In order to truly master any skill you must *practice*. In this section example problems related to the application objectives are discussed. Remember the more you practice, the better your chances of being able to successfully use these objectives to solve problems. Additional practice problems are at the end of the chapter in your textbook and in the online interactive lessons.

Sample Problems

1. Draw the electron dot structure of each atom and the ion it is expected to form.

a. K

b. S

c. Ba

First locate the symbol on the periodic table and draw the dot structure of the atom (See process outlined above). In a, K is drawn as K• since K is in Group I. Since metals lose electrons, the ion is the element's symbol with a positive charge to indicate the number of electrons removed. **Answer is K⁺.**

In b, S is drawn as $\overset{\cdot\cdot}{\cdot}\overset{\cdot\cdot}{S}{:}$ since S is in Group VI. For nonmetals, electrons are gained to make a complete set of eight dots. A negative charge is placed on the ion that indicates the number of electrons (dots) needed to make eight. **Answer is** $:\!\overset{\cdot\cdot}{S}\!:^{2-}$.

In c, Ba is drawn as Ba: since Ba is in Group II. Since metals lose electrons, the ion is the element's symbol with a positive charge to indicate the number of electrons removed. **Answer is Ba ²⁺.**

2. Give the charge for the monatomic ions below:

a. $:\!\overset{\cdot\cdot}{\underset{\cdot\cdot}{As}}\!:$

b. $:\!\overset{\cdot\cdot}{\underset{\cdot\cdot}{Te}}\!\cdot$

c. $:\!\overset{\cdot\cdot}{\underset{\cdot\cdot}{I}}\!:$

Remember that all nonmetals (except for hydrogen which needs only 2) gain electrons to achieve 8 outer electrons. The charge the monatomic ion has is equal to the number of electrons gained.

a. As, arsenic, is in group V which means it starts out with 5 outer electrons and has to gain 3 electrons to have 8 electrons. **Answer is -3.**

b.Te, tellurium, is in group VI which means it starts out with 6 outer electrons and has to gain 2 electrons to have 8 electrons. **Answer is -2.**

c. I, iodine, is in group VII which means it starts out with 7 outer electrons and has to gain 1 electron to have 8 electrons. **Answer is -1.**

When naming ionic compounds, follow these basic rules:
a. Metals are named first followed by the nonmetal name.
b. The nonmetal is changed to "ide" Example: chlorine becomes chloride
or oxygen becomes oxide.

41

c. When using metals from group I, II or Al, Zn, or Ag, the ion name is the same as the metal name.

d. When using a metal capable of forming more than one type of ion (like the transition metals) the name of the metal must be followed by a Roman numeral that indicates the charge of the metal ion (See Table 3.1 in the text).

e. If there is a polyatomic ion, just give the name of that ion (See Table 3.2 in the text).

3. Name the following compounds:

      a. $SrBr_2$ **Answer is strontium bromide**

      b. $CaCO_3$ **Answer is calcium carbonate**

      c. Pb_3N_4 **Answer is lead IV nitride**

First you must note that Pb is not one of the metals from group I, II nor is it Al, Zn, or Ag; Pb therefore requires a Roman numeral. To calculate the charge, multiply the subscript 4 on the N by the charge of the N ion (see above if you have forgotten how to calculate that charge): 4 x -3 = -12 This number is divided by the subscript of the metal to get the charge on the metal. 12/3 = 4, thus the Roman numeral is IV.

Notice that for ionic compounds it is not necessary to include the number of atoms in the compound name.
When writing ionic compound formulas follow these basic rules:
    a. The metal or positive ion symbol is written first.
    b. The nonmetal or negative ion symbol is written last.
    c. Total positive charge must equal the total negative charge.

4. Write the formulas for the following ionic compounds:

      a. sodium hydroxide

Sodium ion is Na^+ and hydroxide is OH^-. Since the metal has a +1 and the nonmetal is -1 it takes one of each to equal zero. **Answer is NaOH.**

      b. barium iodide

Barium ion is Ba^{2+} and iodide is I^- , Since the metal has a +2 and the nonmetal is -1 it takes 1(+2) and 2(-1) to equal zero. **Answer is BaI_2.**

      c. iron III oxide

Iron ion is Fe^{3+} (note the Roman numeral in the name) and oxide is O^{2-}; since the metal has a +3 and the negative ion is -2 it takes 2(+3) and 3(-2) to equal zero. **Answer is Fe_2O_3.**

When writing and naming binary molecules, remember that molecules are mainly formed between two nonmetals. Therefore, it is important to use the following rules only when naming binary molecules (two nonmetals). This rule for writing and naming compounds is not used if a metal or polyatomic ion is in the formula.

When naming binary molecules follow these basic rules:
  a. The nonmetal that is the farthest to the left in a period or farthest down a group is written first.
  b. The ending on the nonmetal written first is NOT changed.
  c. The ending on the nonmetal written last is changed to "ide."
  d. The number of each atom is indicated using the prefixes:
      mono = 1   di = 2   tri = 3   tetra = 4   and penta = 5
      hexa = 6

5. Write the names of the following binary molecules.

   a. N_2O_5 **Answer is dinitrogen pentoxide**

   b. SF_6 **Answer is sulfur hexafluoride**

Note that in the first molecule, penta is used as "pent" when placed in front of an element whose name starts with a vowel. When used with an element whose name starts with a consonant as in hexafluoride the vowel ending in hexa remains. Also note that mono is *not* used on the first element written.

6. Write the formula for the following binary molecules.

   a. silicon tetrabromide   **Answer is $SiBr_4$**

   b. dinitrogen trichloride   **Answer is N_2Cl_3**

When writing formulas of binary molecules follow these basic rules:
  a. Write the symbols of the elements in the order given in the name.
  b. The subscript that goes on each symbol is the number that corresponds to the prefix in front of the element name.

Drawing dot or line structures representing molecules allows a scientist to give more structure to the molecule. A dot drawing allows you to see how the electrons are shared but can be time consuming to draw and can often get confusing when many atoms are present in a molecule. To simplify the drawing a single line is used to replace a pair of dots in the drawing so that: becomes –.

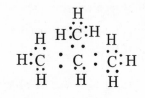

7. Convert into a line drawing.

Place a single line for every two dots that appear between two atoms. Note that in cases where dots are not shared you do not replace those with lines.

Answer is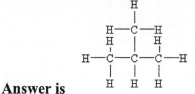

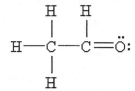

8. Convert into a dot drawing.

Place a pair of dots for every line that appears between two atoms. Note that in cases where dots are already written you simply write them down as is.

$$H : C : C :: Ö:$$

Answer is

Before they can be used as conversion factors, you must first be able to calculate formula weights (mass of ionic compounds) or molecular weights (mass of covalently bonded molecules). You may also see these terms referred to formula mass, molecular mass, and more recently, molar mass. No matter which of these quantities you are asked to calculate, the math works the same way as the following problems illustrate.

9. Calculate the formula weight of $Ca_3(PO_4)_2$.

Determine the number of each atom present, multiply the number of atoms by the atomic weight of that atom to find the weight contributed by that element. Add the total contributed weights of the elements to get the total weight, referred to as the formula weight.

Atom	number of atoms	atomic weight	total weight
Ca	3	40.1	120.3

44

P	2	31.0	62.0
O	8	16.0	<u>128.0</u>
		Formula weight =	**310.3**

10. Calculate the molecular weight of $C_6H_{11}O_6$

Atom	number of atoms	atomic weight	total weight
C	6	12.0	72.0
H	11	1.0	11.0
O	6	16.0	<u>96.0</u>
		Molecular weight =	**179.0**

The second factor to understand in order to do mass mole conversions is the relationship between the formula or molecular weight and the mole. Recall that the mole is a counting term equal to 6.02 x 10²³ particles (like atoms, formula units, or molecules). However this counting term also has a special relationship with the atomic, formula, and molecular weights. If you measure out the atomic, formula, or molecular weight of a substance in grams, you will have 6.02 x 10²³ atoms, formula units, or molecules of that substance. In other words, the amu value in grams is equal to one mole. This relationship gives scientists a very useful set of conversion factors that help them calculate how many moles (a number far too large to actually count out in the lab) of a compound or element they have by measuring its mass in the lab. Likewise if the scientist knows the number of moles required to make an experiment work, this relationship allows the scientist to calculate the mass needed in the lab.

11. How many moles of KCl have a mass of 10.5 g?

First determine your equivalency to be used in the conversion by calculating the formula weight of KCl.

Atom	number of atoms	atomic weight	total weight
K	1	39.0	39.0
Cl	1	35.5	<u>35.5</u>
		Formula weight =	74.5

This means that 1 mole KCl = 74.5 g. Now we can set up the problem and convert from grams to moles just as we have done conversions many times before.

$$10.5 \text{ g KCl} \quad \text{x} \quad \frac{1 \text{ mol}}{74.5 \text{ g}} \quad = \quad \textbf{0.141 mol KCl}$$

12. Determine the mass in grams of 2.33 mol $BaSO_4$.

First determine the equivalency to be used in the conversion by calculating the formula weight of $BaSO_4$

Atom	number of atoms	atomic weight	total weight
Ba	1	137.3	137.3
S	1	32.0	32.0
O	4	16.0	64.0
		Formula weight =	233.3

This means that 1 mole $BaSO_4$ = 233.3 g. Now we can set up the problem and convert from moles to grams.

$$2.33 \; \cancel{mol} \; BaSO_4 \quad x \quad \frac{233.3 \; g}{1 \; \cancel{mol}} \quad = \quad \textbf{544 g BaSO}_4$$

Remember this conversion skill is used often in chemistry. Be sure to practice it until you are very competent at doing it.

Chapter 3 Practice Test

Now that you've had the chance to study and review Chapter 3's practice problems provided in your textbook, you are ready to practice problems as you might see them in a testing situation. The questions and problems below are randomly selected in much the same way your instructor will select them; that is, not necessarily ordered in the same way as they appear in the textbook. To make best use of this practice, have your paper, pencil and calculator ready as you would for a class exam. Use a clock and give yourself 45 minutes to complete this activity. Note that taking this test under similar conditions as a class exam provides two benefits:

1. Testing your knowledge of the chapter's contents and your mastery of the required skills.
2. Practicing working under the stress of a timed exam.

Most instructors have a type of testing format that they prefer and will most likely tell you about it before the test. For practice, this self-test uses a variety of test question formats such as matching, fill-in-the-blank, listing, multiple-choice, and essay questions. *Be sure to have a periodic table!*

Note the time and begin.

1. How does an atom form an ion?

A. It gains electrons
B. It loses electrons
C. It gains or loses protons
D. It gains or loses electrons

2. When an atom has a positive charge it is called _____.

A. Anion
B. Cation
C. Dogion
D. Ion

3. What is the name of the Mg^{+2} ion?

A. Magmite ion
B. Manganese ion
C. Magnesium (II) ion
D. Magnesium ion

4. What is the name of the F^- ion?

A. Fluorine ion
B. Iron (I) ion
C. Fluoride ion
D. Fosphorus ion

5. What is the name of the Cu^+ ion?

A. Copper ion
B. Copper (II) ion
C. Cupric ion
D. Cuprous ion

6. What is the name of Na^{+1} ion?

A. Sodium
B. Sodium ion
C. Sodium (I) ion
D. Salt

7. How many valence electrons does F have?

A. 8
B. 7
C. 6
D. 5

8. How many valence electrons does Li^+ have?

A. 1
B. 0
C. 2
D. 8

9. How many valence electrons does Boron have?

A. 2
B. 7
C. 3
D. 0

10. How many total electrons does $^{16}_{8}O^{-2}$ have?

A. 8
B. 10
C. 12
D. 6

11. Cations have a _____ charge while anions have a _____ charge.

A. positive; negative
B. negative; positive
C. neutral; positive
D. positive; neutral

12. Monatomic anions' names are formed by changing the last syllable of the element name to _____.

A. –ic
B. –ous
C. –ide
D. none of the above

13. In H_3O^+ ion, how many protons are not matched by an electron charge?

A. 1
B. 2
C. 0
D. 3

14. If the sulfide ion has a -2 charge, how many electrons are present?

A. 16
B. 17
C. 15
D. 18

15. How many total electrons are present in the carbon atom?

A. 4
B. 6
C. 5
D. 7

16. The formula name for HCO_3^- is:

A. Carbonate ion
B. Hydrogen carbonate ion
C. Oxycarbonate ion
D. Bicarboxyl ion

17. Give the total number of electrons and valence electrons of arsenic.

A. 5 ; 33
B. 75 ; 5
C. 7; 75
D. 33; 5

18. Give the total number of electrons and valence electrons of tin.

A. 50; 4
B. 119; 4
C. 4; 50
D. 54; 50

19. According to the octet rule, atoms _____ to end up with 8 valence electrons.

A. gain electrons
B. lose electrons
C. share electrons
D. all of the above

20. Nonmetals _____ electrons while metals _____ electrons.

A. Lose; gain
B. Gain; lose
C. Share; gain
D. Gain; share

21. Two or more elements that have interacting opposite charges are called:

A. covalent compounds
B. ionic compounds
C. binary compounds
D. monatomic compounds

22. What is the formula name for the compound Na_2SO_3?

A. Disodium sulfate
B. Sodium sulfate
C. Sodium trisulfate
D. Sodium sulfite

23. The combination of lithium and bromide forms:
A. lithium lromide
B. lithium bromine
C. bromine lithide
D. lithius bromus

24. When two elements share electrons it is called a (an):

A. ionic bond
B. sharing bond
C. covalent bond
D. none of the above

25. An uncharged group of atoms connected to one another by covalent bonds is called a (an):

A. cation
B. anion
C. ion
D. molecule

26. How many bonds can one carbon atom form?

A. 1
B. 2
C. 3
D. 4

27. How many different elements are contained in a binary molecule?

A. 1
B. 2
C. 3
D. 4

28. What is the prefix for four?

A. Tetra
B. Penta
C. Di
D. Quartenary

29. What is the name of the compound SO_2?

A. Sulfur oxide
B. Sulfur dioxide
C. Disulfur oxide
D. Disulfur dioxide

30. What is the name of the substance N_2O_5?

A. Nitrogen Pentoxide
B. Dinitrogen oxide
C. Dinitrogen Pentoxide
D. Nitrogen oxide

31. What is the molecular weight of the compound N_2O_5?

A. 230
B. 35.5
C. 108
D. 158.5

32. What is the formula weight of copper II nitrate $Cu(NO_3)_2$?

A. 63.5
B. 187.5
C. 75.6
D. 184.9

33. How many moles are contained in 25.0 g of copper nitrate?

34. What would be the mass in grams of 3.5 mol of N_2O_5?

35. Draw the correct dot structure of sulfur.

36. Draw the line-bond drawing for the molecule below.

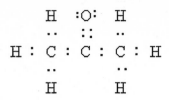

37. Draw the dot structure for the line-bond drawing below.

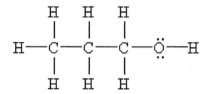

38. Draw the correct electron dot structure of arsenic and of the ion it is expected to form.

39. Draw the correct electron dot structure of bromine and of the ion it is expected to form.

40. How many atoms would be found in 12.0 g of carbon?

PRACTICE TEST Answers

1. D
2. B
3. D
4. C
5. D
6. B
7. B
8. B
9. C
10. B
11. A
12. C
13. A
14. D
15. B
16. B
17. D
18. A
19. D
20. B
21. B
22. D
23. A
24. C
25. D
26. D
27. B
28. A
29. B
30. C
31. 108 amu
32. 187.5 amu
33. 0.133 mol of $Cu(NO_3)_2$
34. 378 grams of N_2O_5

35.

36.

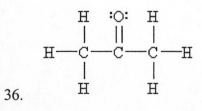

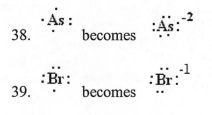

37.

38. $\cdot \overset{\displaystyle \cdot}{\underset{\displaystyle \cdot}{As}} :$ becomes $: \overset{\displaystyle \cdot \cdot}{\underset{\displaystyle \cdot \cdot}{As}} :^{-2}$

39. $: \overset{\displaystyle \cdot \cdot}{\underset{\displaystyle \cdot}{Br}} :$ becomes $: \overset{\displaystyle \cdot \cdot}{\underset{\displaystyle \cdot \cdot}{Br}} :^{-1}$

40. 6.02×10^{23} carbon atoms

ANSWER PAGE FOR BUILDING KNOWLEDGE

Fill-in the term or terms needed to complete the statement.

1. **Octet** is the stable arrangement of eight valence electrons.

2. **Bohr model** describes electrons as circling the nucleus in fixed orbits.

3. **Quantum mechanical model** describes electrons as in orbitals that surround the nucleus in energy levels with a maximum of electrons allowed per energy level.

4. **Orbitals** are energy zones where electrons are believed to reside; each holds two electrons.

5. **Valence electrons** are in the highest numbered, occupied energy level of an atom.

6. **Cations** are ions with a positive charge that are formed when an atom loses electrons.

7. **Electron dot structure** uses dots placed around the symbol of an element to represent the valence electrons of that element.

8. **Anions** are negatively charged ions that are formed when an atom gains electrons.

9. **Ions** are atoms or groups of atoms with an unequal number of protons and electrons caused by the atom or atoms either gaining or losing electrons.

10. **Monatomic** means formed from a single atom.

11. **Molecular weight** is the sum of the weights of the elements in the formula of a molecule.

12. **Molecules** are an uncharged group of atoms joined by covalent bonds.

13. **Formula weight** is the sum of the weights of the elements in the formula of an ionic compound.

14. **Ionic bond** is the attraction between ions of opposite charges.

15. **Covalent bond** is the sharing of electrons as a way of achieving eight outer electrons; typically occurs between nonmetals.

16. **Ionic compounds** are compounds formed with ionic bonds and named by combining the name of the cation with the name of the anion.

17. **Binary compounds** are compounds that contain just two different elements.

Chapter 4
An Introduction to Organic Compounds

Chapter Overview

Now that you have learned to manipulate elements and to discover what is required to predict the product of a chemical reaction (Chapter 3), it is time to investigate a large group of molecules that cannot be so easily predicted. You were introduced to a small sample of this group of compounds in Chapter 3 when learning about dot and line structures. In Chapter 4 you will investigate organic compounds. These molecules have the element carbon as the primary building block. Carbon has two characteristics that make it a very good building block for molecules:

1. Its ability to form four covalent bonds.
2. The ability of one carbon atom to form covalent bonds with other carbon atoms as well as being able to bond with other nonmetals.

These properties allow carbon to form an infinite number of molecules with different numbers of atoms and/or different structures. In Chapter 4 you will first learn how to recognize organic molecular structures and how to draw them. The next part of the chapter clarifies how a chemist can interpret these molecular shapes for predicting properties and classifying the molecules into families. Just like family siblings, the chemist identifies characteristics that make the molecules similar and characteristics that make them different. Within the organic family you will discover how differences in structure account for isomers, conformations, and stereoisomers (also known as geometric isomers). Finally you will learn how to recognize each family group by its one dominant feature, called a functional group.

Knowledge/Comprehension

Typically, there are new terms or phrases that you must memorize in order for the lesson to make sense to you. Look up this information and memorize their explanations and definitions:

condensed structural formula
skeletal structure
conformations
noncovalent interactions
cis
dipole-dipole interaction
alkanes constitutional isomers
functional groups
tetrahedral, pyramidal, trigonal planar, bent, and linear

coordinate-covalent interaction
electronegativity
polar covalent bonds
hydrogen bond
trans
London forces
geometric isomers

alkynes
aromatic
stereoisomers
salt bridge
ion-dipole
hydrocarbons
alkenes

- Describe how condensed structural formulas and skeletal structures differ from the electron dot and line structures introduced in Chapter 3.
- Explain the relationship between electronegativity and polar covalent bonds.
- Explain how shape plays a role in determining overall polarity.
- Describe the noncovalent interactions that attract one compound to another.
- Describe the four families of hydrocarbons.
- Explain the differences between constitutional isomers, conformations, and the stereoisomers known as geometric isomers.
- Describe the features that distinguish hydrocarbons, alcohols, carboxylic acids, and esters from one another.

The following activity is designed to help you master the objectives outlined in the beginning of Chapter 4 and then summarized at the end of the chapter. To make the best use of your study time, *read* Chapter 4 before continuing. Next, work the matching activity provided as though you are taking a test *without first* referring to the answer page. Put your answers on a separate sheet of paper so that you can make more than one attempt without seeing your previous answers. Once you have completed the activity, check your answers. Make *flash cards* for those you missed. Have a study buddy drill you on them if you missed more than half of the terms. Once you have reviewed the ones you missed for at least 15 minutes try the activity again. If you miss fewer than four, write in the answers so that you will have them to review just before taking the test.

Match the terms or phrases to the correct meaning

_____1. Condensed structural formula

_____2. Alkanes

_____3. Skeletal structure

_____4. Electronegativity

_____5. Tetrahedral, pyramidal, trigonal planar, bent, and linear

_____6. Polar covalent bonds

_____7. Functional groups

_____8. Noncovalent interactions

_____9. Hydrogen bond

_____10. Salt bridge

_____11. Alkenes

_____12. Dipole-dipole interaction

a. are hydrocarbons that contain only single covalent bonds

b. are hydrocarbons that contain at least one double covalent bond

c. involves the attraction between a cation and the nonbonding electrons of a nonmetal

d. compounds consist of rings of carbon atoms joined by alternating double and single covalent bonds

e. are molecules that have the same molecular formula but different atomic connections

f. occur when there is unequal sharing of electrons between atoms

g. is a strong dipole attraction found only when hydrogen is bonded to nitrogen, oxygen, or fluorine

h. describes the attachment of atoms to one another

i. used to describe an ionic bond

_____13. *Cis*

_____14. *Trans*

_____15. Ion-dipole

_____16.Coordinate-covalent interaction

_____17. Aromatic

_____18. Alkynes

_____19. London forces

_____20. Hydrocarbons

_____21. Conformations

_____22. Geometric isomers

_____23. Stereoisomers

_____24. Constitutional isomers

_____25. Formal charge

j. are the different three-dimensional shapes that a molecule can assume through rotation of single bonds

k. are molecules that have the same molecular formula and the same atomic connections but different three-dimensional shapes that can be interchanged only by bond breaking.

l. is the ability of an atom to attract electrons in a covalent bond.

m. is an attraction of an ion for a polar group

n. is when the substituents are on opposite sides of a cycloalkane or alkenes

o. are forces that attract one compound to another

p. are stereoisomers that result due to limited rotation about covalent bonds

q. arise from the temporary dipoles that exist in nonpolar compounds

r. represents covalent bonds with lines; carbon atoms are not shown and hydrogen atoms are only shown when attached to an element other than carbon

s. is the attraction between opposite partial charges

t. is when the substituents are on the same side of a cycloalkane or alkenes

u. are molecules that contain only carbon and hydrogen atoms.

v. are the basic shapes for molecules

w. are atoms, groups of atoms, or bonds that give a particular set of chemical properties

x. are hydrocarbons that contain at least one triple covalent bond

y. determined by comparing the number of electrons that surround an atom in a compound to the number of valence electrons that it carries as a neutral atom

Once you know the appropriate tools for Chapter 4, you are ready to evaluate your understanding of those tools and their relationships. In this type of concept building, it is necessary to compare or contrast terms or concepts. In some cases you may be required

to give examples or recognize a term or concept when it is depicted in a picture or sample scenario.

Sample Problems

As you saw in Chapter 3, a chemist draws molecular structures dependent upon what type of information the chemist is trying to convey. Dot structures are used to show how electrons are arranged around an atom and line structures provide the same basic structural information without keeping track of so many dots. You probably noticed very quickly that if you had to draw very many of either of these two types of structures it would become tiresome. When chemists are trying to convey just the essential structural differences, they will sometimes "condense" the structures by showing only the position of the carbons and other non-hydrogen elements, without showing hydrogen bonded to the carbon atom(s). This is referred to as a condensed structural formula; the following structure is the condensed structural formula of 2-butanol:

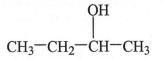

The International Union of Pure and Applied Chemist (IUPAC) developed the naming system used in this chapter. The naming system depends on the number and position of the carbons and other constituents attached to the carbons, like the –OH above, and not the number or position of the hydrogens on the carbon. Therefore, it is sometimes more convenient to represent the structure with just the core or basic skeletal structure of the molecule. In the structure below lines are used to represent the bonds between the carbons; the end of each line represents a carbon position and only the substituent groups are drawn, as the 2-butanol depicted above is seen below:

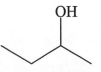

Notice that without all the element symbols cluttering up the drawing, more details about the shape of the molecule can be given.

1. From the dot structure given below draw the condensed and skeletal structures.

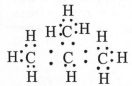 For the condensed structure, simply collapse the hydrogens in next to the carbon they are attached to but leave the carbons as they are.

CH₃
|
CH₃—CH—CH₃   Then for the skeletal structure, the hydrogens are not shown
at all. Think of each carbon as points you will connect with lines. Also in this structure,
the fact that carbons actually bond at angles to each other can be shown.

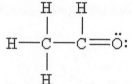

   *Note how the center carbon still has three carbons radiating out but they
are more evenly distributed around the atom as they would be in a real molecule.*

2. From the line structure given below, draw the condensed and skeletal structures.

   Condensing the line-structure is no different than condensing
the electron dot structure. Again, collapse only the hydrogens into the attached carbon
and leave the structure as is for carbon and oxygen. The oxygen is moved up so that it
does not appear to be bonded to the collapsed hydrogen.

  Then for the skeletal structure remove the hydrogens and rotate the
drawing slightly so that the angle between is more distinctive.

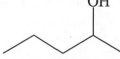

   *Note that this drawing only shows the bare essentials of the
structure which is why they are called **skeletal structures**.*

3. From the skeletal structure given below, draw the condensed structure.

OH

  First, go through and replace each point with a C for carbon
(Remember the condensed structure is drawn without angles.).

$$\text{OH}$$

(structure) becomes

$$\overset{\text{OH}}{\underset{|}{C-C-C-C-C}}$$

. Then count the number of bonds on each carbon and place in front of it the number of hydrogen atoms needed to give each carbon a total of four bonds; this means that the two carbons on the end need three hydrogens each. The two carbons in the middle without the oxygen each need two hydrogens; the carbon with the oxygen only needs one hydrogen atom in order for it to have four bonds. When the hydrogens are put in place you form the condensed drawing below.

$$\overset{\text{OH}}{\underset{|}{CH_3-CH_2-CH_2-CH-CH_3}}$$

Recall that electronegativity is the ability of an atom to attract bonding electrons. Electronegativity tends to increase as you move to the right on a period or move up a group of elements. If there is an equal sharing of electrons, that is, a purely covalent bond, the electronegativity values must be the same. Since these typically only happen between atoms of the same element, most bonds have unequal sharing. This causes the electron to spend more time on one side of the molecule than on the other, resulting in a slight negative charge on one side and a slight positive charge on the other. This type of bond is called a polar bond with two distinct ends.

4. Which of the following bonds would *not* exhibit polar bonding?

a. B—C   b. O—F   c. Br—Br   d. H—O

The only way to form a non-polar bond is to combine the atoms of the same element. **The answer is c.**

5. Which of the following would exhibit a polar bond?

a. C—C   b. F—F   c. Br—Br   d. H—O

A polar bond only occurs when the electronegativity is different. The only way for that to be true is when two different atoms are bonding. **Answer is d.**

Although unequal sharing (indicated by an arrow pointed to the more negative side of the bond), as in C→O, results in a polar bond, this does not mean that a molecule with this type of bond must be polar. If the same type of bond is placed on the opposite side, O←C→O, this makes the molecule negative on both sides and therefore the molecule does not have two distinct ends and is not polar.

If the structural drawing of the molecule is not symmetrical (same atoms or group of atoms in every bond position around a central atom) then the molecule will be polar.

6. Explain why the following molecule is *not* polar.

$$CH_3-CH_2-CH_2-O-CH_2-CH_2-CH_3$$

The molecule would not be polar because it has the same number of carbons and hydrogens on both sides of the oxygen atom.

7. Which of the following molecules is a polar molecule?

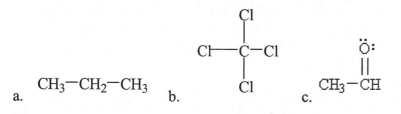

a. $CH_3-CH_2-CH_3$ b.   c. CH_3-CH

In choice a, the number of carbons and hydrogens are the same on both sides of the center carbon, making this molecule non-polar. In choice b, chlorine is bonded in every possible position around the carbon atom making this molecule non-polar. In choice c, there is a CH_3 on one side of the center carbon and a $=O$ on the other side so that this molecule *is not symmetrical*. **Answer is c.**

Learn the definition of each type of noncovalent interaction above. For this objective, you should be able to look at the structures of the molecules (or ions) that are interacting and be able to describe the interaction taking place. The interactions can be of one type such as dipole-dipole if both molecules are polar, or may be a combination of interactions. For example, a dipole is formed if one substance is polar covalent but the other is an ion.

8. For each of the substance combinations below name the type of noncovalent interaction and describe why that interaction exists.

a. Na^+ and H—Cl

The Na^+ is an ion and the H—Cl is dipole but does not have hydrogen bonding.

Interaction is ion-dipole.

b. H—Cl and H—Cl

The H—Cl bond is polar which gives the molecule a dipole. The interaction is dipole-dipole because both interacting substances are dipole.

Interaction is dipole-dipole.

c. S=C=S and O=C=O

Both molecules are symmetrical and therefore do not have a dipole and only exhibit London forces.

Interaction is London-London.

9. Draw two molecules that would have a dipole-hydrogen bond interaction.

Your choice of molecules can vary for this type of question but remember the substances *cannot* be symmetrical, have only one type of atom, or have ions. One of the substances must be polar but *cannot* have a hydrogen atom that is bonded to an oxygen, nitrogen, or fluorine atom. The other substance must be polar and *must* have at least one hydrogen bonded to an oxygen, nitrogen, or fluorine atom.

Example answer: 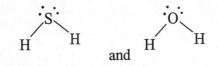 and

The four families of hydrocarbons are classified by their bond types. The alkane family has only single bonds. The alkene family has at least one double bond. The alkyne family has at least one triple bond. The aromatics have alternating double and single bonds within a cyclic structure.

10. Which of the following is an aromatic hydrocarbon?

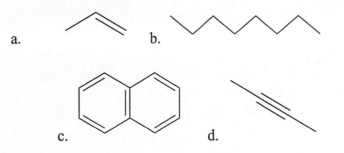

Look for the compound that has alternating double and single bonds. In this case, only choice c has both single and double bonds. **Answer is c.**

11. Which of the following is an alkane?

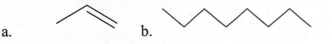

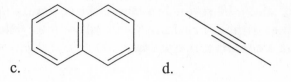

c.   d.

Alkanes have single bonds only. Choice b is the only molecule that does not have double or single bonds. **Answer is b.**

*It is important that you understand the concept of isomer. Remember that carbons can bond with each other in up to four single bonds. This means that as the number of carbons increase the possible combinations creating different structures increase as well. **Isomers** are molecules that have the same molecular formulas (same number of carbons, hydrogens and substituents) but have different structures. A quick review of how these three types of isomer classifications can be identified follows:*

Constitutional	*Conformations*	*Geometric*
Atoms connected in a different pattern	*Atoms connected in same pattern*	*Atoms connected in same pattern*
	Different 3D shape	*Different 3D shape*
	Interchanged by rotating around a single bond	*Interchanged by rotating by bond breaking*

Note: in order to determine if molecules with the same molecular formulas are conformational or geometric isomers, some type of three-dimensional drawing must be given.

12. Label each molecular pair given below as constitutional, conformation, geometric or not-isomers. Explain your answer.

a.

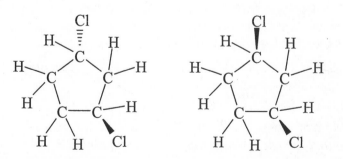

These are stereoisomers (geometric). They are isomers since they have the number of each type of atom (same molecular formula). They would not be constitutional since the Cl atoms are located in the same place. Due to the cyclic nature of the molecule, the chlorines of the molecule on the left (one protrudes out of the plane

and one going into the plane) cannot be rotated to match the structure on the right (both protrude out of the plane) without breaking the bonds of the cyclic structure. This is a geometric isomer not a conformation isomer.

b.

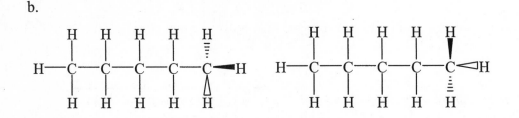

Answer: These are conformation isomers. They are isomers since they have the same number of each type of atom (same molecular formula). They are not constitutional isomers since the atoms are in the same place. The three-dimensional structure shows the H atom on the end as protruding out of the plane in the molecule on the left and on the plane in the molecule on the right. Since the second drawing could be made to match (interchange) with the first by rotating the carbon around the single bond, this is a conformation isomer.

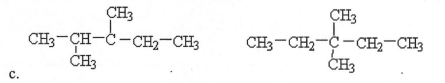

c.

Answer: These are constitutional isomers. They are isomers since they have the same number of each type of atom (same molecular formula). Conformational and geometric isomers are not possible choices since by definition the atoms have to be in the same place, which is not the case in either of these two isomers. The molecule on the left has methyl groups in positions 2 and 3 while the molecule on the right has them both in position 3.

You should have learned how to distinguish the four hydrocarbon families above. The other three family groups are recognized by the characteristic atoms attached as substituents, changing the way the molecule functions. These families are referred to as functional groups. For the alcohol family, the group is the –OH . For the carboxylic

$$\begin{array}{c} \quad O \\ \quad \parallel \\ -C-O-H \end{array}$$

acids family, the functional group is . *In the ester family the functional group is similar to that in carboxylic acids except that the –H is replaced by another hydrocarbon group. Since that hydrocarbon group can vary in number of carbons, this part of the functional group is sometimes symbolized as –R so that the ester*

$$\begin{array}{c} \quad O \\ \quad \parallel \\ -C-O-R \end{array}$$

functional group is .

13. Which of the following molecules is a carboxylic acid?

a. $CH_3-CH_2-CH_2-OH$ b. $CH_3-CH=CH_2$ c. $CH_3-CH_2-CH_2-\overset{\overset{\textstyle O}{\|}}{C}-OH$

Look for the functional group. Choice a has –OH, making it an alcohol.

Choice b has a double bond and is therefore an alkene. Choice c has $-\overset{\overset{\textstyle O}{\|}}{C}-O-H$ which is the functional group for a carboxylic acid. **Answer is c.**

14. Which of the following molecules is an ester?

a. (ring structure) $HC\overset{CH=CH}{\underset{CH-CH}{\diagdown\diagup}}C-\overset{\overset{\textstyle }{\|}}{C}-OH$ b. CH_3-CH_2-OH c. $CH_3-CH_2-CH_2-\overset{\overset{\textstyle O}{\|}}{C}-O-CH_3$

Look at the functional groups; in choice a the functional group is

$-\overset{\overset{\textstyle O}{\|}}{C}-O-H$

making it a carboxylic acid. In choice b the functional group is –

OH, making it an alcohol. In choice c the functional group is $-\overset{\overset{\textstyle O}{\|}}{C}-O-CH_3$ with the CH_3 serving as the –R group, making this molecule an ester. **Answer is c.**

Application

To apply what you've already learned and master the concepts you must utilize all of the knowledge, understanding, facts, and/or techniques that you have gathered up to this point. Simply knowing an equation or process is not enough to master the concept. These you develop by *practice*. Solve as many different types of problems with as many different variations as you *strive to master the skill*. Your textbook provides a very good assortment of problems at the end of the chapter. Work as many as needed for skill mastery, not just those assigned.

- Predict whether or not a covalent bond is polar.
- Use the rules to predict the five basic shapes about an atom in a molecule.
- Give examples of two different families of hydrocarbons that can exist as geometric isomers.
- Calculate the formal charge on a molecule or polyatomic ion.

In order to truly master any skill you must *practice*. In this section example problems related to the application objectives are discussed. Remember, the more you practice, the better your chances of being able to successfully use these objectives to solve problems. Additional practice problems are at the end of the chapter in your textbook and in the online interactive lessons.

Sample Problems

If there are no unshared pairs of electrons on the central atom, the atoms assume an equidistant position around the central atom with the following shapes; 2 atoms only (no central atom) are linear, as are 2 atoms on a central atom with no unshared pairs of electrons, 3 atoms around the central atom forms a trigonal planar and 4 atoms around the central atom forms a tetrahedral. As atoms are replaced by unshared pairs of electrons, the remaining atoms are pushed downward. When one atom is replaced by an unshared pair of electrons in a tetrahedral, the remaining shape is pyramidal. When a second atom is replaced, all that remains is an angle (called bent shape). Replacing one atom with an unshared pair of electrons in a trigonal planar results in a bent shape (see Table 4.2 in the textbook).

1. Give the shape that would be predicted for the following molecule.

Note first that there are four positions which would make it a tetrahedral if it had four surrounding atoms. However, one atom has been replaced by a pair of unshared electrons. **Answer is pyramidal**.

2. Give the shape that would be predicted for the following molecule.

$$
\begin{array}{c}
\text{H} \\
| \\
\text{H}-\text{C}-\text{H} \\
| \\
\text{H}
\end{array}
$$

Note first that there are four positions which would make it a tetrahedral if it had four surrounding atoms. Since it has does have four atoms and no unshared pairs of electrons, the predicted shape is tetrahedral. **Answer is tetrahedral.**

3. Give the shape that would be predicted for the following molecule.

Note first that there are four positions which would make it a tetrahedral if it had four surrounding atoms. Two atoms have been replaced by a pair of unshared electrons. **Answer is bent shape.**

In order to provide examples of different families of hydrocarbons that exist as geometric isomers, you must first understand geometric isomers (see above review). You must also know which families of hydrocarbons can meet the requirements for having geometric isomers (see figures 4.16-4.19 in your textbook).

4. 1-pentene has a double bond structure but does not have geometric *cis* and *trans* isomers, like 2-pentene. What accounts for this difference?

Sample Answer: In order for alkenes to have geometric isomers, the carbons on either side of the double bond cannot have the same atoms attached to them. In the case of the 1- pentene, $CH_3-CH_2-CH_2-CH=CH_2$ **, the carbon on the right has two hydrogens attached to it, which makes its structure look the same no matter which hydrogen is facing up or down. In the case of 2-pentene,**

$CH_3-CH_2-CH=CH-CH_3$ **the carbon on the left side of the double bond has one carbon group,** CH_3-CH_2 **and one hydrogen atom attached to it, while the carbon on the right has a different group, —CH_3, and one hydrogen atom attached. Therefore drawing the two carbon structures on the same side (*cis*) of the double bond is not the same structure as drawing them on opposite sides (*trans*) of the double bond. This allows 2-pentene to have geometric isomers.**

5. Which of the following hydrocarbon families *never* have geometric isomers?

      a. cycloalkanes

      b. alkenes

      c. alkanes

      d. aromatics

      e. alkynes

First think about the basic structure of each family. In order to have geometric isomers it must be possible to create identical molecules, except for certain atoms that are placed with different orientation to each other. The only way to rearrange these atoms so that identical molecules can be created is through some type of bond breaking process. Since alkanes have only single bonds which can rotate freely, they do not have a special orientation that is "locked in" by bonds. *Don't forget the criteria!* There must be atoms attached on opposite sides of the bond that orient in different directions in order for geometric isomers to be possible. In the case of alkynes, the triple bond eliminates two additional atoms from the carbons in the bond which only allows for one type of atom

group on each side. Therefore, alkynes also do not form geometric isomers either.
Answer: c and e.

A systematic method for naming and drawing structures was developed by a group of scientists in an organization called the International Union of Pure and Applied Chemistry (IUPAC). It was necessary to have a systematic way of naming hydrocarbons since there are so many different types of molecules all made from the same two elements, carbon and hydrogen. Although hydrogen atoms fulfill the electron structure of the carbons, they do not dictate the structure of the hydrocarbons since they have only one bond position. The carbon atom's four bonds and its ability to bond to itself determine the structure of the molecules. Therefore, the naming system is designed to describe the number of carbons present in a molecule and their relationship to each other in the molecule. Be sure that you know the prefixes shown in Table 4.5 of you textbook, as well as the alkyl names and structures in Table 4.6.

6. Name the following hydrocarbon.

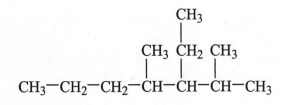

Check the structure to see how many contiguous carbons can be counted without having to pick up your pencil or backtrack. Some of the possibilities are shown below.

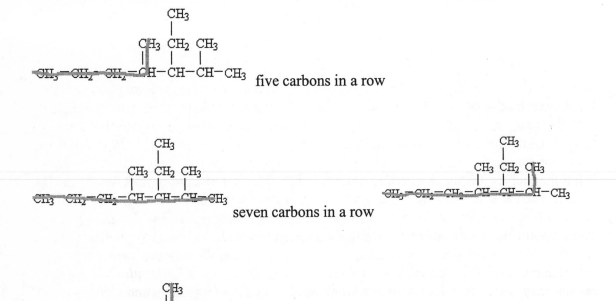

Since seven contiguous carbons is the longest chain, use the simplest of the possibilities (the one that includes the smallest amount of branching) to continue the naming process. This longest chain is the parent or family name for the structure, which in this case is heptane.

Next, identify the substituent groups.

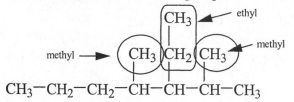

The substituent names are placed in front of the parent chain name in alphabetic order: ethyl methyl methyl heptane. Notice, however, that the methyl name is repetitious (just think what it would look like for a compound with five methyl side chains). To eliminate this, use those counting prefixes (mon, di, tri, etc.) to write ethyl dimethylheptane.

Now there is one last detail!! Remember that in hydrocarbons, atoms in different locations make them different molecules. Therefore, as you write the hydrocarbon's name, it is critical that the next person reading the name is able to draw the substituents in the correct location(s). To ensure this, the carbons of the parent chain are assigned numbers. Since the molecule is three dimensional, it can be counted in either direction. Recall that the rule is to make the number location of the substituents as small as possible.

$$CH_3$$
$$|$$
$$CH_3 \quad CH_2 \quad CH_3$$
$$|4 \quad |3 \quad |2 \quad 1$$
$$\overset{7}{CH_3}-\overset{6}{CH_2}-\overset{5}{CH_2}-\overset{4}{CH}-\overset{3}{CH}-\overset{2}{CH}-\overset{1}{CH_3}$$
$$1 \qquad 2 \qquad 3 \qquad 4 \quad 5 \quad 6 \quad 7$$

Numbering the carbons from left to right, the first methyl substituent is located on carbon 4, followed by the ethyl substituent on carbon 5 and the second methyl on carbon 6. If the carbons are numbered from right to left, the first methyl substituent is located on carbon 2, the ethyl substituent is on carbon 3 and the second methyl is on carbon 4. Note that a 2, 3, carbon number sequence is smaller than a 4, 5, 6 carbon number sequence, so the right to left numbering is used. The numbers are inserted into the name using commas to separate numbers and hyphens are used to separate the numbers from the letters.

The name of this molecule is 3-ethyl-2,4-dimethylheptane.

Remember that alkenes and alkynes are named following the same basic rules as outlined above, except that the multiple bond must be located in the parent chain and a number position must be assigned to the location of that bond, for example, 2-butene. The numbering of the parent chain is always done from the end closest to the double bond.

71

Chapter 4 Practice Test

Now that you've had the chance to study and review Chapter 4's practice problems provided in your textbook, you are ready to practice problems as you might see them in a testing situation. The questions and problems below are randomly selected in much the same way your instructor will select them; that is, not necessarily ordered in the same way as they appear in the textbook. To make best use of this practice, have your paper, pencil and calculator ready as you would for a class exam. Use a clock and give yourself 45 minutes to complete this activity. Note that taking this test under similar conditions as a class exam provides two benefits:

1. Testing your knowledge of the chapter's contents and your mastery of the required skills.
2. Practicing working under the stress of a timed exam.

Most instructors have a type of testing format that they prefer and will most likely tell you about it before the test. For practice, this self-test uses a variety of test question formats such as matching, fill-in-the-blank, listing, multiple-choice, and essay questions. *Be sure to have a periodic table!*

Note the time and begin.

1. Which is the correct line bond structure for methane?

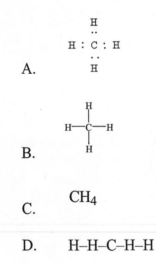

A.

B.

C. CH_4

D.   H–H–C–H–H

2. Which is the correct line bond structure for nonane?

A. $CH_3-(CH_2)_7-CH_3$

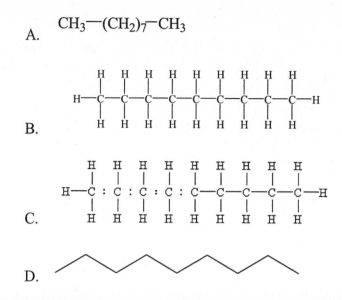

B.

C.

D.

3. What is the shape of a water molecule?

   A. Tetrahedral
   B. Pyramid
   C. Linear
   D. Bent

4. What is the formal charge on nitrogen in nitrite?

   A. 3
   B. 2
   C. 4
   D. 5

5. In calcium carbonate, what is the formal charge on the carbon atom?

   A. 3
   B. 2
   C. 0
   D. 1
   E. none of the above

6. What is the shape of the ammonium ion?

   A. Tetrahedral
   B. Pyramid
   C. Linear
   D. Bent

7. Do hydrogen bonds from between formaldehyde molecules?

    A. Yes
    B. No
    C. Depends on environmental conditions
    D. If we make them

8. What type of noncovalent interactions would be expected to occur between two methanol molecules?

    A. hydrogen bonding
    B. ionic
    C. London forces
    D. dipole-dipole

9. In the following sets of molecules, which two share stronger London force interactions? Two $CH_3CH_2CH_2CH_2CH_3$ or two $CH_3CH(CH_3)CH_2CH_3$?

    A. two $CH_3CH_2CH_2CH_2CH_3$
    B. two $CH_3CH(CH_3)CH_2CH_3$
    C. They are the same
    D. Neither have London force interactions

10. Which would be predicted to have a higher boiling point, pentane or octane?

    A. pentane
    B. octane
    C. they are the same
    D. neither will boil

11. From lowest boiling point to highest boiling point which is the correct order for decane, propane, butane?

    A. Decane, propane, butane
    B. Propane, decane, butane
    C. Propane, butane, decane
    D. Decane, propane, butane
    E. Butane, decane, propane

12. Which is the correct skeletal structure of the alkane CH₃CH₂C(CH₃)₃?

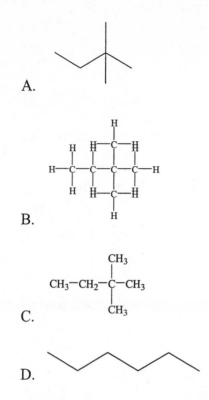

A.

B.

C.

D.

13. Draw the correct line bond structure of the alkane
    CH₃C(CH₂CH₃)₂CH(CH₂CH₃)CH₃

14. Provide the complete IUPAC name for the following compound.

_____

$$CH_3-CH_2-CH_2-\overset{\overset{\displaystyle CH_3}{|}}{\underset{\underset{\displaystyle CH_3}{|}}{C}}-CH_3$$

15. Provide the complete IUPAC name for the following compound:

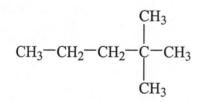

75

16. Based on its functional group, the following compound belongs to the _____ family.

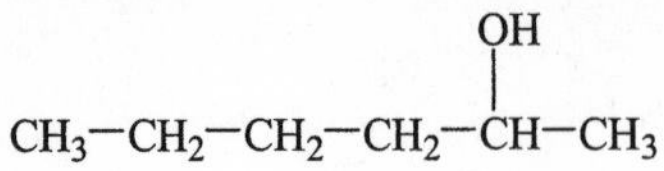

$$CH_3-CH_2-CH_2-CH_2-\overset{\overset{\displaystyle OH}{|}}{CH}-CH_3$$

    A. aldehyde
    B. carboxylic acid
    C. alcohol
    D. ester

17. Based on its functional group, the following compound belongs to the

_____ family.

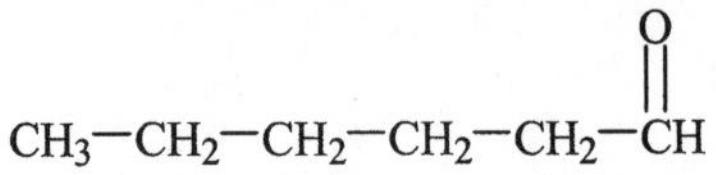

$$CH_3-CH_2-CH_2-CH_2-CH_2-\overset{\overset{\displaystyle O}{\|}}{CH}$$

    A. aldehyde
    B. carboxylic acid
    C. alcohol
    D. ester

18. Which of the following molecules has the longest parent chain?

A. 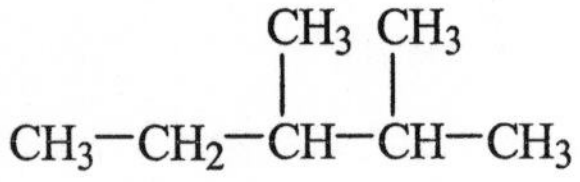
$$CH_3-CH_2-\overset{\overset{\displaystyle CH_3}{|}}{CH}-\overset{\overset{\displaystyle CH_3}{|}}{CH}-CH_3$$

B. 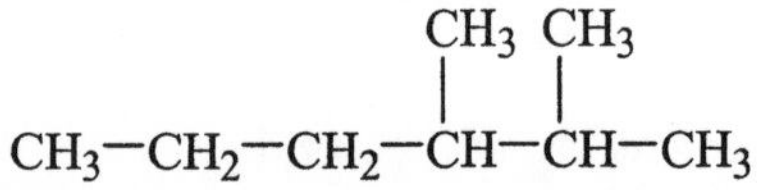
$$CH_3-CH_2-CH_2-\overset{\overset{\displaystyle CH_3}{|}}{CH}-\overset{\overset{\displaystyle CH_3}{|}}{CH}-CH_3$$

C. 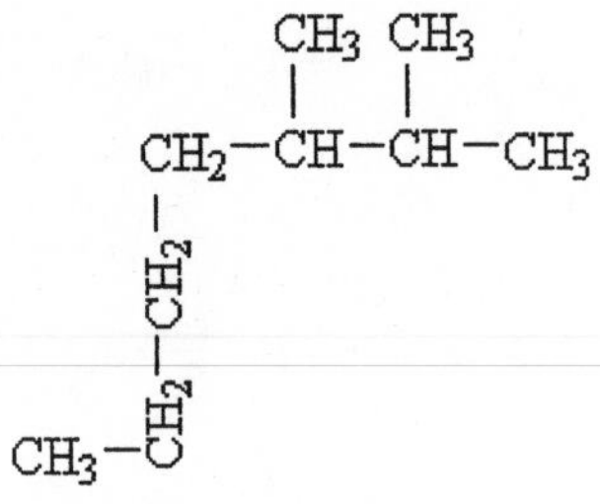
$$CH_3-CH_2-CH_2-\overset{\overset{\displaystyle CH_3}{|}}{CH_2}-\overset{\overset{\displaystyle CH_3}{|}}{CH}-\overset{\overset{\displaystyle CH_3}{|}}{CH}-CH_3$$

D. 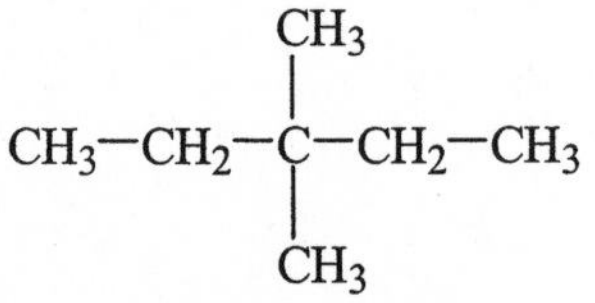
$$CH_3-CH_2-\overset{\overset{\displaystyle CH_3}{|}}{\underset{\underset{\displaystyle CH_3}{|}}{C}}-CH_2-CH_3$$

19. What is the name of the substituents in the following compound?

_____ and _____

20. Which of the hydrocarbons below are constitutional isomers?

    A. pentane and 2-methylpentane
    B. 2-methylpentane and 3-methylpentane
    C. 2,2-dimethylbutane and pentane
    D. 2,2-dimethylpropane and cyclopentane

21. Which pairs of molecules are geometric isomers?

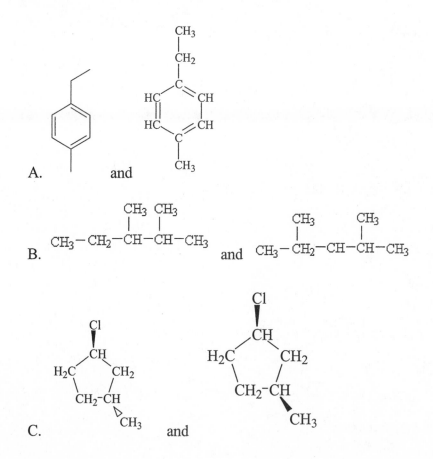

22. Constitutional isomers have the same IUPAC name.

    A. True
    B. False

23. The different conformations of an alkane have the same IUPAC name.

    A. True
    B. False

24. Which of the following molecules can have *cis* and *trans* isomers?

    A. alkanes
    B. alkenes
    C. alkynes
    D. both B and C

25. Only cyclic compounds can have geometric isomers.

    A. True
    B. False

26. In the following structure, which bonds are polar covalent bonds? Explain.

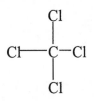

27. Which of the following molecules are polar?

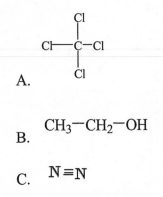

    A.

    CH_3-CH_2-OH
    B.

    $N \equiv N$
    C.

28. Draw 1,3-diethylcyclohexane.

29. Draw 3-propyl-2,5-dimethyloctane.

30. Draw o-dimethylbenzene.

PRACTICE TEST Answers

1. B
2. B
3. D
4. B
5. C
6. B
7. B
8. A
9. A
10. B
11. C
12. A

13.

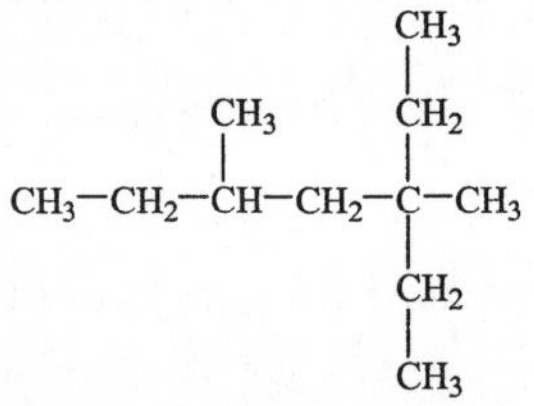

14. 2,2-dimethylpentane
15. 3-ethyl-4-methyl-2-heptene
16. C
17. A
18. C
19. ethyl and methyl
20. B
21. C
22. B
23. A
24. B
25. B
26. All of the bonds are polar since Cl and C do not have the same ability to attract electrons in a bond.
27. B

28. 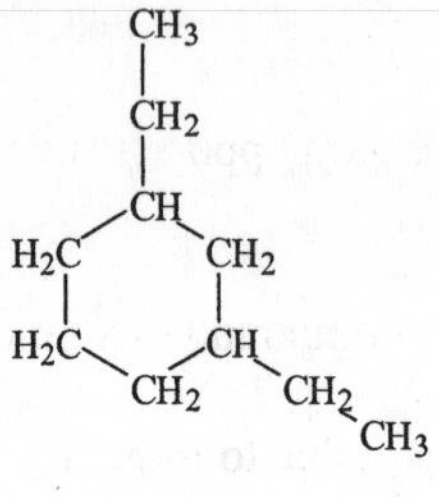   29. 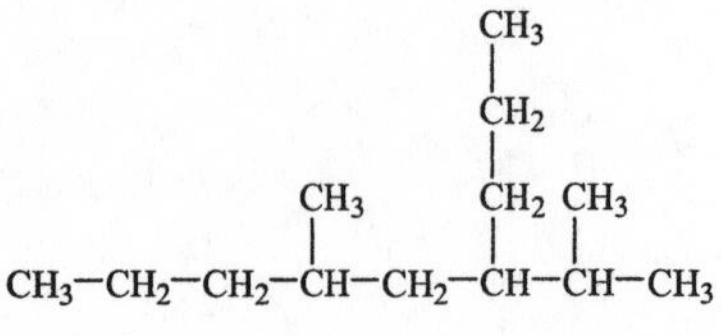   30.

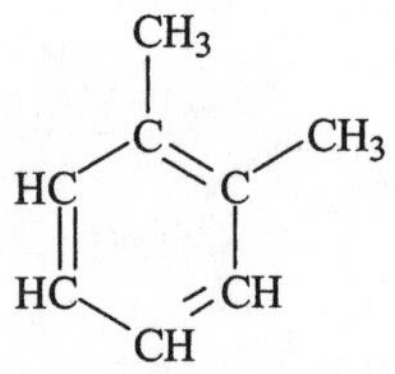

ANSWER PAGE FOR BUILDING KNOWLEDGE

Match the terms or phrases to the correct meaning

___h___ 1. Condensed structural formula

___a___ 2. Alkanes

___r___ 3. Skeletal structure

___l___ 4. Electronegativity

___v___ 5. Tetrahedral, pyramidal, trigonal planar, bent, and linear

___f___ 6. Polar covalent bonds

___w___ 7. Functional groups

___o___ 8. Noncovalent interactions

___g___ 9. Hydrogen bond

___i___ 10. Salt bridge

___b___ 11. Alkenes

___s___ 12. Dipole-dipole interaction

___t___ 13. Cis

___n___ 14. Trans

___m___ 15. Ion-dipole

___c___ 16. Coordinate-covalent interaction

___d___ 17. Aromatic

___x___ 18. Alkynes

___q___ 19. London forces

___u___ 20. Hydrocarbons

___j___ 21. Conformations

___p___ 22. Geometric isomers

___k___ 23. Stereoisomers

___e___ 24. Constitutional isomers

a. are hydrocarbons that contain only single covalent bonds

b. are hydrocarbons that contain at least one double covalent bond

c. involves the attraction between a cation and the nonbonding electrons of a nonmetal

d. compounds consist of rings of carbon atoms joined by alternating double and single covalent bonds

e. are molecules that have the same molecular formula but different atomic connections

f. occur when there is unequal sharing of electrons between atoms

g. is a strong dipole attraction found only when hydrogen is bonded to nitrogen, oxygen, or fluorine

h. describes the attachment of atoms to one another

i. used to describe an ionic bond

j. are the different three dimensional shapes that a molecule can assume through rotation of single bonds

k. are molecules that have the same molecular formula and the same atomic connections but different three-dimensional shapes that can be interchanged only by bond breaking.

l. is the ability of an atom to attract electrons in a covalent bond.

m. is an attraction of an ion for a polar group

n. is when the substituents are on opposite sides of a cycloalkane or alkenes

o. are forces that attract one compound to another

p. are stereoisomers that result due to limited rotation about covalent bonds

q. arise from the temporary dipoles that exist in nonpolar compounds

___y___ 25. Formal charge

r. represents covalent bonds with lines;, carbon atoms are not shown and hydrogen atoms are only shown when attached to an element other than carbon

s. is the attraction between opposite partial charges

t. is when the substituents are on the same side of a cycloalkane or alkenes

u. are molecules that contain only carbon and hydrogen atoms

v. are the basic shapes for molecules

w. are atoms, groups of atoms, or bonds that give a particular set of chemical properties

x. are hydrocarbons that contain at least one triple covalent bond

y. determined by comparing the number of electrons that surround an atom in a compound to the number of valence electrons that it carries as a neutral atom

Chapter 5
Gases, Liquids, and Solids

Chapter Overview

By now you have recognized that matter comes in many shapes and sizes. Atoms can be combined to create a wide variety of substances. Each of these combinations of atoms interacts in different ways depending upon the type of atoms and the structures of the molecules. These differences in interactive forces, structures, and masses make each substance respond uniquely to the energy imparted to it from its surroundings.

Chapter 5 explores the manner in which substances exchange energy and how the energy changes the physical properties of the substances. First you'll learn how to differentiate the states of matter: solids, liquids, and gases, and then calculate the energy needed to change matter from one state to another. In nature, all states of matter balance between the natural tendencies to lose energy with the drive toward random behavior. After exploring how these forces of nature work, you will take a closer look at each state of matter. As you study gases, you'll find out that volume, temperature, pressure, and moles help the chemist control and account for gaseous molecules. For liquids, you will investigate the basic physical properties such as viscosity, density, and vapor pressure. Finally a brief discussion of the different kinds of solid structures that can exist in nature is presented.

Knowledge/Comprehension

Typically, there are new terms or phrases that you must memorize in order for the lesson to make sense to you. Look up this information and memorize their explanations and definitions:

specific heat	heat of fusion	heat of vaporization
enthalpy change	entropy change	free energy
spontaneous	nonspontaneous	pressure
standard pressure units	Boyle's law	Charles' law
Gay-Lussac's law	Avogadro's law	combined gas law
ideal gas law	partial pressure	density
specific gravity	vapor pressure	crystalline solid
amorphous	ionic solid	covalent solid
molecular solids	Metallic solids	
Dalton's law of partial pressure		

The following activity is designed to help you master the objectives outlined in the beginning of Chapter 5 and then summarized at the end of the chapter. To make the best use of your study time, *read* Chapter 5 before continuing. Next, work the matching activity provided as though you are taking a test *without first* referring to the answer page. Put your answers on a separate sheet of paper so that you can make more than one attempt without seeing your previous answers. Once you have completed the activity, check your answers. Make *flash cards* for those you missed. Have a study buddy drill you on them if you missed more than half of the terms. Once you have reviewed the ones you missed for at least 15 minutes try the activity again. If you miss fewer than four, write in the answers so that you will have them to review just before taking the test.

Match the terms or phrases to the correct meaning

1._____ Metallic solid

a. molecules held together by noncovalent forces

2._____ Specific heat

b. arrangement of particles in an ordered fashion

3._____ Heat of vaporization

c. mass/volume

4._____ Entropy change

d. maximum pressure above a liquid

5._____ Pressure

e. is an unordered structure

6._____ Heat of fusion

f. $P_1V_1/ n_1T_1 = P_2V_2 / n_2T_2$

7._____ Spontaneous

g. $V_1/n_1 = V_2 /n_2$

8._____ Boyle's law

h. energy required to vaporize a liquid

9._____ Enthalpy change

i. array of metal cations immersed in a cloud of electrons that spans the entire crystalline structure

10._____ Nonspontaneous

j. $P_1/T_1 = P_2 /T_2$

11._____ Standard pressure units

k. $P_{total} = P_1 + P_2 + P_3...$

12._____ Avogadro's law

l. density of a substance divided by the density of water

13._____ Charles' law

m. $V_1/T_1 = V_2 /T_2$

14._____ Free energy

n. the pressure that each gas in a mixture would exert if it were the only gas present

15._____ Partial pressure

o. energy required to melt a solid

16._____ Dalton's law of partial pressure

p. due to the collision of the gas particles with the walls of the container

17._____ Gay-Lussac's law

18._____ Combined gas law

q. atoms are held together by an array of covalent bonds

19._____ Density

r. the amount of heat required to raise the temperature of one gram of the substance by one degree Celsius

20._____ Specific gravity

21._____ Ideal gas law

s. the heat released or absorbed during a process

22._____ Vapor pressure

84

23.____ Crystalline solid

24.____ Ionic solids

25.____ Amorphous solid

26.____ Covalent solids

27.____ Molecular solids

t. structure held together by oppositely charged ions

u. $PV = nRT$

v. the change in the randomness during a process

x. $P_1V_1 = P_2V_2$

y. will not run by itself

z. a combination of enthalpy change and entropy change and temperature

aa. will continue on its own, once started

bb. 1 atm, 14.7 psi, 760 torr

In addition to knowing the terms or phrases, you should be able to compare, contrast, or list examples that show that you understand the following concepts:

- Explain how the value of free energy change can be used to predict if a process is spontaneous or nonspontaneous.
- Explain Dalton's law of partial pressure.
- Describe the relationship between atmospheric pressure and the boiling point of a liquid.
- Describe the difference between amorphous and crystalline solids.
- Describe the makeup of the four classes of crystalline solids.

Once you know the appropriate tools for Chapter 5, you are ready to evaluate your understanding of those tools and their relationships. In this type of concept building, it is necessary to compare or contrast terms or concepts. In some cases you may be required to give examples or recognize a term or concept when it is depicted in a picture or sample scenario.

Sample Problems

After having studied the definitions of free energy, as well as spontaneous and nonspontaneous energies, the next step is to relate free energy values to the physical process being observed in nature. Free energy, ΔG, is a value that reflects the relationship between enthalpy, ΔH, and entropy, ΔS, at a particular temperature. A negative free energy value indicates that the process provides enough energy to continue on its own once it starts; this process is referred to as spontaneous. If ΔG is positive it means that at the given temperature, the process would stop unless additional energy from another source is added; this process is said to be nonspontaneous. When you are observing a change in physical condition, the ΔG or spontaneity of the system can be predicted. For example, it is 90° F outside and you see a melting block of ice on the sidewalk. The energy present is sufficient for melting the ice without any other action being needed, so you would say the process is spontaneous. This means the reaction has a – ΔG.

1. At 78°C, ethanol begins to boil and turn into a vapor. Does this process favor ΔH or ΔS? Explain?

Since the tendency of enthalpy is to give off energy, the ethanol gaining energy would not favor ΔH and the reaction is nonspontaneous with respect to ΔH. The ethanol is changing from a liquid to a vapor which means it is increasing in randomness. Therefore ΔS is favored and the process is spontaneous with respect to ΔS.

2. Which of the temperatures given below would provide a $- \Delta G$ for the condensation of water?

      a. 50 °C   b. 120 °C   c. 150 °C   d. 200 °C

The boiling point of water is 100°C. This means that condensation would not be expected to occur above this temperature. Therefore choice a. would be the correct answer since it is the only temperature given that is below 100°C. **Answer is a.**

3. Which of the following processes would be nonspontaneous at 100 °C?

      a. ice melting   b. dry ice subliming   c. water boiling   d. water freezing

Ice melting at 100 °C is a process you would expect to occur spontaneously. Dry ice will sublime at room temperature so it certainly would at 100 °C and would also be spontaneous. Water boils at 100 °C, therefore at 100 °C the process is spontaneous. Water normally freezes at temperatures below 0 °C so at 100 °C pressure would be required to reduce the randomness enough for the water to freeze. This process would not be spontaneous. **Answer is d.**

The partial pressure of a gas is the pressure caused by the molecules of that gas. Since molecules of a gas are very small compared to the container size, the pressure exerted by a gas is dependent on how many moles of the gas are present, not what kind of gas it is. This means that if you put 1 mole of hydrogen in a container and add 1 mole of nitrogen to the container, you will have the same effect as if putting2 moles of hydrogen in the container, since there are the same number of particles in each. Because the number of moles present and the pressure in the container are directly related, Dalton expressed the relationship of added moles of gases in terms of the pressure exerted by each gas. So just like adding moles gave total moles, adding pressures caused by those moles gives the total pressure in the container.

4. Which of the following container combinations at the same temperature and volume would have different pressures?

      a. 1 mol of N_2 ; and 1 mol Ar
      b. 1 mol H_2, 1 mol O_2 ; and 2 mol Ne

c. $P_{N2} = .5$ atm, $P_{H2} = .5$ atm ; and $P_{F2} = 2.0$ atm

d. $P_{O2} = 1$ atm, $P_{H2} = .5$ atm;   and $P_{Cl2} = 1.5$ atm

In choice a, the two containers have equal numbers of moles and would therefore have equal pressures. In choice b, the number of moles in the first container, $1 + 1 = 2$, is equal to the number of moles in the second container, making the pressures the same. In choice d, the pressures of the first container, $1 + .5 = 1.5$, making it the same as the pressure of the second container. In c, the first container's pressure adds up to be $.5 + .5 = 1$ which is not the same as 2 atm. **Answer is c.**

5. In the drawings below, identify the container where molecule X has the largest partial pressure.

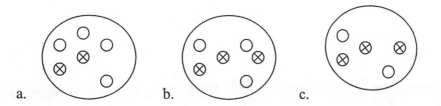

a.   b.   c.

In figure a, there are more O than X molecules, so the partial pressure of O would be the greatest. In figure b, the numbers are the same so the partial pressures would be the same. In figure c, there are more X molecules than O molecules so the partial pressure of the X molecules would be the greatest. **Answer is c.**

Boiling point is defined as the temperature at which the vapor pressure of the liquid is equal to the pressure above the liquid. In an open container, the pressure above the liquid is typically the atmosphere. Therefore the pressure above the container would be the atmospheric pressure. This is why the author of your textbook defines boiling point as the temperature at which the vapor pressure of the liquid equals the atmospheric pressure.

6. At which height would the boiling point be the greatest?

a. 2 km   b. 10 km   c. 5 km   d. 8 km

Recall that air becomes less dense and the pressure decreases as you go up in altitude. Less pressure means less energy is needed to get the vapor pressure to that point. The lowest height of 2 km would have the greatest atmospheric pressure and that would require the highest temperature to start it to boil. **Answer is a.**

7. Looking at the graph below, at what elevation would a liquid with a boiling point of 90°C begin to boil?

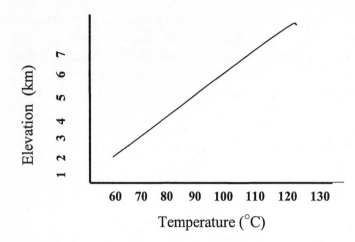

To determine the answer, move along the x-axis until you get to 90°C then mark that point on the line of the graph. Next move parallel to the x-axis over to the y-axis and read the elevation as shown below.

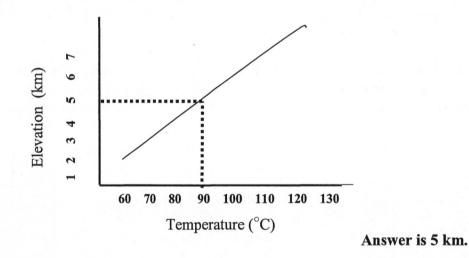

Answer is 5 km.

The definitions studied above tell you that crystalline solids have an ordered structure while amorphous solids do not. This means that connected nuclei of the atoms of crystals form specific shapes such as cubes, rectangles, rhombics or other distinct geometric shapes. Amorphous solids form patterns that are random and connecting the nuclei of the atoms does not give any type of specific geometric shape.

8. If you were to look at particles of an amorphous solid under a magnifying glass, which of the following shapes would *not* be likely to be observed?

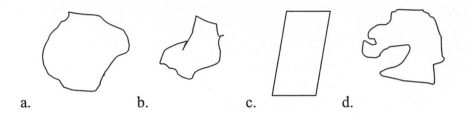

a.   b.   c.   d.

Since amorphous solids do not have an orderly geometric shape it is unlikely you would see shape c. **Answer is c.**

9. Which drawing does *not* represent a crystalline solid?

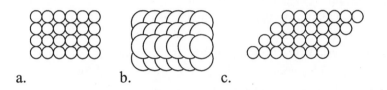

a.   b.   c.

Solid b does not have a distinct geometric shape therefore it would not represent a crystalline solid. **Answer is b.**

The four classes of crystalline solids are based on the types of forces that hold them together. Ionic crystals are solids composed of positive and negative ions held together by the attraction of opposite charges. Some crystalline solids are held together by covalent bonds that occur between connected molecules. The third class, molecular solids, is composed of molecules held together by the noncovalent forces discussed previously in Chapter 3. The fourth type of crystalline solid is metal. The metallic crystal is held together by a cloud of electrons that are passed around the structure. It is this cloud of electrons that make metals good conductors of electricity and allow metals to be bent or hammered into shapes without breaking apart.

10. A crystal is found during an archeological dig. It is determined to be very hard and has a very high melting point but is not a good conductor of electricity. What type of crystalline solid is it?

      a. ionic   b. covalent   c. molecular   d. metallic

Ionic compounds are poor conductors of electricity in the crystal state and can have very high melting points, but they are not very hard. Molecular crystals generally have low melting points and are not very hard. Metallic solids can be very hard and have high melting points but they are good conductors of electricity. Covalent solids have all three properties listed. **Answer is b.**

11. A chemist analyzing a crystalline solid found that the crystal had a moderate melting point, dissolved in water, and that the water solution did not conduct electricity. What class of crystalline solid was he most likely analyzing?

a. ionic   b. covalent   c. molecular   d. metallic

Metallic and covalent solids have high melting points and are not soluble in water. Ionic compounds can have moderate melting points and are soluble in water but since dissolving them in water frees the ions to move, they do conduct an electric current in water. Molecular solids have all of the characteristics listed in the analysis. **Answer is c.**

Application

To apply what you've already learned and master the concepts, you must utilize all of the knowledge, understanding, facts, and/or techniques that you have gathered up to this point. Simply knowing an equation or process is not enough to master the concept. These you develop by *practice*. Solve as many different types of problems with as many different variations as you *strive to master the skill*. Your textbook provides a very good assortment of problems at the end of the chapter. Work as many as needed for skill mastery, not just those assigned. Use specific heat, heat of fusion, and heat of vaporization to calculate energy changes in matter.

- Be able to convert between common pressure units.
- From the units of measured values given in a gas problem, be able to list the variables and select the appropriate gas law for determining the missing variable.

In order to truly master any skill you must *practice*. In this section example problems related to the application objectives are discussed. Remember, the more you practice, the better your chances of being able to successfully use these objectives to solve problems. Additional practice problems are at the end of the chapter in your textbook and in the online interactive lessons.

Sample Problems

Once you have learned the basic definitions of specific heat, heat of fusion and heat of vaporization, the next level of learning involves applying them in calculations. Since each term can be expressed as a ratio value (specific heat = cal/g °C, heat of fusion = cal/g, and heat of vaporization = cal/g) they can be used as conversion factors. Specific heat is used when temperature is changing within the same state of matter. If the substance is going to change states, heat of fusion is used for the melting/freezing process and the heat of vaporization is used for the boiling (vaporization)/condensation process. If it is necessary to calculate energy for heating a substance and changing its state, the heating calculation is done, followed by the state change calculation; the two values are then added together. It is unlikely that your instructor will require you to memorize these values so your task is to know when to apply each value.

1. Calculate the energy required to warm 125 g of ice from -5.0°C to 0°C

Note that the ice only heated to the melting point so that the temperature is raised but the ice has not started to change states. To calculate the energy needed for this process, the mass is multiplied by the temperature change and then converted to energy using the factor label method from Chapter 1 and the specific heat as the conversion factor. The specific heat of ice is 0.500 cal/g°C.

$$125 \text{ g } \times \quad 5.0°\cancel{C} \quad \times \quad \frac{0.500 \text{ cal}}{\text{g·}°\cancel{C}} = 312.5 \text{ cal}$$

Answer: 312.5 cal

2. Calculate the heat required to vaporize 215 g of water.

Note that the water is heated so that the temperature is not being raised but the water is changing states from liquid to gas. To calculate the energy needed for this process, the mass is converted to energy using the factor label method from Chapter 1 and the heat of vaporization as the conversion factor. The heat of vaporization for water is 540 cal/g.

$$215 \text{ g } \times \quad \frac{540 \text{ cal}}{\text{g}} = 11.6 \times 10^4 \text{ cal}$$

Answer: 11.6 x 10^4 cal

3. An ice pack used to prevent joint swelling contains 411 g of ice. How much energy will it absorb if all of the ice melts?

Note that the ice is melting so that the temperature is not being raised; rather the ice is changing states from solid to liquid. To calculate the energy needed for this process, the mass is converted to energy using the factor label method from Chapter 1 and the heat of fusion as the conversion factor. The heat of fusion for water is 80 cal/g. Note that energy absorbed by the ice is the same as the amount required to melt it.

$$411 \text{ g } \times \quad \frac{80 \text{ cal}}{\text{g}} = 3.29 \times 10^4 \text{ cal}$$

Answer: 3.29 x 10^4 cal

4. A 7.5 g piece of gold has a temperature of 33°C. What is the new temperature of the gold if 145 cal of heat are added?

Note that the gold is only heated so that the temperature is raised but has not started to change states. To calculate the temperature for the heating process, the mass is multiplied by the energy used. That is then converted to temperature using the factor label method from Chapter 1 and the specific heat as the conversion factor. The temperature value calculated is equal to the temperature increase. To find the new temperature, add the temperature increase to the original temperature. The specific heat of gold is 0.0310 cal/g°C.

7.5 g   x $\dfrac{0.0310 \text{ cal}}{\text{g·}^{\circ}\text{C}}$ = 0.232 cal/°C then this number can be used to convert from cal to °C

145 ~~cal~~   x $\dfrac{1\,^{\circ}\text{C}}{0.232 \text{ ~~cal~~}}$ = 625 °C and   33°C + 625°C = 658 °C

Answer: 658 °C

Depending on your instructor, you may or may not be required to know the common pressure units. At sea level and 273, K 1.00 atm = 760 mmHg = 101.3 kPa = 14.7 psi. These equivalencies can be used to convert from one pressure unit used to another. Remember, when working gas law calculations, dividing or multiplying measurements with different units will result in an incorrect answer.

5. How many atmospheres are equivalent to a pressure of 40.6 psi?

As with any conversion, the first step is to know the equivalency between the different units used. Here, 1 atm = 14.7 psi. Use this as a conversion factor and convert from psi to atm.

40.6 ~~psi~~   x $\dfrac{1\text{ atm}}{14.7 \text{ ~~psi~~}}$ = 2.76 atm

Answer: 2.76 atm

6. How many torr is equivalent to a pressure of 22.5 psi?

As with any conversion, the first step is to know the equivalency between the different units used. Here, 760 torr = 14.7 psi. Use this as a conversion factor and convert from psi to torr.

22.5 ~~psi~~   x $\dfrac{760 \text{ torr}}{14.7 \text{ ~~psi~~}}$ = 1163 torr

Answer: 1163 torr

7. At a pressure of 560 torr a balloon has a volume of 2.50 L. If the balloon is put into a container and the pressure is increased to 1500 torr (at constant temperature), what is the new volume of the balloon?

Identify your variables by the units given.

$P_1 = 560$ torr $P_2 = 1500$ torr
$V_1 = 2.50$ L $V_2 = ?$ (this is what you are asked to solve for)

From the gas laws studied select the one that has *only* these variables. In this case it will be **Boyle's law: $P_1V_1 = P_2V_2$** . Replace the variables in the equation with the ones identified above and algebraically solve for the missing variable.

$P_1V_1 = P_2V_2$
$P_1V_1 / P_2 = V_2$
(560 ~~torr~~ x 2.50 L) / 1500 ~~torr~~ = V_2
0.933 L = V_2

Answer: 0.933 L

8. At a temperature of 50°C, a balloon has a volume of 3.50 L. If the temperature is increased to 100°C (at constant pressure), what is the new volume of the balloon?

Make sure that you convert °C to Kelvin!!! Remember °C does not work in gas law problems:

50°C + 273 = 323 K
100° C + 273 = 373 K

Identify your variables by the units given.

$T_1 = 323$ K $T_2 = 373$ K
$V_1 = 3.50$ L $V_2 = ?$ (this is what you are asked to solve for)

From the gas laws studied select the one that has *only* these variables. In this case it will be **Charles' law:; $V_1/ T_1 = V_2/ T_2$** . Replace the variables in the equation with the ones identified above and algebraically solve for the missing variable.

$V_1/ T_1 = V_2/ T_2$
(V_1/ T_1) x $T_2 = V_2$
(3.50 L / 323 ~~K~~) x 373 ~~K~~ = V_2
            4.04 L = V_2

Answer: 4.04 L

9. At a temperature of 20°C, a gas inside a 1.25 L metal canister has a pressure of 860 torr. If the temperature is increased to 50°C (at constant volume), what is the new pressure of the gas?

Make sure that you convert °C to Kelvin!!! Remember °C does not work in gas law problems.

20°C + 273 = 293 K
50° C + 273 = 323 K

Identify your variables by the units given.

$T_1 = 293$ K $T_2 = 323$ K

$$P_1 = 860 \text{ torr} \qquad P_2 = ? \text{ (this is what you are asked to solve)}$$

Note that even though the volume was stated it is not included as a variable because the problem states that it remained constant.

From the gas laws studied select the one that has *only* these variables. In this case it will be **Gay-Lussac's law: $P_1/T_1 = P_2/T_2$**. Replace the variables in the equation with the ones identified above and algebraically solve for the missing variable.

$$P_1/T_1 = P_2/T_2$$
$$(P_1/T_1) \times T_2 = P_2$$
$$(860 \text{ torr} / 293 \text{ K}) \times 323 \text{ K} = P_2$$
$$948 \text{ torr} = P_2$$

Answer: 948 torr

10. A 5.0 L balloon contains 0.50 mol of $Cl_2(g)$. At constant pressure and temperature, what is the new volume of the balloon if 0.30 mol of gas is removed?

Identify your variables by the units given.
First note 0.30 is the amount removed not final number of moles.
$n_2 = 0.50 - 0.30 = 0.20 \text{ mol}$

$$n_1 = 0.50 \text{ mol} \qquad n_2 = 0.20 \text{ mol}$$
$$V_1 = 5.0 \text{ L} \qquad V_2 = ? \text{ (this is what you are asked to solve for)}$$

From the gas laws studied select the one that has *only* these variables. In this case it will be **Avogadro's law: $V_1/n_1 = V_2/n_2$**. Replace the variables in the equation with the ones identified above and algebraically solve for the missing variable.

$$V_1/n_1 = V_2/n_2$$
$$(V_1/n_1) \times n_2 = V_2$$
$$(5.0 \text{ L} / 0.50 \text{ mol}) \times 0.20 \text{ mol} = V_2$$
$$2.0 \text{ L} = V_2$$

Answer:2.0 L

11. A balloon with a volume of 4.50 L is at a pressure of 660 torr and a temperature of $10°C$. The balloon is put into a container, the pressure is increased to 1500 torr, and the temperature is raised to $50°C$. What is the new volume of the balloon?

Make sure that you convert $°C$ To Kelvin!!! Remember $°C$ does not work in gas law problems.

$$10°C + 273 = 283 \text{ K}$$
$$50°C + 273 = 323 \text{ K}$$

Identify your variables by the units given.

$$T_1 = 283 \text{ K} \qquad T_2 = 323 \text{ K}$$

$$P_1 = 660 \text{ torr} \qquad P_2 = 1500$$
$$V_1 = 4.50 \text{ L} \qquad V_2 = ? \text{ (this is what you are asked to solve for)}$$

From the gas laws studied select the one that has all of these variables. In this case it will be the **combined gas law: $P_1V_1/n_1T_1 = P_2V_2/n_2T_2$** . Replace the variables in the equation with the ones identified above and algebraically solve for the one missing variable. Note it is not necessary to include n_1 and n_2 since moles are assumed to be constant.

$$P_1V_1/T_1 = P_2V_2/T_2$$
$$(P_1V_1) \times T_2 / T_1 = V_2$$
$$[(P_1V_1) \times T_2] / (T_1 \times P_2) = V_2$$
$$[(660 \text{ torr}) (4.50 \text{ L}) \times 323 \text{ K} /] / (283 \text{ K} \times 1500 \text{ torr}) = V_2$$
$$2.26 \text{ L} = V_2$$

Answer: 2.26 L

12. A 275 mL metal can contains 1.50×10^{-3} mol of He at a temperature of 398K. What is the pressure (in atm) inside the can?

Identify your variables by the units given.
$$T = 398 \text{ K} \qquad P = ?$$
$$V = 0.275 \text{ L} \qquad n = 1.50 \times 10^{-3} \text{ mol}$$

Note: since the R constant given has the units atm and L in it, it is necessary to convert the volume from mL to L.   275 mL x 1 L/ 1000 mL = .275 L

From the gas laws studied select the one that has just one of each variable. In this case it will be the **ideal gas law: $PV = nRT$,** where R is the gas constant defined as 0.0821 L atm /mol K. Replace the variables in the equation with the ones identified above and algebraically solve for the one missing.

$$PV = nRT$$
$$P = nRT/V$$
$$P = (1.50 \times 10^{-3} \text{ mol})(0.0821 \text{ L atm /mol K})(398 \text{ K})/(0.275 \text{ L})$$
$$P = 2.17 \text{ atm}$$

Answer:2.17 atm

Chapter 5 Practice Test

Now that you've had the chance to study and review Chapter 5's practice problems provided in your textbook, you are ready to practice problems as you might see them in a testing situation. The questions and problems below are randomly selected in much the same way your instructor will select them; that is, not necessarily ordered in the same way as they appear in the textbook. To make best use of this practice, have your paper, pencil and calculator ready as you would for a class exam. Use a clock and give yourself 45 minutes to complete this activity. Note that taking this test under similar conditions as a class exam provides two benefits:

1. Testing your knowledge of the chapter's contents and your mastery of the required skills.
2. Practicing working under the stress of a timed exam.

Most instructors have a type of testing format that they prefer and will most likely tell you about it before the test. For practice, this self-test uses a variety of test question formats such as matching, fill-in-the-blank, listing, multiple-choice, and essay questions. ***Be sure to have a periodic table!***

Note the time and begin.

1. Which of the following processes involve only a phase change with no change in temperature?

A. warming water
B. ethanol cooling down
C. boiling water
D. metal heating in the sun

2. Calculate the energy needed to heat 25g of water from 25 ^{o}C to 50 ^{o}C. (Specific heat of water = 1.00 cal/g^{o}C)

A. 25 cal
B. 50 cal
C. 1250 cal
D. 625 cal

3. Calculate the amount of heat needed to vaporize 20 g of 2-propanol. (heat of vaporization = 159 cal)

A. 3180 cal
B. 20 cal
C. 159 cal
D. 7.95 cal

4. Calculate the amount of energy that must be removed in order to freeze 30 g of ethanol (heat of fusion = 26.05)

A. 26.05 cal
B. 30 cal
C. 782 cal
D. 0.87 cal

5. Ice is observed melting on a cold day with a temperature of -10°C. Which of the following statements is *not* true?

A. The process is spontaneous with respect to entropy.
B. ΔG is positive for the process.
C. The process is nonspontaneous with respect to enthalpy.
D. ΔH is nonspontaneous for the process.

6. Which of the following variables is not part of the free energy calculation?

A. ΔG
B. ΔH
C. ΔS
D. ΔP

7. According to the graph below, what is the estimated atmospheric pressure at 7.5 km?

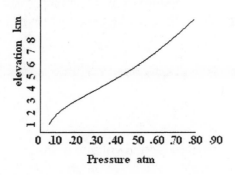

A. 0.10 atm
B. 0.40 atm
C. 0.70 atm
D. 0.90 atm

8. Which of the following is not a gas law?

A. Charles'
B. Avogadro's
C. Boyle's
D. Lewis'

9. Which of the following equations is the mathematical representation of the Gay-Lussac's gas law?

A. $PV = nRT$
B. $P_1/T_1 = P_2/T_2$
C. $V_1/T_1 = V_2/T_2$
D. $P_1V_1 = P_2V_2$

10. If a can with a volume of 2.0 L contains 2 mol of a gas, how many moles of gas will a 1.0 L can contain at the same temperature and pressure as the first can?

A. 1 mol
B. 0.5 mol
C. 2 mol
D. 4 mol

11. If a balloon has a volume of 3.5 L at $20^{\circ}C$, what is its new volume if the temperature is increased to $100^{\circ}C$?

A. 17.5 L
B. 350 L
C. 4.5 L
D. 0.012 L

12. A 2.5 L container has a temperature of $30^{\circ}C$ and a pressure of 1.2 atm. How many moles of gas particles are in the container? ($R = 0.0821$ atm L/ mol K)

A. 0.12 mol
B. 1.2 mol
C. 0.45 mol
D. 2.5 mol

13. A bubble leaves a diver's mask with a volume of 20 mL and pressure of 1.5 atm. If the temperature is constant, what is the new volume of the bubble when it reaches the surface with a pressure of 0.955 atm?

A. 200 mL
B. 31 mL
C. 42 mL
D. 2.0 mL

14. A bubble leaves a diver's mask with a volume of 20 mL, pressure of 2.5 atm, and a temperature of 10°C. If the temperature is 33°C at the surface, what is the new volume of the bubble when it reaches the surface with a pressure of 0.955 atm?

A. 31 mL
B. 400 mL
C. 56 mL
D. 173 mL

15. A canister with a pressure of 0.950 atm at 25°C is placed in a burning brush pile with a temperature of 350°C. What is the new pressure in the canister?

A. 23.7 atm
B. 1.00 atm
C. 0.454 atm
D. 1.98 atm

16. How many grams of He are present in a 250 mL bottle that has a pressure of 780 torr and a temperature of 22°C?

A. 0.0106 g
B. 0.0425 g
C. 0.143 g
D. 0.570 g

17. If the total pressure in a flask is 2 atm and the partial pressure of oxygen is 1.5 atm, what is the partial pressure of the remaining gas?

A. 0.5 atm
B. 2 atm
C. 1.5 atm
D. 3.0 atm

18. What volume is occupied by 15.0 grams of bone with a density of 1.90 g/mL?

A. 1.00 mL
B. 7.89 mL
C. 15 mL
D. 8.45 mL

19. Which of the following has the smallest density?

A. 1.56 g/mL
B. a specific gravity of 1.56
C. a mass of 12.0 g and 7.69 mL
D. a mass of 7.69g and a volume of 12.0 mL

20. As the elevation where you live decreases, the temperature required to boil water will:

A. increase
B. decrease
C. stay the same
D. decrease and then increase

21. Draw a pattern that could be used to describe an amorphous crystal.

22. Diamonds are known for their hardness and extremely high melting points. What type of force holds the particles together? Explain why this makes the melting point so high.

23. When your diaphragm goes down it relieves the pressure on the lungs, and air rushes into the lungs. Which gas law explains this? Explain why.

24. _____, _____, _____ and _____ are the four forces that hold particles together in a solid.

25. True or False: A substance that is a gas at STP would be a liquid at room temperature (25°C).

PRACTICE TEST Answers

1. C
2. D
3. A
4. C
5. B
6. D
7. C
8. D
9. B
10. A
11. C
12. A
13. A
14. C
15. D
16. B
17. A
18. B
19. D
20. B
21.

22. Covalent. Covalent forces are very strong interactive forces and the stronger the interactive forces that hold the solid particles together, the higher the melting point.

23. Boyle's law. As the pressure on the lungs decreases, the volume of the lungs will increase. This increase in volume makes the air pressure inside the lungs less than atmospheric pressure as air is pushed into the lungs until the two pressures are equal again.

24. Ionic, covalent, molecular, and metallic

25. False

ANSWER PAGE FOR BUILDING KNOWLEDGE

Match the terms or phrases to the correct meaning

1. __i__ Metallic solid

2. __r__ Specific heat

3. __h__ Heat of vaporization

4. __v__ Entropy change

5. __p__ Pressure

6. __o__ Heat of fusion

7. __aa__ Spontaneous

8. __x__ Boyle's law

9. __s__ Enthalpy change

10. __y__ Nonspontaneous

11. __bb__ Standard pressure units

12. __g__ Avogadro's law

13. __m__ Charles' law

14. __z__ Free Energy

15. __n__ Partial pressure

16. __k__ Dalton's law of partial pressure

17. __j__ Gay-Lussac's law

18. __f__ Combined gas law

19. __c__ Density

20. __l__ Specific gravity

21. __u__ Ideal gas law

22. __d__ Vapor pressure

23. __b__ Crystalline solid

24. __t__ Ionic solids

a. molecules held together by noncovalent forces

b. arrangement of particles in an ordered fashion

c. mass/volume

d. maximum pressure above a liquid

e. is an unordered structure

f. $P_1V_1/n_1T_1 = P_2V_2/n_2T_2$

g. $V_1/n_1 = V_2/n_2$

h. energy required to vaporize a liquid

i. array of metal cations immersed in a cloud of electrons that spans the entire crystalline structure

j. $P_1/T_1 = P_2/T_2$

k. $P_{total} = P_1 + P_2 + P_3...$

l. density of a substance divided by the density of water

m. $V_1/T_1 = V_2/T_2$

n. the pressure that each gas in a mixture would exert if it were the only gas present

o. energy required to melt a solid

p. due to the collision of the gas particles with the walls of the container

q. atoms are held together by an array of covalent bonds

r. the amount of heat required to raise the temperature of one gram of the substance by one degree Celsius

s. the heat released or absorbed during a process

t. structure held together by oppositely charged ions

u. $PV = nRT$

v. the change in the randomness during a process

103

25.__e__ Amorphous solid

26.___q_ Covalent solids

27.__a__ Molecular solids

x. $P_1V_1 = P_2V_2$

y. will not run by itself

z. a combination of enthalpy change and entropy change and temperature

aa. will continue on its own, once started

bb. 1 atm, 14.7 psi, 760 torr

Chapter 6
Reactions

Chapter Overview

In Chapter 5, the manner in which substances exchange energy and how the energy changes the physical properties of the substances were explored. However, just being able to recognize that a chemical reaction is taking place and that energy is involved is not enough. For example, in industry when a company produces a product, it would be financially foolish for them to pay more for the starting materials than what they would be able to sell the finished product for at a later time. Therefore, it is important to be able to look at the amount of starting materials and be able to predict the amount of product that can be produced.

In this chapter you will learn how chemists represent chemical reactions and the proper proportions that allow chemicals to react. You will be investigating some common types of reactions that can occur in both nature and in industry. You will then learn how to quantify the reactions in order to be able to predict the amount of product produced when given a starting amount. Finally, a chemist must control production so that the product doesn't take too long to produce or is produced so fast that the energy is released in an explosion. This is accomplished by understanding the factors that can affect the rate at which the chemical reaction occurs.

Knowledge/Comprehension

Typically, there are new terms or phrases that you must memorize in order for the lesson to make sense to you. Look up this information and memorize their explanations and definitions:

chemical equation	balanced equation	coefficients
oxidation	reduction	combustion reaction
catalytic hydrogenation	hydrolysis reaction	hydration reaction
dehydration reaction	limiting reactant	theoretical yield
actual yield	percent yield	activation energy
catalysts	rate of reaction	temperature
reactants	products	oxidizing agent
reducing agent	enzyme	
reaction energy diagram		

The following activity is designed to help you master the objectives outlined in the beginning of Chapter 6 and then summarized at the end of the chapter. To make the best use of your study time, *read* Chapter 6 before continuing. Next, work the matching activity provided as though you are taking a test *without first* referring to the answer page. Put your answers on a separate sheet of paper so that you can make more than one attempt without seeing your previous answers. Once you have completed the activity, check your answers. Make *flash cards* for those you missed. Have a study buddy drill you on them if you missed more than half of the terms. Once you have reviewed the ones you missed for at least 15 minutes try the activity again. If you miss fewer than four, write in the answers so that you will have them to review just before taking the test.

Match the terms or phrases to the correct meaning

1._____ Chemical equation

2._____ Balanced equation

3._____ Coefficients

4._____ Oxidation

5._____ Reduction

6._____ Combustion reaction

7._____ Oxidizing agent

8._____ Catalytic hydrogenation

9._____ Hydrolysis reaction

10._____ Hydration reaction

11._____ Theoretical yield

12._____ Activation energy

13._____ Limiting reactant

14._____ Percent yield

15._____ Rate of reaction

16._____ Reaction energy diagram

17._____ Enzyme

18._____ Catalysts

19._____ Products

20._____ Dehydration reaction

21._____ Actual yield

22._____ Temperature

a. is the average kinetic energy of the molecular forces

b. can lower the activation energy of a reaction and speed up a reaction

c. is the measure of how fast a product is being produced

d. are the elements or compounds present at the start of a reaction

e. are the new substances formed

f. is the substance that is removing the electrons

g. is the substance that is providing electrons

h. is a biochemical catalyst

i. is a drawing that represents the energy change that takes place during a chemical reaction

j. found by dividing the actual yield by the theoretical yield and then multiplying by 100%

k. the amount of product actually collected in the laboratory

l. the calculated amount of product expected when the limiting reactant is used up

m. the reactant that runs out first

n. water being removed from an alcohol to make an alkene

o. adds water across the double bond of an alkene to make an alcohol

p. uses water to split another molecule

q. involves the reduction of a double bond by H_2 in the presence of Pt

106

23._____ Reactants

24._____ Reducing agent

r. the gaining of electrons

s. the loss of electrons

t. used as multipliers in front of formulae in balancing equation

u. has the same number and kind of each atom on both sides of the equation

v. uses an arrow to separate reactants from products

w. is the energy barrier that must be crossed to go from reactants to products

x. reactant is oxidized (burned by O_2)

In addition to knowing the terms or phrases, you should be able to compare, contrast, or list examples that show that you understand the following concepts:

- Identify oxidation, reduction, combustion, and hydrogenation reactions.
- Identify hydrolysis, hydration, and dehydration reactions of organic compounds.
- Describe the difference in energy changes for spontaneous and nonspontaneous reactions.
- List the factors that affect the rate of a chemical reaction.

Once you know the appropriate tools for Chapter 6, you are ready to evaluate your understanding of those tools and their relationships. In this type of concept building, it is necessary to compare or contrast terms or concepts. In some cases you may be required to give examples or recognize a term or concept when it is depicted in a picture or sample scenario.

Sample Problems

When identifying the type of reaction that is taking place, it is important to look for particular features of the reaction. Oxidation and reduction occur in any reaction that involves the exchange of electrons; one does not occur without the other. Therefore, for these two terms, you are not trying to identify the type of reaction but instead trying to identify the species that is oxidized and the species that is reduced within a reaction equation.

1. In the following reaction, which element is oxidized and which is reduced?

$$ZnSO_4 \ + \ Mg(s) \ \rightarrow \ Zn(s) \ + \ MgSO_4$$

To identify the occurrence of oxidation and reduction you must be able to recognize which species is losing electrons (oxidation) and which species is gaining electrons (reduction). To do this, first identify the charge on each species by going through the equation and predicting possible charges from their periodic position (in the case of Zn you have already learned that its charge is always +2 as an ion). Pure elements have an assigned charge of zero. If there is a metal that typically has a Roman numeral in its name or a nonmetal not in the rightmost position, find the charge by determining the total negative charge of the rightmost nonmetal and then adding the positive charge contributed by another metal (if any). The positive charge assigned to the element is equal to the remaining negative charge, divided by the subscript of the species to which you are assigning the charge:

$$\overset{+2\ +6\ -2}{ZnSO_4} + \overset{0}{Mg_{(s)}} \rightarrow \overset{0}{Zn_{(s)}} + \overset{+2\ +6\ -2}{MgSO_4}$$

On the left side of the equation, the S atom's charge is found by multiplying the -2 of oxygen by 4 (there are four oxygen atoms in the compound) for a total negative charge of -8. Adding the +2 of the Zn atom to the negative charge leaves a charge of -6. Since there is only one S atom its charge must be +6. Because Mg has the same charge in the $MgSO_4$ on the right side of the equation, the charge on the S atom is the same both on the right and left sides of the equation. When this happens you know that element did not lose or gain electrons and therefore was not oxidized or reduced. Note that the Zn^{+2} ion became $Zn_{(s)}$, a pure element with a charge of zero. In order to do this, the Zn^{+2} ion had to gain two electrons, which lowered *(reduced)* its charge. The $Mg_{(s)}$ pure element with a charge of zero became Mg^{+2} ion, which required it to lose two electrons *(oxidized)*. **Answer: $Mg_{(s)}$ is oxidized and Zn^{+2} ion is reduced.**

A question that often accompanies the type of problem above is "Which species is the oxidizing agent and which is the reducing agent?" The "agent" is the species causing the action described. Therefore, the oxidizing agent is the species that gained the electrons from the species that was oxidized; the oxidizing agent is the substance that was reduced. The reducing agent is the species that lost electrons to the substance that was gaining electrons; the reducing agent was the substance that was oxidized. In the above equation, the Zn^{+2} ion is the oxidizing agent and $Mg_{(s)}$ is the reducing agent.

Combustion and hydrogenation reactions typically describe a particular type of reaction equation. Combustion is the process of burning something with O_2, therefore adding O_2 to another material is a key feature. Combustion reactions involving organic reactants generate CO_2 and H_2O as products. Hydrogenation reactions are just what the name implies, reactions that hydrogenate a molecule. This typically occurs when alkenes are reduced to alkanes with H atoms in the presence of a catalyst. The key feature of this type of reaction equation is an alkene (double bond) → alkane (single bond) and H_2 as a reactant.

2. Which of the following reactions is a combustion reaction?

a. $HCl + AgNO_3 \rightarrow AgCl + HNO_3$

b. $ZnSO_4 + Mg_{(s)} \rightarrow Zn_{(s)} + MgSO_4$

c. $2C_6H_6 + 15O_2 \rightarrow 12CO_2 + 6H_2O$

d. $H_2SO_4 + 2NaOH \rightarrow Na_2SO_4 + 2H_2O$

Looking for the key feature of O_2 means that equation c. is the only choice that fits a combustion reaction. **Answer is c.**

3. Do any of the reactions in Question 1 represent a hydrogenation reaction?
Answer: No. In order to be a hydrogenation reaction, hydrogen must be one of the reactants.

4. Which of the following is NOT a hydrogenation reaction?

a. $CH_3-CH_2-CH_2-CH=CH_2 + H_2 \longrightarrow CH_3-CH_2-CH_2-CH_2-CH_3$

b. $H_2C=CH_2 + H_2 \longrightarrow CH_3-CH_3$

c. $\underset{\overset{|}{CH_3}}{CH_3-CH-CH_2-CH_3} \longrightarrow \underset{\overset{|}{CH_3}}{CH_3-CH-CH=CH_2} + H_2$

d. $\underset{\overset{|}{CH_3}}{CH_3-C=CH_2} + H_2 \longrightarrow \underset{\overset{|}{CH_3}}{CH_3-CH-CH_3}$

In this case every equation has H_2 as a reactant so that factor is not a clue. Notice that choice c is the only equation that does not have an alkene as a reactant. Since hydrogenation changes alkenes into alkanes, this reaction cannot be hydrogenation. **Answer is c.**

To achieve this objective you must demonstrate that you understand these terms by being able to correctly identify an equation that illustrates the process. For hydrolysis the key to understanding is identifying a reactant that has the ester functional group in it. Hydrolysis is the process of splitting the ester group into two components, an alcohol and a carboxylate ion. In hydration, you need to look for an alkene as a reactant and an alcohol as a product. Hydration is the process of adding the H and OH of water in place of one bond in the double bond. Dehydration is simply the opposite of hydration so the equation will have an alcohol as the reactant and will produce an alkene as a product. Samples of each type of reaction are given below with the key components in bold face type

Hydrolysis:

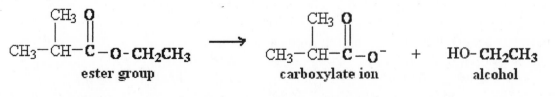

109

Hydration: water in

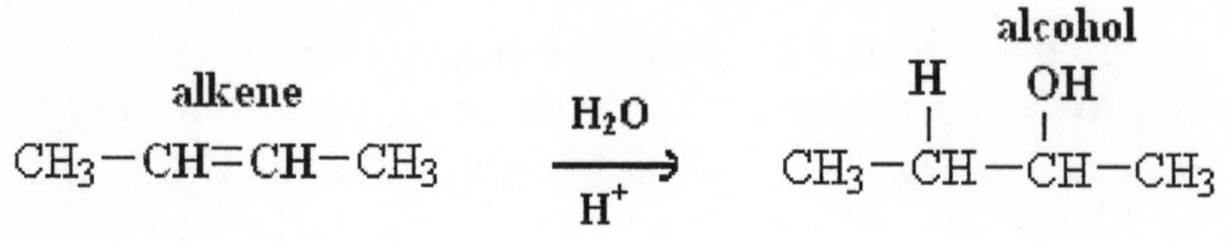

alkene   **alcohol**

$$CH_3-CH=CH-CH_3 \quad \xrightarrow[H^+]{H_2O} \quad CH_3-\underset{\overset{|}{H}}{CH}-\underset{\overset{|}{OH}}{CH}-CH_3$$

Dehydration: water out

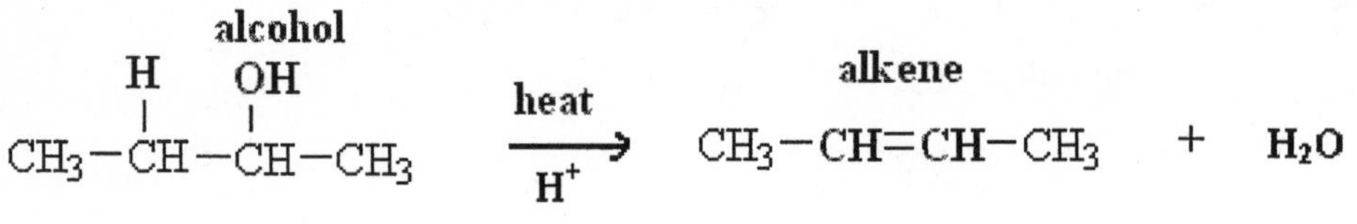

                    alcohol   **alkene**

$$CH_3-\underset{\overset{|}{H}}{CH}-\underset{\overset{|}{OH}}{CH}-CH_3 \quad \xrightarrow[H^+]{heat} \quad CH_3-CH=CH-CH_3 \quad + \quad H_2O$$

Since enthalpy, ΔH, and entropy, ΔS, were discussed in Chapter 5, Chapter 6 presents the interaction of the two in an energy value known as free energy, ΔG.
A negative ΔG value means the reaction is spontaneous and a positive ΔG means the reaction is nonspontaneous.

5. Which of the following reactions would be expected to be spontaneous?

a. $2H_2 + O_2 \rightarrow 2H_2O$ $ΔG = -280kcal/mol$

b. $H_2O_2 \rightarrow H_2 + O_2$ $ΔG = +30kcal/mol$

c. $CO_2 + 2H_2O \rightarrow CH_4 + 2O_2$ $ΔG = +275kcal/mol$

Answer is a since it is the only one with a negative *ΔG*

6. Which of the following reactions would be nonspontaneous?

a. $2H_2 + O_2 \rightarrow 2H_2O$ $ΔG = -280kcal/mol$

b. $H_2O_2 \rightarrow H_2 + O_2$ $ΔG = +30kcal/mol$

c. $CH_4 + 2O_2 \rightarrow CO_2 + 2H_2O$ $ΔG = -275kcal/mol$

Answer is b since it is the only one that has a positive *ΔG*.

The three main ways to change the rate of a reaction are: to change the temperature; change the concentration; or add a catalyst. In addition to listing these changes be sure you know how each one affects the rate. Raising the temperature increases kinetic energy; this increases both the energy of collisions and causes the molecules to collide more frequently, resulting in speeding up the reaction. Concentrations determine the number of collisions in a given time; therefore a higher concentration usually results in a faster reaction. A catalyst will lower the energy of activation and allow more reactions to occur with the energy that is already available, also causing the reaction to speed up.

7. If you have a solid reactant that is destroyed at higher temperatures (like proteins) what would be the best way to increase the speed of the reaction? Explain.

Use a catalyst. Since the reactant is a solid, the concentration cannot be increased; if the temperature is raised it may destroy the reactant. A catalyst is the best choice for increasing the speed of the reaction.

8. Which of the following is not a factor for speeding up a reaction?

a. add a catalyst
b. undergo hydration
c. raise the temperature
d. increase the concentration

Answer is d. Hydration is a chemical reaction process, not a factor for increasing the speed of a reaction.

9. Which of the following is not a factor for speeding up a reaction?

a. add a catalyst
b. lower the temperature
c. increase the concentration

Answer is b. Lowering the temperature would decrease the speed of the reactions.

Application

To apply what you've already learned and master the concepts, you must utilize all of the knowledge, understanding, facts, and/or techniques that you have gathered up to this point. Simply knowing an equation or process is not enough to master the concept. These you develop by *practice*. Solve as many different types of problems with as many different variations as you *strive to master the skill*. Your textbook provides a very good assortment of problems at the end of the chapter. Work as many as needed for skill mastery, not just those assigned.

- Be able to interpret and balance chemical equations.
- Determine the limiting reactant, theoretical yield, and percent yield of a reaction.

In order to truly master any skill you must *practice*. In this section example problems related to the application objectives are discussed. Remember, the more you practice, the better your chances of being able to successfully use these objectives to solve problems. Additional practice problems are at the end of the chapter in your textbook and in the online interactive lessons.

Sample Problems

The balanced equation plays a very important role in chemistry. It uses a shorthand notation that indicates the formulae of both reactants and products. Just as important though, is that the balanced equation gives the mole relationship between reactants and products. It is this mole relationship that provides the conversion factors chemists need to convert moles of one material to moles of another, as you will be doing in the next objective.

Interpreting chemical equations means that you can recognize the basic information provided, such as which atoms are present; how many of each atom are present; and any physical information provided such as (s) for solid; (l) for liquid; (g) for gas; or (aq) for aqueous, which means dissolved in water.

1. In the equation given below which of the following is *not* true?

$$C_2H_5OH(l) \ + \ 3O_2(g) \ \rightarrow \ 2CO_2(g) \ + \ 3H_2O(g)$$

a. 3 moles of O_2 are required to react with one mole of $C_2H_5OH(l)$
b. CO_2 is a gas
c. there are 5 molecules produced as products
d. $3O_2(g)$ is a product
e. there are seven O atoms on both sides of the arrow

Choice a is true since the coefficient 3 in front of the molecule can stand for three moles and there is an understood 1 in front of the $C_2H_5OH(l)$. The coefficients give the mole to mole ratio. Choice b is true because the small letter (g) indicates it is a gas. Choice c is true because there are 2 $CO_2(g)$ and 3$H_2O(g)$ so 2 + 3 = 5 molecules produced. Choice d. is *not* true since reactants are on the left side of a chemical equation and products are on the right. Answer e is true because on the reactant side, $C_2H_5OH(l)$ has one O atom and 3$O_2(g)$ has 3 x 2 = 6 atoms, so 6 + 1 = 7. On the product side, 2$CO_2(g)$ has 2 x 2 = 4 oxygen atoms and 3$H_2O(g)$ has 3 x 1 = 3 oxygen atoms, so 4 + 3 = 7. Be sure to examine all possibilities with this type of question as some professors may have more than one choice that is not true. **Answer is d.**

Often times, the basic reaction will be given so that you do not have to translate names to formulae or predict the products. This does not mean that the equation is finished. By the use of the term "equation" the implication is that the numbers and kinds of atoms on both sides of the arrow are the same. When starting a problem that requires an equation, <u>always</u> check to be sure it is balanced before you continue.

2. Balance the equation: $KClO_3 \rightarrow KCl + O_2$

Note: physical properties do not play a role in balancing the equation and may not be given.
Start by counting each atom.

reactant atoms	product atoms
1 K atom	1 K atom
1 Cl atom	1 Cl atom
3 O atoms	2 O atoms

The equation is not balanced since there are a different number of O atoms on each side. Since you *cannot* change the subscripts, note that the subscript on the O atom in the product is an even number and that any number multiplied by it will be an even number. Start by multiplying the formula with the odd number of reactant oxygens by a coefficient of 2. Then recount the atoms.

$$2KClO_3 \rightarrow KCl + O_2$$

reactant atoms	product atoms
2 K atom	1 K atom
2 Cl atom	1 Cl atom
6 O atoms	2 O atoms

Multiply the O_2 in the product by 3 to provide 6 atoms on the product side and recount.

$$2KClO_3 \rightarrow KCl + 3O_2$$

reactant atoms	product atoms
2 K atom	1 K atom
2 Cl atom	1 Cl atom
6 O atoms	**6** O atoms

Finally, place a coefficient of 2 in front of the KCl to provide 2 K atoms and 2 Cl atoms on the product side and recount.

$$2KClO_3 \rightarrow 2KCl + 3O_2$$

reactant atoms	product atoms
2 K atom	**2** K atom
2 Cl atom	**2** Cl atom
6 O atoms	**6** O atoms

The count is now the same for all atoms on both sides and the equation is balanced.
Answer: 2KClO₃ → 2KCl + 3O₂

3. Balance the equation: $CaCO_3 + HCl \rightarrow CaCl_2 + CO_2 + H_2O$

Note: physical properties do not play a role in balancing the equation and may not be given.

Start by counting each atom.

reactant atoms	product atoms
1 Ca atom	1 Ca atom
1 Cl atom	2 Cl atom
3 O atoms	3 O atoms
1 H atom	2 H atoms

The equation is not balanced since there are a different number of H atoms and Cl atoms on each side. Since you *cannot* change the subscripts, note that the subscript on the H atom in the product is an even number and that any number multiplied by it will be even so start by multiplying the formula with the odd number of reactant hydrogens by a coefficient of 2. Then recount the atoms.

$CaCO_3 + $**2**$HCl \rightarrow CaCl_2 + CO_2 + H_2O$

reactant atoms	product atoms
1 Ca atom	1 Ca atom
2 Cl atom	2 Cl atom
3 O atoms	3 O atoms
2 H atom	2 H atoms

Notice how it also multiplied the Cl atom count by two. Now the number of atoms on both sides are equal and the equation is balanced.
Answer: CaCO₃ + 2HCl → CaCl₂ + CO₂ + H₂O

The theoretical yield calculation combines the previously learned calculation skill of converting from grams to moles or moles to grams with another conversion that changes the formula of the reactant amount to the formula of the product amount. This is accomplished by using the mole to mole ratio between the reactant and the product given by the balanced equation.

There are three possible steps in this type of calculation:
Step 1: if given grams of a reactant convert it to moles of reactant
Step 2: convert the moles of the reactant to moles of the product

Step 3: convert from moles of product to grams of product

Be sure to read the problem carefully to see just which steps you will need to use. Note: Step 2 must be used to convert reactant formula to product formula and will always be included in a theoretical yield. Step 1 is needed when grams of reactant are given instead of moles and Step 3 is needed when grams of product are desired instead of mole of product.

4. How many moles of propane are expected from the complete reaction of 27.5 g of propene?

$$CH_3{-}CH{=}CH_2 \quad \xrightarrow[\text{Pt}]{\text{H}^+} \quad CH_3{-}CH_2{-}CH_3$$

Organic reactions are calculated in the same way as the inorganic reactions studied previously. Use the molar mass of propene to convert from grams to moles. Then, use the mol to mol ratio for C_3H_6 to C_3H_8 given by the equation as a conversion factor to convert from moles of C_3H_6 to moles of C_3H_8.

$$27.5 \ \cancel{g \ C_3H_6} \quad x \quad \frac{mol \ C_3H_6}{42.0 \ \cancel{g \ C_3H_6}} \ = \ 0.655 \ mol \ \ C_3H_6$$

$$0.655 \ \cancel{mol \ C_3H_6} \quad x \quad \frac{1 \ mol \ C_3H_8}{1 \ \cancel{mol \ C_3H_6}} \ = \ \textbf{0.655 mol C}_3\textbf{H}_8$$

5. How many grams of propane are expected from the complete reaction of 155 g of propene?

Follow the first two steps as in problem 4 above. Next, use the molar mass of 2-propanol to convert moles of propane to grams.

$$155 \ \cancel{g \ C_3H_6} \quad x \quad \frac{mol \ C_3H_6}{42.0 \ \cancel{g \ C_3H_6}} \ = \ 3.69 \ mol \ \ C_3H_6$$

$$3.69 \ \cancel{mol \ C_3H_6} \quad x \quad \frac{1 \ mol \ C_3H_8}{1 \ \cancel{mol \ C_3H_6}} \ = \ 3.69 \ mol \ C_3H_8$$

$$3.69 \ \cancel{mol \ C_3H_8} \quad x \quad \frac{44.0 \ g \ C_3H_8}{1 \ \cancel{mol \ C_3H_8}} \ = \ \textbf{162 g C}_3\textbf{H}_8$$

6. Given the equation: $4Al + 3O_2 \rightarrow 2Al_2O_3$
How many grams of Al_2O_3 can be expected to be produced with 2.50 g of Al?

$$2.50 \ \cancel{g \ Al} \quad x \quad \frac{mol \ Al}{27.0 \ \cancel{g \ Al}} \ = \ 0.0926 \ mol \ Al$$

$$0.926 \ \cancel{mol \ Al} \quad x \quad \frac{2 \ mol \ Al_2O_3}{} \ = \ 0.0463 \ mol \ Al_2O_3$$

4 ~~mol Al~~

$$0.0463 \; \text{mol Al}_2\text{O}_3 \quad \text{x} \quad \frac{102 \; \text{g Al}_2\text{O}_3}{1 \; \text{mol Al}_2\text{O}_3} \quad = \quad \textbf{4.72 g Al}_2\textbf{O}_3$$

Note that working this type of theoretical yield problem makes one big assumption. It assumes that there is more than enough oxygen to use up all of the aluminum. When only one reactant is given it is generally assumed to be used up. When both reactant quantities are given, you must first determine which will run out first (limit the reaction). The reaction stops when you run out of the limiting reactant.

To determine the limiting reactant, find the moles of each reactant, use the mole of reactant to mole of product ratio for each one. The one that produces (yields) the smallest number of product moles is the limiting reactant. The product moles produced by the limiting reactant is the amount of product that will be produced by the reaction. If you need to know the grams of product produced, convert the moles of product produced by the limiting reactant to grams.

7. Given the equation: $4\text{Al} + 3\text{O}_2 \rightarrow 2\text{Al}_2\text{O}_3$

How many grams of Al_2O_3 can be expected to be produced with 2.50 g of Al mixed with 3.00 g of O_2?

First calculate the Al moles as before:

$$2.50 \; \text{g Al} \quad \text{x} \quad \frac{\text{mol Al}}{27.0 \; \text{g Al}} \quad = \quad 0.0926 \; \text{mol Al}$$

Convert the moles of O_2:

$$3.00 \; \text{g O}_2 \quad \text{x} \quad \frac{\text{mol O}_2}{32.0 \; \text{g O}_2} \quad = \quad 0.0938 \; \text{mol O}_2$$

Convert the moles of Al to moles of Al_2O_3 as before:

$$0.926 \; \text{mol Al} \quad \text{x} \quad \frac{2 \; \text{mol Al}_2\text{O}_3}{4 \; \text{mol Al}} \quad = \quad 0.0463 \; \text{mol Al}_2\text{O}_3$$

Convert the moles of O_2 to moles of Al_2O_3:

$$0.0938 \; \text{mol O}_2 \quad \text{x} \quad \frac{2 \; \text{mol Al}_2\text{O}_3}{3 \; \text{mol O}_2} \quad = \quad 0.0625 \; \text{mol Al}_2\text{O}_3$$

Compare the moles of Al_2O_3 produced by the Al to the moles of Al_2O_3 produced by the O_2. Al produces 0.0463 mol of Al_2O_3 compared to O_2 that produces 0.0625 mol Al_2O_3. Since the amount produced by the Al is least, it is the limiting reactant.

Use the moles of Al_2O_3 produced by the Al (limiting reactant) to convert to grams of Al_2O_3.

$$0.0463 \text{ mol } Al_2O_3 \quad \text{x} \quad \frac{102 \text{ g } Al_2O_3}{1 \text{ mol } Al_2O_3} \quad = \quad \textbf{4.72 g } Al_2O_3$$

No matter how you calculate the theoretical yield with or without an assumed reactant in excess, it is still theoretical (calculated on paper). This calculation does not take into account the realities of life such as some reactant or product escaping from the container during the reaction, or compounds sticking to glassware or a stirring rod. In some cases the reaction simply will not go to completion. When working in the laboratory the amount of material that actually is produced is referred to as the actual yield. Chemists like to compare the actual yield to the theoretical yield to see a relative comparison between the two values. This is done as a percent calculation called percent yield.

Percent yield = (actual yield ÷ theoretical yield) x 100

8. If for problem 7 above, 4.00 g of Al_2O_3 were actually produced, calculate the percent yield.

$$\% \text{ yield} = \frac{\text{actual yield}}{\text{theoretical yield}} \text{ x } 100 =$$

$$\% \text{ yield} = \frac{4.00 \text{ g}}{4.72 \text{ g}} \text{ x } 100 = \textbf{84.7\%}$$

Chapter 6 Practice Test

Now that you've had the chance to study and review Chapter 6's practice problems provided in your textbook, you are ready to practice problems as you might see them in a testing situation. The questions and problems below are randomly selected in much the same way your instructor will select them; that is, not necessarily ordered in the same way as they appear in the textbook. To make best use of this practice, have your paper, pencil and calculator ready as you would for a class exam. Use a clock and give yourself 45 minutes to complete this activity. Note that taking this test under similar conditions as a class exam provides two benefits:

1. Testing your knowledge of the chapter's contents and your mastery of the required skills.
2. Practicing working under the stress of a timed exam.

Most instructors have a type of testing format that they prefer and will most likely tell you about it before the test. For practice, this self-test uses a variety of test question formats such as matching, fill-in-the-blank, listing, multiple-choice, and essay questions. ***Be sure to have a periodic table!***

Note the time and begin.

1. When the equation, C_6H_{14} + O_2 → CO_2 + H_2O is balanced, what is the coefficient in front of the O_2?

A. 15
B. 19
C. 2
D. 12

2. How many carbon atoms are in the molecule C_7H_7OH?

A. 7
B. 14
C. 1
D. 8

Use the equation below for the next three questions

$$\overset{+2 \ +6 \ -2}{NiSO_4} + \overset{0}{Mg(s)} \rightarrow \overset{0}{Ni(s)} + \overset{+2 \ +6 \ -2}{MgSO_4}$$

3. Which substance is being oxidized?

A. Ni^{2+}
B. $Mg(s)$
C. S^{6+}
D. $Ni(s)$

4. Which substance is being reduced?

A. Ni^{2+}
B. $Mg(s)$
C. S^{6+}
D. $Ni(s)$

5. Which substance is the oxidizing agent?

A. Ni^{2+}
B. $Mg(s)$
C. S^{6+}
D. $Ni(s)$

6. Which of the following best describes a balanced equation?

A. The loss of electrons
B Used as multipliers in front of formulae in balancing equation
C. Has the same number and kind of each atom on both sides of the equation
D. Uses an arrow to separate reactants from products

7. According to the equation below what would be the coefficient needed for the O_2 molecule?

$$2KClO_3 \rightarrow 2KCl + \_?\_ O_2$$

reactant atoms	product atoms
2 K atom	2 K atom
2 Cl atom	2 Cl atom
6 O atoms	2 O atoms

A. 1
B. 2
C. 3
D. 4

8. In the following equation, what should you expect to observe that would indicate a chemical reaction is occurring?

$$CaCO_3(s) + 2HCl(aq) \rightarrow CaCl_2(aq) + CO_2(g) + H_2O(l)$$

A. nothing without the use of an indicator
B. a solid precipitate
C. the solution becoming wet
D. bubbling from the production of a gas

9. Which of the following products best completes the reaction below?

$$H_2C{=\!=}CH_2 + H_2 \longrightarrow$$

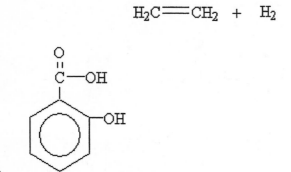

A.

B. $CH_3-CH_2-CH_2-CH{=}CH_2$

C. CH_3-CH_3

D. $CH_3-\overset{\overset{\displaystyle O}{\|}}{C}-H$

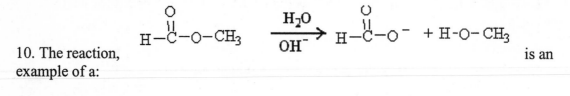

10. The reaction, is an example of a:

A. hydration reaction
B. dehydration reaction
C. hydrogenation reaction
D. hydrolysis

11. The reaction

$$CH_3-CH_2-CH=CH_2 \quad + \quad H_2O \quad \xrightarrow{H^+} \quad CH_3-\overset{\overset{\displaystyle OH}{\displaystyle |}}{CH}-CH_2-CH_3$$

is an example of a:

A. hydration reaction
B. dehydration reaction
C. hydrogenation reaction
D. hydrolysis

12. The reaction

$$\overset{\text{alcohol}}{\underset{\displaystyle CH_3-\overset{\overset{\displaystyle H}{\displaystyle |}}{CH}-\overset{\overset{\displaystyle OH}{\displaystyle |}}{CH}-CH_3}{}} \quad \xrightarrow[H^+]{\text{heat}} \quad \overset{\text{alkene}}{CH_3-CH=CH-CH_3} \quad + \quad H_2O$$

is an example of a

A. hydration reaction
B. dehydration reaction
C. hydrogenation reaction
D. hydrolysis

13. Which of the following is *not* a factor that increases the speed of a reaction?

A. increasing the temperature
B. adding a catalyst
C. increasing the particle size
D. increasing concentration

14. A reaction is most likely to be spontaneous when:

A. ΔS is negative
B. ΔG is negative
C. ΔH is positive
D. temperature is low

15. In the reaction, $2C_4H_2 \ + \ 13O_2 \ \rightarrow \ 8CO_2 \ + \ 10H_2O$
Which of the following statements is true?

A. the equation is not balanced
B. there are 18 oxygen atoms on the product side
C. there are 10 hydrogen atoms on the reactant side
D. there are 2 moles of C_4H_2 required for 13 moles of O_2 for a complete reaction

16. How many moles of C_2H_5OH are required to produce 1.5 moles of $2CO_2(g)$?

$$C_2H_5OH(l) + 3O_2(g) \rightarrow 2CO_2(g) + 3H_2O(g)$$

A. 0.25 mol
B. 0.50 mol
C. 0.75 mol
D. 1.00 mol

17. In a chloride ion determination, 0.075 g of AgNO3 were required to react completely with the chloride ions in the solution. How many moles of chloride ions were present?

$$Cl^-(aq) + AgNO_3(aq) \rightarrow AgCl(s) + NO_3(aq)$$

A. 4.4×10^{-4} mol
B. 0.075 mol
C. 1.5×10^{-2} mol
D. 1.28×10^1 mol

18. According to the equation below, how many grams of H_2 would be needed to produce 50.0 g of C_5H_{12}?

$$CH_3-CH_2-CH_2-CH=CH_2 + H_2 \longrightarrow CH_3-CH_2-CH_2-CH_2-CH_3$$

A. 0.644 g
B. 1.39 g
C. 50.0 g
D. 3.11 g

19. A student reacted 15.0 g of C_4H_8 with 5.00 g of H_2O according to the reaction below. What is the theoretical yield of C_4H_7OH?

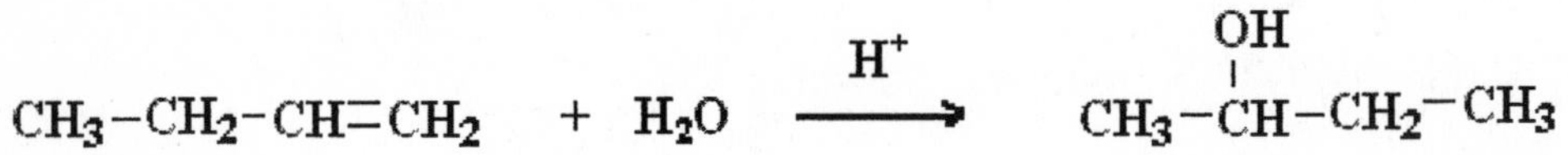

$$CH_3-CH_2-CH=CH_2 + H_2O \xrightarrow{\text{H}^+} CH_3-\underset{\underset{\text{OH}}{|}}{CH}-CH_2-CH_3$$

A. 19.3 g
B. 0.268 g
C. 0.278 g
D. 20.0 g

20. The theoretical yield for a reaction was calculated to be 25.0 g but when the student weighed the final product all they actually had was 23.4 g. What is the percent yield of their reaction?

A. 6.0%
B. 85%

C. 94%
D. 25%

21. Calculate the oxidation number for the S atom in $CaSO_3$.

Oxidation number = _____

22. Draw in the product for the following catalyst reaction.

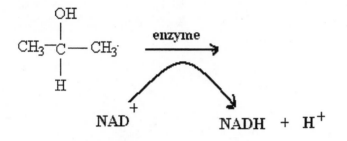

23. Circle the ester group on the molecule below.

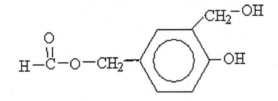

24. How many hydrogen atoms would be required to hydrogenate all of the double bonds in the molecule below?

$$CH_3-CH_2-CH=C=CH-CH_2-CH=C=CH-CH_3$$

25. True or False: Heating a reaction should increase the rate of reaction.

PRACTICE TEST Answers

1. B
2. A
3. B
4. A
5. A
6. C
7. C
8. D
9. C
10. D
11. A
12. B
13. C
14. B
15. D
16. C
17. A
18. B
19. A
20. C
21. +4

22.

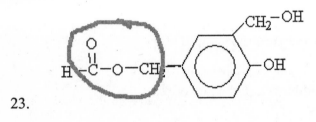

23.

24. 16 H atoms are needed

25. False

ANSWER PAGE FOR BUILDING KNOWLEDGE

Match the terms or phrases to the correct meaning.

1.___v___ Chemical equation

2.___u___ Balanced equation

3.___t___ Coefficients

4.___s___ Oxidation

5.___r___ Reduction

6.___x___ Combustion reaction

7.___f___ Oxidizing agent

8.___q___ Catalytic hydrogenation

9.___p___ Hydrolysis reaction

10.__o___ Hydration reaction

11.__l___ Theoretical yield

12.__w___ Activation energy

13.__m___ Limiting reactant

14.___J___ Percent yield

15.___c___ Rate of reaction

16.___i___ Reaction energy diagram

17.___h___ Enzyme

18.___b___ Catalysts

19.___e___ Products

20.___n___ Dehydration reaction

21.___k___ Actual yield

22.___a___ Temperature

23.___d___ Reactants

24.___g___ Reducing agent

a. is the average kinetic energy of the molecule forces

b. can lower the activation energy of a reaction and speeds up a reaction

c. is the measure of how fast a product is being produced

d. are the elements or compounds present at the start of a reaction

e. are the new substances formed

f. is the substance that is removing the electrons

g. is the substance that is providing electrons

h. is a biochemical catalyst

i. is a drawing that represents the energy change that takes place during a chemical reaction

j. found by dividing the actual yield by the theoretical yield and then multiplying by 100%

k. the amount of product actually collected in the laboratory

l. the calculated amount of product expected when the limiting reactant is used up

m. the reactant that runs out first

n. water being removed from an alcohol to make an alkene

o. adds water across the double bond of an alkene to make an alcohol

p. uses water to split another molecule

q. involves the reduction of a double bond by H_2 in the presence of Pt

r. the gaining of electrons

s. the loss of electrons

125

t. used as multipliers in front of formulae in balancing equation

u. has the same number and kind of each atom on both side of the equation

v. uses an arrow to separate reactants from products

w. is the energy barrier that must be crossed to go from reactants to products

x. reactant is oxidized (burned by O_2)

Chapter 7
Solutions, Colloids, and Suspensions

Chapter Overview

Up until this point you've been investigating substances made up of only one type of element or compound. You realize, of course, that in the real world substances are generally mixed together. In fact, in the chemical world substances mixed at the molecular level are called solutions. These mixtures appear pure even when viewed under a microscope. Many chemicals are shipped as solutions because the time involved in separating them is not cost effective. This means that in order for a chemist to work with these materials, it is necessary to understand that solutions may affect both the physical and chemical properties of its component materials.

In Chapter 7 you will investigate the parts of a solution and how they are identified. You will then investigate the different types of reactions that solutions can undergo. Next, methods used for determining how much of a substance is in a solution and how chemists create solutions of a particular concentration is discussed. Finally, the chapter explains how to distinguish solutions from other mixtures with similar properties by looking at the properties of suspensions and colloids, and the behavior of molecules or ions in a solution.

Knowledge/Comprehension

Typically, there are new terms or phrases that you must memorize in order for the lesson to make sense to you. Look up this information and memorize their explanations and definitions:

pure substance	homogeneous	heterogeneous
solvent	solute	solution
solubility	precipitate	Henry's law
hydrophilic	hydrophobic	amphipathic
weight/volume percent	volume/volume percent	weight/weight percent
parts per thousand (ppt)	parts per million (ppm)	parts per billion (ppb)
molarity (M)	dilution equation	suspension
colloid	diffusion	osmosis
hypotonic	hypertonic	isotonic

The following activity is designed to help you master the objectives outlined in the beginning of Chapter 7 and then summarized at the end of the chapter. To make the best use of your study time, *read* Chapter 7 before continuing. Next, work the matching activity provided as though you are taking a test *without first* referring to the answer page. Put your answers on a separate sheet of paper so that you can make more than one attempt without seeing your previous answers. Once you have completed the activity, check your answers. Make *flash cards* for those you missed. Have a study buddy drill you on them if you missed more than half of the terms. Once you have reviewed the ones you missed for at least 15 minutes try the activity again. If you miss fewer than four, write in the answers so that you will have them to review just before taking the test.

Match the terms or phrases to the correct meaning

1. _____ Pure substance

2. _____ Homogeneous

3. _____ Heterogeneous

4. _____ Solvent

5. _____ Solute

6. _____ Solution

7. _____ Solubility

8. _____ Precipitate

9. _____ Henry's law

10. _____ Hydrophilic

11. _____ Hydrophobic

12. _____ Amphipathic

13. _____ Weight/volume percent

14. _____ Volume/volume percent

15. _____ Weight/weight percent

16. _____ Parts per thousand (ppt)

17. _____ Parts per million (ppm)

18. _____ Parts per billion (ppb)

19. _____ Molarity (M)

20. _____ Dilution equation

21. _____ Suspension

22. _____ Colloid

23. _____ Diffusion

a. the solubility of a gas is proportional to the pressure of the gas over the liquid

b. $V_{original} \times C_{oringinal} = V_{final} \times C_{final}$

c. weight of solute (g)/volume of solution (mL) x 10^6

d. material present in greatest amount

e. movement of substances from areas of higher concentration to areas of lower concentration

f. weight of solute (g)/weight of solution (g) x 100

g. water fearing (insoluble in water)

h. made up of only one element or compound

j. volume of solute (mL)/volume of solution (mL) x 100

i. water-insoluble products

k. amount of solute that will dissolve in a solvent at a given temperature

l. weight of solute (g)/volume of solution (mL) x 10^3

m. moles of solute/liters of solution

n. water loving (soluble in water)

o. homogeneous mixture of a solvent and a solute

p. has both a water-soluble part and a water-insoluble part unaffected

q. concentration is such that water does not move and cells are not changed

r. particles are too large to dissolve and will settle out

s. weight of solute (g)/volume of solution (mL) x 100

t. only one visible material present

u. weight of solute (g)/volume of solution (mL) x 10^9

24.____Osmosis

v. movement of water across semipermeable membranes from areas of lower concentration to areas of higher concentration

25.____Hypotonic

w. material present in a lesser amount

26.____Hypertonic

x. intermediate size particles that do not settle out

27.____Isotonic

y. concentration causes water to move from the cells into the solution

z. concentration causes water to move from the solution into cells

aa. more than one visible material present

In addition to knowing the terms or phrases, you should be able to compare, contrast, or list examples that show that you understand the following concepts:

- Explain what differentiates solutes from solvents.
- Describe the effect of temperature on the solubility of gases, liquids, and solids in water.
- Explain Henry's law.
- Provide examples of compounds that are hydrophilic, hydrophobic, and amphipathic.
- List some commonly encountered concentration units.
- Describe the principles of diffusion and osmosis.

Once you know the appropriate tools for Chapter 7, you are ready to evaluate your understanding of those tools and their relationships. In this type of concept building, it is necessary to compare or contrast terms or concepts. In some cases you may be required to give examples or recognize a term or concept when it is depicted in a picture or sample scenario.

Sample Problems

Most students think of a solid dissolved in a liquid when asked to define a solution. As a result, they tend to think that the solute is always the solid and the liquid is the solvent. A solution is not actually defined in terms of "dissolving." It is a homogeneous mixture of atomic/molecular size particles and dissolving is a way of describing the mixing process. Solutions can be composed of particles in any of the three states of matter such as liquid in liquid, gas in liquid, gas in gas, solid in solid, etc. This is why the solute and solvent are distinguished by their amounts rather than by their physical states.

1. Identify the solvent and the solute in each of the mixtures below. Explain.

a. $NaCl_{(s)}$ 50.0 g $H_2O_{(l)}$ 300.0 g
b. $CO_{2(g)}$ 20.0 g $O_{2(g)}$ 5.0 g

c. $C_2H_5OH_{(l)}$ 15 mL $H_2O_{(l)}$ 350.0 g

In a. the mass of NaCl is 50.0 g compared to the 300.0 grams of H_2O.
The answer is: NaCl would be the solute and water would be the solvent.

In b, the mass of O_2 is 5.0 g compared to the 20.0 g of CO_2.
The answer is: O_2 would be the solute and the CO_2 would be the solvent.

In c, the volume of $C_2H_5OH(l)$ is 15 mL compared to the 350.0 mL of H_2O.
The answer is: $C_2H_5OH(l)$ would be the solute and the H_2O would be the solvent.

Remember that dissolving is a way of describing the mixing process. Temperature is the average kinetic energy (related to how fast molecules move and how much energy they have). As the amount of solute increases, the ability of the solvent particles to knock them apart and keep them apart becomes more difficult as illustrated below.

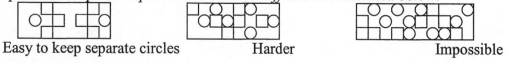

Easy to keep separate circles   Harder   Impossible

Raising the temperature of the solvent allows the solvent particles to hit the solute particles more often and keep them moving away from each other, allowing more solute to be dissolved.

For gases, the gas particles are kept in the mixture because they are surrounded by the solvent particles (they do not need to be moved apart). Raising the temperature will move more gas solute to the surface where they are free to move away from the solvent. Raising the temperature of the solvent decreases the solubility of gases.

2. If you wanted to make rock candy with a hot solution of sugar water would you want to raise the temperature higher or cool the solution? Explain.

Cool the solution. Rock candy is solid sugar; as the solution cools, there would be more sugar in the solution than is soluble at room temperature and the solid sugar (rock candy) would form.

3. You want to tie-dye some T-shirts but can't get all of the dye to dissolve into the amount of water you are using. Would it be best to cool the solution or heat it up? Explain.

Heat the solution; heating the solution should increase the solubility of the dye so that more of it can be mixed with the solvent.

4. Demonstrate your understanding of Henry's law by relating it to a similar event.

Sample Answer: Henry's law states that the amount of gas (concentration) that dissolves in a solvent is proportional to the pressure of the gas over the liquid. This means that if you double the pressure of the gas over a liquid, twice as much gas will dissolve in the solvent. Recall from above that the gas molecules will move away

when they reach the surface of the solvent. As the pressure of the gas above the liquid increases it becomes more difficult for the solute particles to escape. Think of it as you and friend (solute particles) trying to leave the classroom as students (additional solute particles) for the next class period come pushing (pressure above) into the classroom (liquid). You (a solute gas particle) may eventually get out but by the time you do, the number of students (additional solute gas particles) in the room has increased more than it has decreased, causing an increase in the concentration of students in the classroom. If there were no students outside the room (no pressure) you could leave the room quickly and the concentration in the room goes down.

This set of terms is a solubility classification scheme used primarily for biochemical compounds. It separates the compounds according to their ability to mix (dissolve) with water. Recall that the solubility of a solute increases the more its noncovalent forces are like those of the solvent (like dissolve like). The hydrophilic character increases as a biochemical molecule has more OH groups or NH groups that exhibit hydrogen bonding. The hydrogen bonding is similar to that found in water, making these molecules water-soluble (water-loving). Compounds that have a dominating carbon base have very little in common with water and therefore are water-insoluble (water-fearing). Some biochemical molecules are large enough that one end can have a large carbon base while another end is highly polar or ionic. These molecules are soluble in both water and organics and are called amphipathic.

5. Identify the molecules below as hydrophilic, hydrophobic, or amphipathic. Explain your choice.

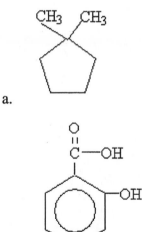

a.

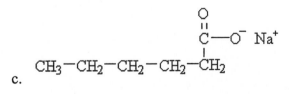

b.

c.

Structure a is *hydrophobic* because it is completely organic and has only dispersion forces as noncovalent forces.

Structure b is *hydrophilic* because it has multiple OH and =O sites for hydrogen bonding with water.

Structure c is *amphipathic* because one is purely organic (dispersion forces) while the other end is ionic, which could be attracted to the water.

6. In the box below, circle as many of the commonly used concentration units as you can find. Answer given on page 140.

```
dL        g      Al      ppm        Al              mg        Al      Al

     % (V/V)      g        Na       ppt      % (W/V)    Na      dL
 dL
          ppm    Na     ppb        mg         Al   Al         mg   ppm
  % (V/V)                          ppt    % (W/W)            % (W/V)
     g      ppb      ppt    g   % (V/V)   Al      g

          Al     mg    M                         mol/L   % (W/W)      ppb
  % (W/V)  ppm          ppt       dL
              % (W/W)                     Na    M    g      % (W/V)
  M                       mg      ppb
      ppb     ppt    Na      ppt   Na    mg    % (W/V)   Na      dL
          M                             M              M
    Na    mol      g       ppm    Al  % (V/V)    % (W/W)      dL
```

7. Examine the drawings below and describe how diffusion and osmosis are similar and different.

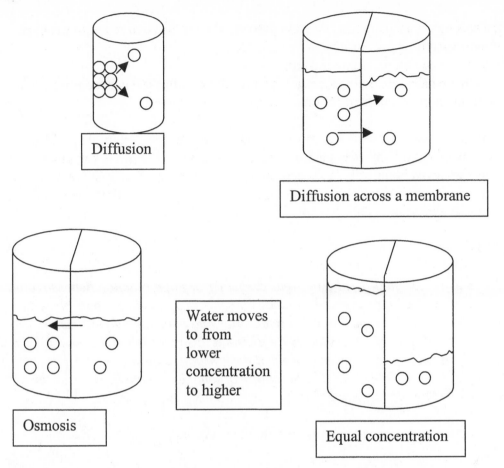

Diffusion

Diffusion across a membrane

Water moves to from lower concentration to higher

Osmosis

Equal concentration

The two processes differ because in diffusion, the *solute* particles move from a higher concentration to a lower concentration while in osmosis, the *solvent* moves from the lower concentration to the higher concentration. Osmosis also requires a semipermeable membrane between the concentrations, allowing only the solvent to pass through. Diffusion does not require a semipermeable membrane to occur. Students often confuse the two processes since the end result is similar –lowering the concentration of the solute in the solution.

Application

In order to apply what you've already learned and master the concepts, you must utilize all of the knowledge, understanding, facts, and/or techniques that you have gathered up to this point. Simply knowing an equation or process is not enough to master the concept. These you develop by *practice*. Solve as many different types of problems with as many different variations as you *strive to master the skill*. Your textbook provides a very good assortment of problems at the end of the chapter. Work as many as needed for skill mastery, not just those assigned.

- Be able to use the solubility rules to predict whether or not a reaction between ionic compounds will produce a precipitate.
- Perform calculations involving concentration.
- Be able to calculate the concentration of a solution after it has been diluted, or the volume required for the dilution.

In order to truly master any skill you must *practice*. In this section example problems related to the application objectives are discussed. Remember, the more you practice, the better your chances of being able to successfully use these objectives to solve problems. Additional practice problems are at the end of the chapter in your textbook and in the online interactive lessons.

Sample Problems

To be able to use the solubility rules to predict whether or not a reaction between ionic compounds will produce a precipitate, you must be able to apply three concepts. You have studied two of the concepts, predicting formulas and balancing equations, in previous chapters. The third concept is using the solubility rules to determine if products are aqueous (soluble in water) or solid precipitates (insoluble in water). For this review of the concepts, refer to Table 7.1 "The Solubility of Ionic Compounds" in your textbook. Some instructors may require you to know these rules for testing purposes.

1. Predict whether each ionic compound is soluble or insoluble in water. Give your reason.

a. Na_2SO_4 **Soluble—sulfates are soluble except those with Pb^{2+}, Sr^{2+}, Ba^{2+}, and Hg_2^{2+}**

b. KCl **Soluble—chlorides are soluble except those with Pb^{2+}, Ag^+, and Hg_2^{2+}**

c. $CaCO_3$ **Insoluble—carbonates are insoluble except those with alkali metals and NH^+**

d. $LiNO_3$ **Soluble—all nitrates are soluble**

2. When aqueous nickel (II) sulfate is mixed with aqueous lithium sulfide, a precipitate forms. Write a balanced equation for this reaction.

Before attempting to balance the equation make sure you write down the correct formula. Nickel (II) sulfate is composed of Ni^{2+} and SO_4^{2-}; since they have the same charge, their ratio is one to one and the formula is $NiSO_4(aq)$. Lithium sulfide is composed of Li^+ and

S^{2-}; its formula is Li$_2$S*(aq)*. Now that you have identified the reactants you need to predict the products. Previously you've learned that metals do not react with metals nor do nonmetals react with nonmetals. The only possible combination in this case is for the Ni^{2+} ion to react with the S^{2-} ion and form NiS. The Li$^+$ ion reacts with the SO$_4$ ion to form Li$_2$SO$_4$. Since you are told a precipitate forms and Table 7.1 in the text indicates that sulfates are soluble, the NiS must be the solid formed. Now we can write the equation and balance it.

$$\text{NiSO}_4(aq) \ + \ \text{Li}_2\text{S}(aq) \ \rightarrow \ \text{NiS}(s) \ + \ \text{Li}_2\text{SO}_4(aq)$$

Start by counting each atom.

reactant atoms	product atoms
1 Ni atom	1 Ni atom
2 S atoms	2 S atoms
4 O atoms	4 O atoms
2 Li atoms	2 Li atoms

In this case, all the atoms are the same on both sides of the equation. The equation is already balanced as is.

$$\textbf{NiSO}_4\textbf{\textit{(aq)}} \ + \ \textbf{Li}_2\textbf{S}\textbf{\textit{(aq)}} \rightarrow \ \textbf{NiS}\textbf{\textit{(s)}} \ + \ \textbf{Li}_2\textbf{SO}_4\textbf{\textit{(aq)}}$$

3. Complete and balance the following reaction: SrBr$_2$*(aq)* + Rb$_3$PO$_4$*(aq)* →

First, predict the products as in problem 2 above: Sr^{2+} reacts with PO$_4$$^{3-}$ to form Sr$_3$(PO$_4$)$_2$ and Rb$^+$ reacts with Br$^-$ to form RbBr. Table 7.1 indicates that phosphates are insoluble, but rubidium bromide is soluble—you can now complete and balance the equation.

$$\text{SrBr}_2(aq) + \text{Rb}_3\text{PO}_4(aq) \ \rightarrow \ \text{Sr}_3(\text{PO}_4)_2(s) \ + \ \text{RbBr}(aq)$$

Start by counting each atom.

reactant atoms	product atoms
1 Sr atom	3 Sr atoms
2 Br atoms	1 Br atom
4 O atoms	8 O atoms
3 Rb atoms	1 Rb atom
1 P atom	2 P atoms

When starting with a more complicated equation like this, it is usually best to start by balancing the atoms with the largest subscripts. Multiply Rb$_3$PO$_4$ by 2 as in 2 Rb$_3$PO$_4$ and recount the atoms.

reactant atoms	product atoms
1 Sr atom	3 Sr atom
2 Br atoms	1 Br atom
8 O atoms	8 O atoms
6 Rb atoms	1 Rb atom
2 P atoms	2 P atoms

Note that this balances the O atoms and the P atoms but not the Rb atoms. Next place a 6 in front of the RbBr as in 6RbBr and recount the atoms.

reactant atoms	product atoms
1 Sr atom	3 Sr atoms
2 Br atoms	6 Br atoms
8 O atoms	8 O atoms
6 Rb atoms	6 Rb atoms
2 P atoms	2 P atoms

This balances the Rb atoms but makes 6 Br on the left and 2Br on the right. To balance the Br, place a 3 in front of the SrBr$_2$ (3SrBr$_2$) and recount the atoms.

reactant atoms	product atoms
3 Sr atoms	3 Sr atoms
6 Br atoms	6 Br atoms
8 O atoms	8 O atoms
6 Rb atoms	6 Rb atoms
2 P atoms	2 P atoms

Now the number of atoms is the same on both sides of the arrow, creating a balanced equation.

3SrBr$_2$(aq) + 2Rb$_3$PO$_4$(aq) → Sr$_3$(PO$_4$)$_2$(s) + 6RbBr(aq)

Drill and practice problems like those given below until you can recall the process for each type of conversion.

4. You carefully measure 5 drops of ethanol into a test tube and add exactly 5 drops of water. What is the % (v/v) of ethanol in this aqueous solution?

To find the volume of the solution, add the volume of the solute to the volume of the solvent.

$$5 \text{ drops} + 5 \text{ drops} = 10 \text{ drops of solution}$$

To find the percent by volume use the equation

$$\% \text{ (v/v)} = \frac{\text{volume of solute}}{\text{volume of solution}} \times 100 = \%$$

$$\frac{5 \text{ drops solute}}{10 \text{ drops of solution}} \times 100 = \textbf{50 \%}$$

5. The normal concentration range of testosterone in blood serum is 300-1000 ng/mL. Express this concentration range in parts per million and parts per billion.

Since parts per million = (g of solute/mL) x 10^6 of solution and parts per billion = (g of solute/mL) x 10^9, convert ng to grams.

$$300\text{-}1000 \text{ ng} \quad \times \quad \frac{1 \text{ g}}{10^9 \text{ ng}} = 3 \times 10^{-7}\text{-} 1 \times 10^{-6} \text{ g/mL}$$

Parts per million = $(3 \times 10^{-7} \text{ g/mL}) \times 10^6 = 3 \times 10^{-1}$ ppm and $(1 \times 10^{-6} \text{ g/mL}) \times 10^6 = 1$ ppm

Parts per billion = $(1.6 \times 10^{-7} \text{ g/mL}) \times 10^9 = 1.6 \times 10^2$ ppb and $(1.7 \times 10^{-6} \text{ g/mL}) \times 10^9 = 1.7 \times 10^3$ ppb

Range in ppm = 3×10^{-1}-1 ppm
Range in ppb = 3×10^2-1 $\times 10^2$ ppb

6. If 30.0 g of $BaCl_2$ are present in 500 mL of aqueous solution, what is the concentration in terms of the following?

a. Molarity

Molarity is equal to the moles of the solute divided by liters of solution (M = mol/L). Convert grams of $BaCl_2$ to moles and 500 mL to L.

$$\frac{30.0 \text{ g } BaCl_2}{500 \text{ mL}} \quad \times \quad \frac{\text{mol } BaCl_2}{137.3 \text{ g } BaCl_2} \quad \times \quad \frac{1000 \text{ mL}}{1 \text{ L}} = 0.437 \text{ mol/L } BaCl_2$$

0.437 mol/L $BaCl_2$

b. Weight/volume percent

Since percent by weight/volume is the weight of the solute (in grams) divided by the volume of the solution (in mL) x 100, no conversions are necessary.

$$\frac{30.0 \text{ g BaCl}_2}{500 \text{ mL}} \quad \text{x } 100 \quad = 6.0\%$$

6.0 % BaCl$_2$

c. Parts per thousand

Parts per thousand is the weight volume ratio multiplied by a 10^3 instead of 100 as used in finding percent.

$$\frac{30.0 \text{ g BaCl}_2}{500 \text{ mL}} \quad \text{x } 10^3 \quad = 60 \text{ ppt}$$

60 ppt BaCl$_2$

d. Parts per million

Parts per million is the weight volume ratio multiplied by a 10^6 instead of 100 as used in finding percent.

$$\frac{30.0 \text{ g BaCl}_2}{500 \text{ mL}} \quad \text{x } 10^6 \quad = 6 \text{ x } 10^4 \text{ ppm}$$

6 x 10^4 ppm BaCl$_2$

e. Parts per billion

Parts per billion is the weight volume ratio multiplied by a 10^9 instead of 100 as used in finding percent.

$$\frac{30.0 \text{ g BaCl}_2}{500 \text{ mL}} \quad \text{x } 10^9 \quad = 6 \text{ x } 10^7 \text{ ppb}$$

6 x 10^7 ppb BaCl$_2$

7. Calculate the molarity of each.

a. 2.75 mole of LiCl in 25.0 L of solution

> Molarity (M) = mole of solute divided by liters of solution

> $M = \dfrac{2.75 \text{mol LiCl}}{25.0 \text{ L}} = 0.11 \text{ mol/L LiCl}$

0.11 mol/L LiCl

b. 170 mg of Na_2O in 2.00 mL of solution

> Since molarity (M) = mole of solute divided by liters of solution, convert mg to g, then g to mol, and then convert mL to L.

> $\dfrac{170 \text{ mg } Na_2O}{2.00 \text{ mL}} \times \dfrac{1 \text{ g } Na_2O}{10^3 \text{ mg } Na_2O} \times \dfrac{\text{mol } Na_2O}{62 \text{ g } Na_2O} \times \dfrac{1000 \text{ mL}}{1 \text{ L}} = 1.37 \text{ mol/L } Na_2O$

1.37 mol/L Na_2O

For dilution equations it is usually just a matter of inputting the given variables into the dilution equation, $V_{original} \times C_{original} = V_{final} \times C_{final}$, and algebraically solving for the variable that is missing. Both volume units and concentration units must match each other, but no specific volume units or concentration units are needed unless the problem specifically asks you to solve for a particular unit.

8. If 15 drops of 6.0 M HCl are diluted to a final volume of 100 drops, what is the new concentration?

> First, recall the dilution equation is $V_{original} \times C_{original} = V_{final} \times C_{final}$.
> Then, replace known variables and solve for the one missing.

> 15 drops x   6.0 M = 100 drops x C_{final}

> $\dfrac{(15 \text{ drops} \times 6.0 \text{ M})}{100 \text{ drops}} = C_{final} = 0.90 \text{ M}$

0.90 M

9. A 5.0% (w/v) solution of methanol is diluted from 150.0 mL to 250.0 mL. What is the new weight/volume percent?

Solution: recall the dilution equation is $V_{original}$ x $C_{original}$ = V_{final} x C_{final}.

150.0 mL x 5 % = 250.0 mL x C_{final}

$\dfrac{(150.0 \text{ mL } \text{ x } 5 \text{ \%})}{250.0 \text{ mL}}$ = C_{final} = 3 %

3 %

Answers for Concentration Units on page 132

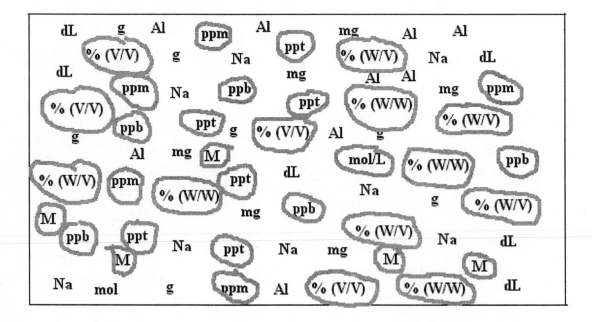

Chapter 7 Practice Test

Now that you have had the chance to study and review Chapter 7's concepts and objectives, and work the practice problems provided in your textbook, you are ready to practice problems as you might see them in a testing situation. The questions and problems below are randomly selected in much the same way your instructor will select them; that is, not necessarily ordered in the same way as they appear in the textbook. To make best use of this practice, have your paper, pencil and calculator ready as you would for a class exam. Use a clock and give yourself 45 minutes to complete this activity. Note that taking this test under similar conditions as a class exam provides two benefits:

1. Testing your knowledge of the chapter's contents and your mastery of the required skills.
2. Practicing working under the stress of a timed exam.

Most instructors have a type of testing format that they prefer and will most likely tell you about it before the test. For practice, this self-test uses a variety of test question formats such as matching, fill in the blank, listing, multiple choice, and essay questions.. **Be sure to have a periodic table!**

Note the time and begin.

1. Which of the following is *not* found in a solution?

A. A homogeneous mixture
B. Visible particles
C. Solvent
D. Solute

2. In a solution, which is found in the largest quantity?

A. Solute
B. Visible particles
C. Solvent
D. very small visible particles

3. When this equation is balanced, what is the coefficient of SrS?
$$SrSO_4(aq) \ + \ Na_2S(aq) \ \rightarrow \ SrS(s) \ + \ Na_2SO_4(aq)$$
A. 1
B. 2
C. 3
D. 4

4. When this equation is balanced, what is the coefficient of KCl?

$$BaCl_2(aq) + K_2CO_3(aq) \rightarrow BaCO_3(s) + KCl(aq)$$

A. 1
B. 2
C. 3
D. 4

5. Which of the following is a product of this reaction?

$$FeCl_2(aq) + Li_2CO_3(aq) \rightarrow$$

A. Li_2Fe
B. Cl_2CO_3
C. $FeCO_3$
D. CLi

6. Which of the following compounds is soluble in water?

A. CaS
B. $NaNO_3$
C. AgCl
D. $BaSO_4$

7. What would be the product formed by the ions Al^{3+} and O^{2-}?

A. AlO
B. Al_3O_2
C. Al_2O_3
D. O_2Al

8. Which of the following would be more soluble if the pressure over the liquid is increased?

A. $H_2O(l)$
B. NaCl(s)
C. HCl(aq)
D. $O_2(g)$

9. If you wanted to be able to dissolve more sugar in a solution after it has become saturated, the temperature should be:

A. increased
B. decreased
C. left alone

10. Soda fizzing or bubbling when the top is removed is best explained by:

A. Boyle's law
B. Henry' law
C. Charles' law
D. Ideal gas law

11. A large organic molecule with only carbons and hydrogens would probably be:

A. hydrophilic
B. hydrophobic
C. amphipathic
D. mutagenic

12. When a solution is saturated, this means:

A. no more solute will go into solution at that temperature
B. hydrogens have been added
C. no more solvent will go into the solution
D. only a small amount of the solute has been added

13. Hydrophilic compounds typically have:

A. hydrocarbons only
B. hydrogens bonded to oxygens
C. nonpolar bonds
D. dispersion forces

14. Amphipathic molecules generally have:

A. a hydrophobic end
B. a hydrophilic end
C. both a hydrophobic and a hydrophilic end

15. Smoke is an example of a:

A. suspension
B. colloid
C. solution
D. aqueous

16. The normal concentration range of vitamin B_{12} in blood serum is 160-170 ng/L. Express this concentration range in parts per million.

17. If 40.0 g of $MgCl_2$ are present in 150 mL of aqueous solution, what is the concentration in terms of molarity?

18. If 50.0 g of KCl are present in 150 mL of aqueous solution, what is the concentration in terms of weight/volume percent?

19. If 150.0 g of KCl are present in 1500 mL of aqueous solution, what is the concentration in terms of parts per thousand?

20. If 75.0 g of NaCl are present in 1000 mL of aqueous solution, what is the concentration in terms of parts per million?

21. If 175.0 g of KBr are present in 500 mL of aqueous solution, what is the concentration in terms of parts per billion?

22. Calculate the molarity of a solution made with 2.75 mole of KCl in 1.2 L of solution.

23. The normal serum concentration of chloride ion (Cl^-) is 95-107 mmol/L and you were found to have a Cl^- concentration of 105 mEq/L. Would you be in the normal range?

24. How many millimoles of $H_2PO_4^-$ are present in a water sample with a concentration of 2.5 mEq/L of $H_2PO_4^-$?

25. If 25.0 mL of 2.0 M HBr are diluted to a final volume of 100.0 mL, what is the new concentration?

PRACTICE TEST Answers

1. B
2. C
3. A
4. B
5. C
6. B
7. C
8. D
9. A
10. B
11. B
12. A
13. B
14. C
15. A
16. $1.6 \times 10^{-1} - 1.7 \times 10^{-1}$ ppm
17. 2.8 M
18. 33%
19. 100ppt
20. 7.5×10^{4} ppm
21. 3.5×10^{8} ppb
22. 2.3 M
23. Yes
24. 2.5 mmol
25. 0.5 M

ANSWER PAGE FOR BUILDING KNOWLEDGE

Match the terms or phrases to the correct meaning

1.__h___ Pure substance

2.__t___ Homogeneous

3.__aa__ Heterogeneous

4.__d__ Solvent

5.__w__ Solute

6.__o_ Solution

7.__k__ Solubility

8.__i___ Precipitate

9.__a___ Henry's law

10.__n___ Hydrophilic

11.__g__ Hydrophobic

12.__p__ Amphipathic

13.__s__ Weight/volume percent

14.__j__ Volume/volume percent

15.__f___ Weight/weight percent

16.__l___ Parts per thousand (ppt)

17.__c___ Parts per million (ppm)

18.__u__ Parts per billion (ppb)

19.__m___ Molarity (M)

20.___b__ Dilution equation

21.___r_ Suspension

22.___x__ Colloid

23.__e__ Diffusion

24.__v__ Osmosis

a. the solubility of a gas is proportional to the pressure of the gas over the liquid

b. $V_{original}$ x $C_{oringinal}$ = V_{final} x C_{final}

c. weight of solute (g)/volume of solution (mL) x 10^6

d. material present in greatest amount

e. movement of substances from areas of higher concentration to areas of lower concentration

f. weight of solute (g)/weight of solution (g) x 100

g. water fearing (insoluble in water)

h. made up of only one element or compound

j. volume of solute (mL)/volume of solution (mL) x 100

i. water-insoluble products

k. amount of solute that will dissolve in a solvent at a given temperature

l. weight of solute (g)/volume of solution (mL) x 10^3

m. moles of solute/liters of solution

n. water loving (soluble in water)

o. homogeneous mixture of a solvent and a solute

p. has both a water-soluble part and a water-insoluble part unaffected

q. concentration is such that water does not move and cells are not changed

r. particles are too large to dissolve and will settle out

s. weight of solute (g)/volume of solution (mL) x 100

t. only one visible material present

u. weight of solute (g)/volume of solution (mL) x 10^9

v. movement of water across semipermeable membranes from areas of lower concentration to areas of higher concentration

146

25.__z__ Hypotonic

26.__y__ Hypertonic

27.___q__ Isotonic

w. material present in a lesser amount

x. intermediate size particles that do not settle out

y. concentration causes water to move from the cells into the solution

z. concentration causes water to move from the solution into cells

aa. more than one visible material present

Chapter 8
Lipids and Membranes

Chapter Overview

In today's health conscious environment there is a lot of talk about good and bad cholesterol, as well as whether or not to eat saturated or unsaturated fats. Many people are influenced by advertising trends. However, the reality is that most of them wouldn't even recognize cholesterol or fats, let alone whether they're saturated or unsaturated. This chapter will serve to remove some of that mystery for you.

This chapter begins with a discussion about determining if a compound is classified as a fat (also called lipids). Then the chapter explains the characteristics of fatty acid structures that make them saturated, unsaturated (mono), or polyunsaturated. Despite what the media may have led you to believe, not all fats are bad. The chapter will help you identify the various types of fats and describe their primary biological functions. One group of fats, called steroids, has gotten a lot of media attention lately; they are discussed in detail, including the steroid cholesterol. Finally, you will learn about the important role that fats play in cell membrane function.

Knowledge/Comprehension

Typically, there are new terms or phrases that you must memorize in order for the lesson to make sense to you. Look up this information and memorize their explanations and definitions:

lipid	fatty acids	saturated
monounsaturated	polyunsaturated	waxes
triglyceride	phospholipids	glycolipids
glycerophospholipids	sphingolipids	sphingosine
steroids	cholesterol	sex hormones
adrenocorticoid	bile salts	eicosanoids
membranes	fluid mosaic model	facilitated diffusion
active transport		

The following activity is designed to help you master the objectives outlined in the beginning of Chapter 8 and then summarized at the end of the chapter. To make the best use of your study time, *read* Chapter 8 before continuing. Next, work the matching activity provided as though you are taking a test *without first* referring to the answer page. Put your answers on a separate sheet of paper so that you can make more than one attempt without seeing your previous answers. Once you have completed the activity, check your answers. Make *flash cards* for those you missed. Have a study buddy drill you on them if you missed more than half of the terms. Once you have reviewed the ones

you missed for at least 15 minutes try the activity again. If you miss fewer than four, write in the answers so that you will have them to review just before taking the test.

Match the terms or phrases to the correct meaning

_____ 1. Membranes   a. water-insoluble biochemical compound

_____ 2. Bile salts   b. consists of long chain carboxylic acids having between 12 and 20 carbon atoms

_____ 3. Active transport   c. containing only single carbon-carbon bonds in their hydrocarbon tail

_____ 4. Fluid mosaic model   d. fatty acids with only one carbon-carbon double bond

_____ 5. Adrenocorticoid   e. fatty acids with two or more carbon-carbon double bonds

_____ 6. Glycolipids   f. esters formed by combining long chain fatty acids with long chain alcohols

_____ 7. Sex hormones   g. a triester formed by combining one glycerol molecule with three fatty acids

_____ 8. Sphingosine   h. fats that have the phosphate ion as one of the components used in their formation.

_____ 8. Waxes   i. lipids that contain a sugar residue

_____ 10. Facilitated diffusion   j. consists of two glycerol, two fatty acids, phosphate, and an alcohol

_____ 11. Cholesterol   k. consists of residues of sphingosine

_____ 12. Phospholipids   l. consists of one fatty acid, phosphate, and an alcohol

_____ 13. Eicosanoids   m. lipids that share the same basic fused ring structure −three 6-carbon atom rings and one 5-carbon atom ring

_____ 14. Sphingolipids   n. steroid used to make other steroids in the body including sex hormones, adrenocorticoid hormones, and bile salts

_____ 15. Triglyceride   o. progesterone, testosterone, and estradiol

_____ 16. Polyunsaturated   p. hormones produced in the adrenal glands and include cortisol and cortisone

_____ 17. Fatty acids   q. produced in the gallbladder and aid in digestion by forming emulsions with dietary lipids.

_____ 18. Steroids   r. hormones derived from arachidonic acid and include prostaglandins, thromboxanes, and leukotrienes.

_____ 19. Lipid   s. bilayers containing phospholipids, glycoplipids, cholesterol, and proteins

_____ 20. Monounsaturated   t. proteins are viewed as being inlayed into a membrane like a mosaic, and membrane flexibility or fluidity is provided by the hydrocarbon tails of unsaturated fatty acid residues

_____ 21. Glycerophospholipids   u. diffusion across a cell membrane assisted by specific proteins

_____ 22. Saturated   v. energy-requiring process in which molecules and ions are moved across a membrane in the opposite direction of diffusion

In addition to knowing the terms or phrases, you should be able to compare, contrast, or list examples that show you understand the following concepts:

- Describe the structure of fatty acids.
- Describe the make-up of triglycerides and list their biological functions.
- Name the three types of eicosanoids and describe their biological function.
- Describe the make-up of a cell membrane.
- Identify the molecules used to make waxes and describe the primary biological function of waxes.
- Identify the basic steroid structure and list important members of this class of lipids.

Once you know the appropriate tools for Chapter 8, you are ready to evaluate your understanding of those tools and their relationships. In this type of concept building, it is necessary to compare or contrast terms or concepts. In some cases you may be required to give examples or recognize a term or concept when it is depicted in a picture or sample scenario.

Sample Problems

Fatty acids are carboxylic acids that contain between 12 and 20 carbons. The most characteristic feature of fatty acids is their carboxyl functional group. Without this group

you know that it cannot possibly be a fatty acid. Once you have determined that the compound is a carboxylic acid, count the carbons to determine if there are between 12 and 20 carbons.

Fatty acids are further subdivided into three groups: saturated, monounsaturated, and polyunsaturated. This subdivision is based on the bonding of the carbons: saturated-single bonds only, monounsaturated-only one double bond, polyunsaturated-more than one double bond.

1. Which of the following is *not* a carboxylic acid?

a.

$$\underset{\text{CH}_3\text{CH}_2}{\overset{\overset{\displaystyle O}{\|}}{\text{C}}}\text{-OH}$$

b.

$$\text{CH}_3-\text{CH}_2-\text{O}-\text{CH}_2-\text{CH}_3$$

c.

$$\text{CH}_3(\text{CH}_2)_{10}\overset{\overset{\displaystyle O}{\|}}{\text{C}}\text{-OH}$$

d.

$$\text{CH}_3(\text{CH}_2)_5\overset{\overset{\displaystyle O}{\|}}{\text{C}}\text{-OH}$$

$$\text{CH}_3\overbrace{(\text{CH}_2-\text{O}-\text{CH}_2}\text{CH}_3$$

Choice b does not have the carboxyl functional group
Answer is b.

2. Which of the molecules listed in question 1 is a fatty acid?

Begin as you did in question 1 by eliminating any choices that are not carboxylic acids. This eliminates choice. b. Next, count the carbons. Choice a. has 3 carbons, choice c has 12 carbons and choice d has 7 carbons. A fatty acid must have between 12 and 20 carbons. **Answer is c.**

3. In the following molecules, identify the functional group and count the carbons for the molecules that could possibly be fatty acids.

a.

$$\text{CH}_3-(\text{CH}_2)_6-\overset{\overset{\displaystyle O}{\|}}{\text{C}}\text{-OH}$$

b.

$$\text{CH}_3(\text{CH}_2)_{15}\overset{\overset{\displaystyle H}{|}}{\text{C}}=\overset{\overset{\displaystyle O}{\|}}{\text{C}}\text{-OH}$$

c.

$$CH_3-CH-CH_3$$
$$CH_3-CH_2-CH_2-CH-CH_2-CH_2-OH$$

d.

$$CH_3-(CH_2)_5-CH-\overset{O}{\overset{||}{C}}-CH_2-CH-CH_3$$

(with CH_3 substituents on the first and third CH groups)

Choice a has a carboxyl functional group and 8 carbons. Choice b has a carboxyl functional group and 18 carbons. Choice c. has an alcohol functional group and therefore no need to count the carbons, since it can't be a fatty acid. Choice d has a ketone functional group and therefore no need to count the carbons because it can't be a fatty acid. **Answer is a and b.**

4. Is the molecule in choice b saturated? Explain.

No. The molecule in choice b has a double bond in its structure which classifies it as a monounsaturated fatty acid.

5. Draw a triglyceride and identify the major components.

Triglycerides (animal fats and vegetable oils) are composed of three fatty acid residues joined to glycerol by ester bonds. Most of them have two or three different fatty acid residues.
Answer:

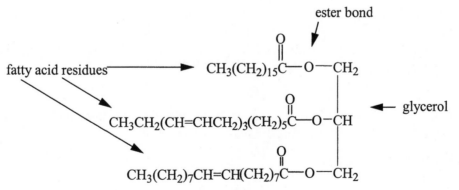

6. Triglycerides have three main biological functions. List them.

1. They provide energy
2. They protect against the cold

3. They prevent injury to internal organs by providing padding around them

7. Give the name of four types of eicosanoids and describe the biological function of each.

Eicosanoids are polyunsaturated fatty acids containing 20 (eico-) carbons.

Prostaglandins: Biological effects include causing pain, inflammation, and fever; affect blood pressure; PGE$_2$ induces labor.

Thromboxanes: Biological effect includes blood clotting.

Leukotrienes: Biological effect includes inducing muscle contractions in the lungs.

8. Describe the make-up of a cell membrane.

The cell membrane is described in terms of its two major components: its *fluid mosaic model* and the *bilayers of* amphipathic lipids and proteins. The lipids are usually *phospholipids, glycolipids, and cholesterol*. The lipids are arranged with the hydrophilic ends together on one side of the membrane and the hydrophobic ends together on the other side of the membrane. The polyunsaturated nature of the lipids with *cis* double bonds makes the *membrane a fluid*, while the more inflexible cholesterol adds rigidity. The *proteins are embedded* in membrane. The function of proteins in membranes will be discussed in Chapter 13.

9. Identify the molecules used to make waxes.

Waxes are esters produced by combining fatty acids with long chain alcohols.

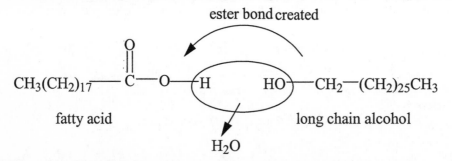

Waxes are esters with long 16 to 30 carbon atom chains attached to the carboxyl carbon, and a long chain of carbons attached to the –O.

10. What is the primary biological function of waxes?

The biological function of a wax is primarily that of protection and generally serves to keep water in or out of an organism. Another biological function of wax includes keeping skin or hair soft. Some organisms use wax as an energy storage molecule.

11. Identify the basic steroid structure.

The basic steroid structure is composed of three 6 carbon rings and one 5-carbon ring fused together.

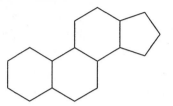

The important members of this class of lipids are differentiated by the arrangement of bonds and side chains on this structure.

12. List important members of the steroid class of lipids.

Answer: The important members of this group include: cholesterol, hormones, and bile salts.

Application

To apply what you've already learned and master the concepts, you must utilize all of the knowledge, understanding, facts, and/or techniques that you have gathered up to this point. Simply knowing an equation or process is not enough to master the concept. These you develop by *practice*. Solve as many different types of problems with as many different variations as you *strive to master the skill.* Your textbook provides a very good assortment of problems at the end of the chapter. Work as many as needed for skill mastery, not just those assigned.

- Distinguish phospholipids from glycolipids.
- Explain how saturated, monounsaturated, and polyunsaturated fatty acids differ from one another.
- Explain how various compounds cross the cell membrane.

In order to truly master any skill you must *practice*. In this section example problems related to the application objectives are discussed. Remember, the more you practice, the better your chances of being able to successfully use these objectives to solve problems. Additional practice problems are at the end of the chapter in your textbook and in the online interactive lessons.

Sample Problems

The distinguishing characteristic of a phospholipid is apparent from its name; it has a phosphate ion forming phosphoester bonds. The phospholipids are further subdivided into two types:

155

1. the glycerophospholipids, which has the glycerol backbone like triglycerides and;

2. the sphingolipids, in which the glycerol backbone is replaced by the alcohol sphingosine

1. Draw a glycerophospholipid and label the major components.

glycerophospholipid

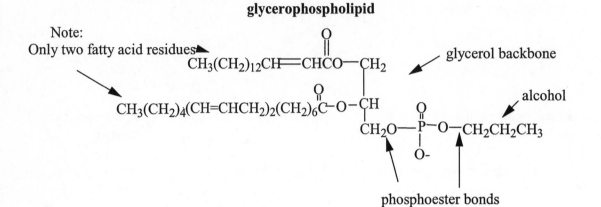

2. Draw a sphingolipid and label the major components.

sphingolipid

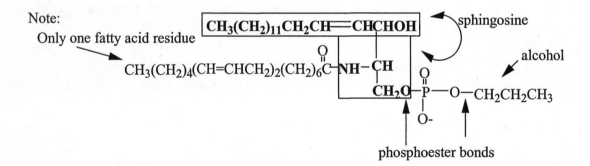

3. Draw a glycolipid and label the major components.

glycolipid

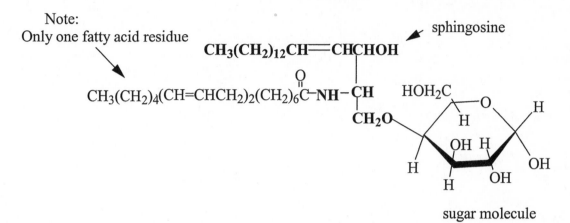

4. What major component distinguishes glycolipids from phospholipids?

Glycolipids have a sugar molecule instead of phosphoester bonds.

5. How do saturated, monounsaturated, and polyunsaturated fatty acids differ from one another?

They differ in the number of carbon-carbon double bonds. Saturated fatty acids have no carbon-carbon double bonds, monounsaturated fatty acids have one carbon-carbon double bond and polyunsaturated fatty acids have two or more carbon-carbon double bonds.

In questions 6 through 10, label the following molecules S for saturated, M for monounsaturated, and P for polyunsaturated.

_____6.

$$CH_3-(CH_2)_8-CH=CH-CH=CH-CH_2-\overset{O}{\overset{\|}{C}}-OH$$

_____7.

$$CH_3-(CH_2)_{10}-CH=CH-CH_2-\overset{O}{\overset{\|}{C}}-OH$$

_____8.

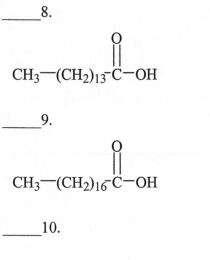

$$CH_3-(CH_2)_{13}-\overset{\overset{\displaystyle O}{\|}}{C}-OH$$

_____9.

$$CH_3-(CH_2)_{16}-\overset{\overset{\displaystyle O}{\|}}{C}-OH$$

_____10.

$$CH_3-(CH_2)_7-CH=CH-CH=CH-CH_2-CH=CH-CH_2-CH_2-\overset{\overset{\displaystyle O}{\|}}{C}-OH$$

6. (P) polyunsaturated—it has two double bonds in the carbon chain
7. (M) monounsaturated—it has one double bond in the carbon chain
8. (S) saturated—it has no double bonds in the carbon chain
9. (S) saturated—it has no double bonds in the carbon chain
10. (P) polyunsaturated—it has three double bonds in the carbon chain

11. How do various compounds cross a cell membrane?

Water, small nonpolar molecules (including O_2, N_2, and CO_2), and amphipathic steroids are able to diffuse directly through the membranes.

In a process called facilitated diffusion, some small ions (K^+, Na^+, etc.) and glucose are assisted through the membrane by proteins. This diffusion is always from high concentration to low concentration.

Protein-assisted movement through the membrane from low concentrations to high concentrations can move amino acids and small ions (K^+, Na^+, etc.) but requires energy to do so.

Chapter 8 Practice Test

Now that you've had the chance to study and review Chapter 8's practice problems provided in your textbook, you are ready to practice problems as you might see them in a testing situation. The questions and problems below are randomly selected in much the same way your instructor will select them; that is, not necessarily ordered in the same way as they appear in the textbook. To make best use of this practice, have your paper, pencil and calculator ready as you would for a class exam. Use a clock and give yourself 45 minutes to complete this activity. Note that taking this test under similar conditions as a class exam provides two benefits:

1. Testing your knowledge of the chapter's contents and your mastery of the required skills.
2. Practicing working under the stress of a timed exam.

Most instructors have a type of testing format that they prefer and will most likely tell you about it before the test. For practice, this self-test uses a variety of test question formats such as matching, fill-in-the-blank, listing, multiple-choice, and essay questions. *Be sure to have a periodic table!*

Note the time and begin.

For questions 1 through 4, identify the major components of the phospholipid below.

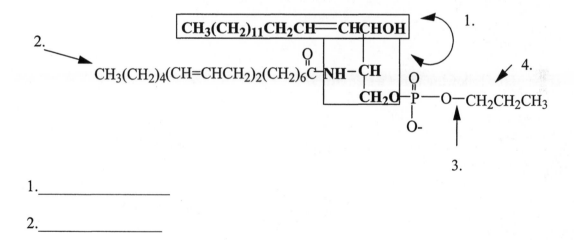

1._____

2._____

3_____

4._____

5. The molecule used in questions 1-4 would be classified as a:

A. glycerophopholipid
B. sphingolipid
C. glycolipids
D. triglyceride

6. In order to be an eicosanoid, the molecule must have:

A. 14 carbons
B. 30 carbons
C. 20 carbons
D. 6 carbons

7. Which of the following is not an eicosanoid?

A. leukotrienes
B. prostaglandins
C. thromboxanes
D. cholesterol

8. Bile salts are released from the gallbladder and:

A. aid digestion
B. lower blood pressure
C. control urine release
D. deepen the voice

9. Which of the following is not a steroid?

A. cholesterol
B. progesterone
C. leukotrienes
D. bile salts

10. Cell membranes can move selected ions from low concentration to high concentration in a process called:

A. facilitated diffusion
B. active transport
C. diffusion
D. filtration

11. The two classes of phospholipids are _____ and _____.

12. The molecule $CH_3(CH_2)_{14}$-CH=CH-$CH_2(CH_2)_5COOH$ would be classified as:

A. monounsaturated
B. polyunsaturated
C. saturated
D. concentrated

13. Describe the fluid mosaic model of a cell membrane.

For questions 14 through 17 draw the molecules produced when a triglyceride below is saponified.

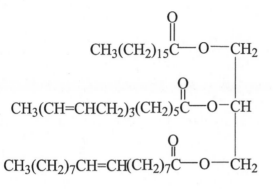

18. _____ is a female hormone and _____ is a male hormone.

19. Draw a line model of a phosphate ion.

20. True or False: A carboxylic acid with nine carbons would be classified as a fatty acid.

21. Would a polyunsaturated fat be found as a liquid or a solid at room temperature? Explain.

22. _____ is the molecule that makes up the backbone of the triglyceride.

23. Draw the basic ring structure associated with steroids.

24. _____ are lipids with the primary function of either keeping water in or out of an organism.

25. _____ consists of one fatty acid, phosphate, and an alcohol.

PRACTICE TEST Answers

1. sphingosine
2. fatty acid residue
3. phosphoester bond
4. alcohol
5. B
6. C
7. D
8. A
9. C
10. B
11. glycerophopholipids
12. A.
13. Bilayers containing phospholipids, glycolipids, cholesterol, and proteins inlayed like a mosaic.
14. through 17.

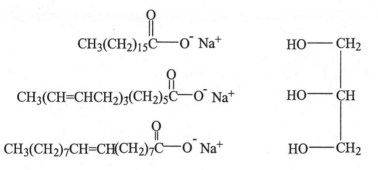

18. progesterone and testosterone
19.

$$
\begin{array}{c}
\overset{\displaystyle O^{-}}{\underset{\displaystyle O^{-}}{O=\overset{\textstyle |}{\underset{\textstyle |}{P}}-O^{-}}}
\end{array}
$$

20. F
21. Liquid. As the number of double bonds increases, the melting point of the fatty acid decreases.
22. glycerol
23.
24. waxes
25. sphingosines

ANSWER PAGE FOR BUILDING KNOWLEDGE

Match the terms or phrases to the correct meaning

___s__ 1. Membranes

a. water-insoluble biochemical compound

___q__ 2. Bile salts

b. consists of long chain carboxylic acids having between 12 and 20 carbon atoms

___v__ 3. Active transport

c. containing only single carbon-carbon bonds in their hydrocarbon tail

___t__ 4. Fluid mosaic model

d. fatty acids with only one carbon-carbon double bond

___p__ 5. Adrenocorticoid

e. fatty acids with two or more carbon-carbon double bonds

___i__ 6. Glycolipids

f. esters formed by combining long chain fatty acids with long chain alcohols

___o__ 7. Sex hormones

g. a triester formed by combining one glycerol molecule with three fatty acids

___l__ 8. Sphingosine

h. fats that have the phosphate ion as one of the components used in their formation.

___f__ 8. Waxes

i. lipids that contain a sugar residue

___u__ 10. Facilitated diffusion

j. consist of two glycerol, two fatty acids, phosphate, and an alcohol

___n__ 11. Cholesterol

k. consist of residues of sphingosine

___h__ 12. Phospholipids

l. consist of one fatty acid, phosphate, and an alcohol

___r__ 13. Eicosanoids

m. lipids that share the same basic fused ring structure —three 6-carbon atom rings and one 5-carbon atom ring

___k__ 14. Sphingolipids

n. steroid used to make other steroids in the body including sex hormones, adrenocorticoid hormones, and bile salts.

___g__ 15. Triglyceride

o. progesterone, testosterone, and estradiol

___e__ 16.Polyunsaturated

p. hormones produced in the adrenal glands and include cortisol and cortisone

___b__ 17. Fatty acids

q. produced in the gallbladder and aid in digestion by forming emulsions with dietary lipids.

___m__ 18. Steroids

r. hormones derived from arachidonic acid and include prostaglandins, thromboxanes, and leukotrienes.

___a__ 19. Lipid

s. bilayers containing phospholipids, glycoplipids, cholesterol, and proteins

___d__ 20. Monounsaturated   t. proteins are viewed as being inlayed into a membrane like a mosaic, and membrane flexibility or fluidity is provided by the hydrocarbon tails of unsaturated fatty acid residues

___j__ 21. Glycerophospholipids   u. diffusion across a cell membrane assisted by specific proteins

___c__ 22. Saturated   v. energy-requiring process in which molecules and ions are moved across a membrane in the opposite direction of diffusion

Chapter 9
Acids, Bases, and Equilibrium

Chapter Overview

By this time in the semester you've probably experienced at least one of those all-nighters with too much caffeine and junk food. When your stomach began to churn and burn you probably thought "Oh no! I have acid indigestion." In this chapter you will learn more about acids.

The chapter begins with a discussion of the common characteristics that distinguish acids from bases. This is followed by a more scientific definition of acids and bases. In order to understand acids, it is necessary to investigate how they behave in water and how their reactions can be controlled. Next, the concept of pH is explained so that you understand how solutions can be checked for acidity. Finally, the processes which allow acids and bases to be changed or quantified are discussed.

Knowledge/Comprehension

Typically, there are new terms or phrases that you must memorize in order for the lesson to make sense to you. Look up this information and memorize their explanations and definitions:

acids	bases	Bronsted-Lowery
conjugate	amphoteric	equilibrium
equilibrium constant	acidity constant	Le Châtelier's principle
catalyst	K_w	ionization
acidic	basic	neutral
pH	dissociates	pK_a
neutralization	titration	buffer

The following activity is designed to help you master the objectives outlined in the beginning of Chapter 9 and then summarized at the end of the chapter. To make the best use of your study time, *read* Chapter 9 before continuing. Next, work the matching activity provided as though you are taking a test *without first* referring to the answer page. Put your answers on a separate sheet of paper so that you can make more than one attempt without seeing your previous answers. Once you have completed the activity, check your answers. Make *flash cards* for those you missed. Have a study buddy drill you on them if you missed more than half of the terms. Once you have reviewed the ones you missed for at least 15 minutes try the activity again. If you miss fewer than four, write in the answers so that you will have them to review just before taking the test.

Match the terms or phrases to the correct meaning

_____ 1. Catalyst

a. dissolve some metals, turn litmus red, release H^+, and those safe to ingest have a sour taste

_____ 2. pH

b. feel slippery or soapy, turn litmus blue, accept H^+, and have a bitter taste

_____ 3. Equilibrium constant

c. definition of acids and bases that states that acids release H^+ and bases accept H^+

_____ 4. Buffer

d. the corresponding acid or base that is produced when a H^+ is gained or lost

_____ 5. Titration

e. can act as either an acid or a base

_____ 6. Equilibrium

f. the rate of the forward reaction and the rate of the reverse reaction are equal

_____ 7. Bronsted-Lowery

g. a value that expresses the relationship between the forward rate and the reverse rate of a reaction

_____ 8. Neutralization

h. the constant expressed for an acid in water

_____ 9. Dissociates

i. when a reversible reaction is pushed out of equilibrium, the reaction responds to reestablish equilibrium

_____ 10. Acids

j. increase reaction rates by lowering activation energy

_____ 11. pK_a

l. equilibrium constant for water equal to 1.0×10^{-14}

_____ 12. Basic

m. the formation of ions from molecules

_____ 13. Neutral

n. $[H_3O^+]$ is greater than 1×10^{-7} and the pH is less that 7

_____ 14. Ionization

o. $[H_3O^+]$ is less than 1×10^{-7} and the pH is greater that 7

_____ 15. Acidic

p. $[H_3O^+]$ is equal to 1×10^{-7} and the pH is equal to 7

_____ 16. Bases

q. $= -\log[H_3O^+]$

_____ 17. Le Châtelier's principle

r. ions falling apart

_____ 18. K_w

s. $= -\log K_a$

_____ 19. Acidity constant

t. an acid and a base react to form water and a salt

_____ 20. Amphoteric

u. technique used to determine the concentration of an acid or base

_____ 21. Conjugate

v. a solution prepared from weak acids and their conjugate bases that resists changes in pH

In addition to knowing the terms or phrases, you should be able to compare, contrast, or list examples that show that you understand the following concepts:

- List the common characteristics of acids and bases.
- Describe the processes of neutralization and titration.
- Explain how the pH of a solution can affect the relative concentrations of an acid and its conjugate base.
- Describe buffers.
- Explain how Bronsted-Lowry acids and bases differ from their conjugates.

Once you know the appropriate tools for Chapter 9, you are ready to evaluate your understanding of those tools and their relationships. In this type of concept building, it is necessary to compare or contrast terms or concepts. In some cases you may be required to give example or recognize a term or concept when it is depicted in a picture or sample scenario.

Sample Problems

1. Which of the following would not be part of a neutralization reaction?

    a. production of a salt

    b. formation of water

    c. peroxide decaying to water

    d. H^+ ions are transferred

A neutralization reaction is the reaction between an acid and a base. As the process proceeds, the two typical products are a salt (the negative anion of the acid combined with the positive cation of the base) and water. Therefore, choices a and b are not the answer. Peroxide decaying to water is a chemical reaction that does not involve an acid and base; it is not a neutralization reaction. The way acids react with bases is by donating H^+ ions to the cation (often OH^-) of the base. Therefore, choice d would be part of a neutralization process. **Answer is c.**

2. Which of the following would not be needed in a titration process?

    a. acid and base

    b. thermometer

    c. indicator

    d. buret

The titration process is a well-controlled neutralization reaction that allows you to determine the concentration of an acid or base by using a known amount of one to neutralize the other. To determine the acid concentration, a known amount is pipetted into an Erlenmeyer flask and the base of known concentration is added until the acid is neutralized; choice b cannot be the correct answer. The base is usually delivered through a special measuring device called a buret; choice d cannot be the answer. The point of neutralization, referred to as the *endpoint*, is detected by an indicator; choice c is not the correct answer. Although temperature may affect reaction rates, it does not have a significant effect on concentration and therefore a thermometer is not needed. **Answer is b.**

3. Which of the following change the pH of a solution?

    a. adding H^+ ions

    b. adding OH^- ions

    c. adding more conjugate base

    d. adding water

    e. all of the above

The pH of a solution is a logarithmic-based representation of the hydrogen ion concentration in a solution (also represented in solution as H_3O^+) so naturally choice a, adding H^+ ions, is correct. On the other hand, adding OH^- ions would result in H^+ ions reacting to form water, therefore reducing the H^+ ion concentration and changing the pH; choice b is also correct. For reversible acid solutions, adding more conjugate base pushes the reaction back toward the weak acid and also removes H^+ ions, lowering the pH; choice c is correct. Since pH is based on the H^+ ion concentration, adding more water will dilute the solution and make the H^+ ion concentration lower, thus changing the pH, which means choice d is also correct. **Answer is e.**

4. For the following reversible weak acid solution, which of the following would make the reaction produce more of the conjugate base?

$$HC_2H_3O_2 \quad \rightleftharpoons \quad H_3O^+ \ + \ C_2H_3O_2^-$$

    a. lowering the pH

    b. raising the pH

    c. leaving the pH constant

The reaction is currently at equilibrium, leaving the pH, the H_3O^+ ion concentration, and the conjugate base, $C_2H_3O_2^-$ concentration all constant. Choice c is not correct–the pH of the solution is based on the $[H_3O^+]$; raising the pH increases $[H_3O^+]$. This results in the production of more weak acid and the removal of the conjugate base so choice b is not correct. Lowering the pH means a lower $[H_3O^+]$–H_3O^+ is being removed and the reaction shifts to produce more $[H_3O^+]$ which means it also produces more conjugate base. **Answer is a.**

5. Which of the following would not be an important property of a buffer?

    a. a weak acid

    b. the conjugate base

    c. weak acid salt

    d. a strong acid

A buffer must have the ability to react in more than one direction so that if a small amount of acid or base is added to the solution, the pH will not change. In order for this to happen, the reaction must be a reversible reaction like that of a weak acid; choice a is not correct. To control the input of acid there must also be a substantial amount of the conjugate base present; choice b is not correct. Since a conjugate base is an anion, it cannot exist alone; it is introduced into the buffer in the form of the weak acid salt, therefore, choice c is not correct. A strong acid would ionize completely and not be reversible to any significant degree. Choice d is correct because you would not want a strong acid in a buffer solution. **Answer is d.**

6. In the equations below, identify both the Bronsted-Lowry acid and base as well as their conjugates.

    A. $HC_2H_3O_2 + H_2O \rightleftharpoons H_3O^+ + C_2H_3O_2^-$

    B. $H_2S + NH_3 \rightleftharpoons NH_4^+ + HS^-$

According to the Bronsted-Lowry definition, acids give off H^+ ions. In equation A the $HC_2H_3O_2$ is becoming $C_2H_3O_2^-$ by giving off H^+, making it the acid. The ion created as the H^+ is given off is the conjugate base of the acid, making the $C_2H_3O_2^-$ ion the conjugate base. Since the H_2O became H_3O^+ it gained an H^+ and according to the Bronsted-Lowry definition of acids and bases, the H_2O is the base. The ion created by the base accepting the H^+ is the conjugate acid, making the H_3O^+ the conjugate acid.

acid	base		conjugate acid		conjugate base
$HC_2H_3O_2$ +	H_2O	$\rightleftharpoons$	H_3O^+	+	$C_2H_3O_2^-$

According to the Bronsted-Lowry definition, acids give off H^+ ions, so in equation B the H_2S is becoming HS^- by giving off H^+, making it the acid. The ion created as the H^+ is given off is the conjugate base of the acid, making the HS⁻ ion the conjugate base. Since the NH_3 became NH_4^+ it gained an H^+ and according to the Bronsted-Lowry definition, the NH_3 is the base. The ion created by the base accepting the H^+ is the conjugate acid, making the NH_4^+ the conjugate acid.

acid		base		conjugate acid		conjugate base
H_2S	$+$	NH_3	$\rightleftharpoons$	NH_4^+	$+$	HS^-

Application

In order to apply what you've already learned and master the concepts, you must utilize all of the knowledge, understanding, facts, and/or techniques that you have gathered up to this point. Simply knowing an equation or process is not enough to master the concept. These you develop by *practice*. Solve as many different types of problems with as many different variations as you *strive to master the skill*. Your textbook provides a very good assortment of problems at the end of the chapter. Work as many as needed for skill mastery, not just those assigned.

- Relate acid strength to conjugate base strength.
- Write the equilibrium constant for a reversible reaction and use Le Châtelier's principle to explain how equilibrium responds to being disturbed.
- Use $[H_3O^+]$ to identify a solution as being acidic, basic, or neutral.
- Use pH to identify a solution as being acidic, basic, or neutral.

In order to truly master any skill you must *practice*. In this section example problems related to the application objectives are discussed. Remember, the more you practice, the better your chances of being able to successfully use these objectives to solve problems. Additional practice problems are at the end of the chapter in your textbook and in the online interactive lessons.

Sample Problems

1. Use the table below to decide if the acid concentration or conjugate base concentration would be greater when the weak acid, NH^+, is placed in water.

acid	K_a
HF	6.6×10^{-4}
CH_3CO_2H	1.8×10^{-5}
NH^+	5.6×10^{-10}
HCO_3^-	5.6×10^{-11}

K_a is an equilibrium expression that indicates the relationship between products formed and reactants used. Acids which produce more products (more H_3O^+ created) before equilibrium is reached have higher K_a values. For every H_3O^+ created, a conjugate base will also be created. **The extremely low K_a for NH^+ indicates that very little of its**

conjugate base, NH_3, is being produced compared to the initial concentration of the weak acid, NH^+.

2. Which of the following equations is the correct equilibrium constant for

$$HC_2H_3O_2 \rightleftharpoons H_3O^+ + C_2H_3O_2^- \ ?$$

a. $K_a = \dfrac{[HC_2H_3O_2][H_3O^+]}{[C_2H_3O_2^-]}$

b. $K_a = \dfrac{[HC_2H_3O_2]}{[C_2H_3O_2^-][H_3O^+]}$

c. $K_a = \dfrac{[H_3O^+][C_2H_3O_2^-]}{[HC_2H_3O_2]}$

d. $K_a = \dfrac{[H_3O^+][C_2H_3O_2^-]^2}{[HC_2H_3O_2]^2}$

Choice a has one reactant in the numerator and one in the denominator; it is not the correct choice. Choice b has the reactants in the numerator instead of the denominator and is also not the correct choice. Choice c has the products and the reactants as they should be and since the coefficients are 1, each concentration is expressed to the correct power. Choice c is correct. Choice d has the reactants and products in the correct order but there are no coefficients in the equation so the power of 2 is not needed. **Answer is c.**

3. Based on Le Châtelier's principle, indicate in the following equilibrium equation whether a net forward or net reverse reaction will result in order to reestablish equilibrium:

$$NH_3 + H_2O \rightleftharpoons NH_4^+ + OH^-$$

a. NH_3 is added **A net forward reaction will result**

b. OH^- is removed **A net forward reaction will result**

c. H_2O is added **No affect, water is a pure liquid and adding more does not significantly change its concentration**

173

4. Calculate the H_3O^+ concentration present in water when $[OH^-] = 7.8 \times 10^{-5}$ M.

$K_w = 1.0 \times 10^{-14} M = [H_3O^+][OH^-]$

$[H_3O^+] = 1.0 \times 10^{-14} M / [OH^-]$

$[H_3O^+] = 1.0 \times 10^{-14} M / [7.8 \times 10^{-5} M]$

$\mathbf{[H_3O^+] = 1.3 \times 10^{-10} M}$

5. Calculate the H_3O^+ concentration present in water when $[OH^-] = 1.6 \times 10^{-4}$ M.

$K_w = 1.0 \times 10^{-14} M = [H_3O^+][OH^-]$

$[H_3O^+] = 1.0 \times 10^{-14} M / [OH^-]$

$[H_3O^+] = 1.0 \times 10^{-14} M / [1.6 \times 10^{-4} M]$

$\mathbf{[H_3O^+] = 6.3 \times 10^{-11} M}$

6. Indicate whether the solutions in problems 4 and 5 are acidic, basic, or neutral.

Acidic solutions have $[H_3O^+]$ greater than 1×10^{-7} M. Basic solutions have $[H_3O^+]$ less than 1×10^{-7} M. Neutral solutions have $[H_3O^+] = 1 \times 10^{-7}$ M.

In problem 4, the $[H_3O^+] = 1.3 \times 10^{-10}$ M, which is less than 1×10^{-7} M so the solution is basic. **Answer is basic.**

In problem 5, the $[H_3O^+] = 6.3 \times 10^{-11}$ M, which is less than 1×10^{-7} M so the solution is basic. **Answer is basic.**

7. Calculate the OH^- concentration present in water when $[H_3O^+] = 2.2 \times 10^{-3}$ M

$K_w = 1.0 \times 10^{-14} M = [H_3O^+][OH^-]$

$[OH^-] = 1.0 \times 10^{-14} M / [H_3O^+]$

$[OH^-] = 1.0 \times 10^{-14} M / [2.2 \times 10^{-3} M]$

$\mathbf{[OH^-] = 4.5 \times 10^{-12} M}$

8. Calculate the OH⁻ concentration present in water when $[H_3O^+] = 4.5 \times 10^{-9}$ M

$$K_w = 1.0 \times 10^{-14} M = [H_3O^+][OH^-]$$

$$[OH^-] = 1.0 \times 10^{-14} M / [H_3O^+]$$

$$[OH^-] = 1.0 \times 10^{-14} M / [4.5 \times 10^{-9} M]$$

$$\mathbf{OH^-] = 2.2 \times 10^{-6} M}$$

9. Indicate whether the solutions in problems 7 and 8 are acidic, basic, or neutral.

Acidic solutions have [OH⁻] less than 1 x 10⁻⁷ M. Basic solutions have [OH⁻] greater than 1 x 10⁻⁷ M. Neutral solutions have [OH⁻] = 1 x 10⁻⁷ M.

In problem 7, $[OH^-] = 4.5 \times 10^{-12}$ M, which is less than 1×10^{-7} M so the solution is acidic. **Answer is acidic.**

In problem 8, $[OH^-] = 2.2 \times 10^{-6}$ M, which is greater than 1×10^{-7} M so the solution is basic. **Answer is basic.**

10. Calculate the pH of a solution in which $[H_3O^+] = 1 \times 10^{-7}$ M

pH = -log [H₃O⁺]

pH = -log [1 x 10⁻⁷ M]   On a scientific calculator, enter the number 1 x 10⁻⁷ into the calculator, then press the **log** button. This gives -7. Then, - (-7) = 7.

pH = 7.0

11. Calculate the pH of a solution in which $[H_3O^+] = 2.0 \times 10^{-4}$ M

pH = -log [H₃O⁺]

pH = -log [2.0 x 10⁻⁴ M]   On a scientific calculator, enter the number 2.0 x 10⁻⁴ M into the calculator, then press the **log** button. This gives -3.69. Then, - (-3.69) = 3.69.

pH = 3.69

12. Indicate whether the solutions in problems 10 and 11 are acidic, basic or neutral.

Acidic solutions have a pH less than 7. Basic solutions have a pH greater than 7. Neutral solutions have a pH = 7.

In problem 10, pH = 7.0, which is equal to 7, so the solution is neutral. **Answer is neutral.**

In problem 11, pH = 3.69, which is less than 7 so the solution is acidic. **Answer is acidic.**

13. What is $[H_3O^+]$ in a solution if the pH is 3.54?

pH = -log [H₃O⁺]
-pH = log [H₃O⁺]
Antilog (-pH) = [H₃O⁺]

Antilog (-5.54) = $[H_3O^+]$ On a scientific calculator, enter the number -3.54 (remember to put in 3.54 then press the sign change, *do not* subtract) into the calculator, then press the **2ⁿᵈ** button followed by the **log** button. This gives 2.88×10^{-4}.

$[H_3O^+]$ = 2.9 x 10⁻⁴ M

14. What is $[H_3O^+]$ in a solution if the pH is 11.9

Antilog (-pH) = [H₃O⁺]

Antilog (-11.9) = $[H_3O^+]$ On a scientific calculator, enter the number -11.9 (remember to put in 11.9 then press the sign change, *do not* subtract), then press the **2ⁿᵈ** button followed by the **log** button. This gives 1.258×10^{-12}.

$[H_3O^+]$ = 1 x 10⁻¹² M

Chapter 9 Practice Test

Now that you have had the chance to study and review Chapter 9's concepts and objectives, and work the practice problems provided in your textbook, you are ready to practice problems as you might see them in a testing situation. The questions and problems below are randomly selected in much the same way your instructor will select them; that is, not necessarily ordered in the same way as they appear in the textbook. To make best use of this practice, have your paper, pencil and calculator ready as you would for a class exam. Use a clock and give yourself 45 minutes to complete this activity. Note that taking this test under similar conditions as a class exam provides two benefits:

1. Testing your knowledge of the chapter's contents and your mastery of the required skills.
2. Practicing working under the stress of a timed exam.

Most instructors have a type of testing format that they prefer and will most likely tell you about it before the test. For practice, this self-test uses a variety of test question formats such as matching, fill in the blank, listing, multiple choice, and essay questions.. **_Be sure to have a periodic table!_**

Note the time and begin.

1. Which of the following is not a property of an acid?

A. sour taste
B. turns litmus red
C. gives off OH^-
D. has a pH less than 7

2. Which of the following ions is the conjugate base for HCl?

A. Cl^-
B. H^+
C. HCl^-
D. HCl^+

_____3. For the equation, $2H_2(g) + O_2(g) \rightleftharpoons 2H_2O(l)$,
$K_{eq} =$

    A. $\dfrac{[H_2][O_2]}{[H_2O]}$

    B. $\dfrac{[H_2]^2[O_2]}{[H_2O]^2}$

C. $\dfrac{[H_2O]^2}{[H_2]^2[O_2]}$

D. $[H_2]^2[O_2]$

4. Which of the following could not be a Bronsted-Lowry acid?

A. HCl
B. Na_2CO_3
C. HNO_3
D. HSO_4^-

5. For the reversible reaction,

$$H_2S + NH_3 \rightleftharpoons NH_4^+ + HS^-,$$

which of the following would not favor the forward reaction?

A. adding H_2S
B. adding a catalyst
C. adding NH_3
D. removing NH_4^+

6. If the $[OH^-] = 5.2 \times 10^{-9}$ what is the $[H_3O^+]$?

A. 1.9×10^{-6}
B. 5.2×10^{-9}
C. 1×10^{-6}
D. 8.8×10^{-9}

7. The solution in problem 6 would be considered to be:

A. acidic
B. basic
C. neutral
D. salty

8. If the $[H_3O^+] = 3.2 \times 10^{-11}$ what is the $[OH^-]$?

A. 3.9×10^{-6}
B. 5.2×10^{-9}
C. 3.1×10^{-4}
D. 1.8×10^{-9}

9. The solution in problem 8 would be considered to be:

A. acidic
B. basic
C. neutral
D. salty

10. If the $[H_3O] = 3.20 \times 10^{-3}$ what is the pH?

A. 11.5
B. 3.2
C. 3
D. 2.5

11. The solution of problem 10 would be considered to be:

A. acidic
B. basic
C. neutral
D. salty

12. If the pH is 8.5 what is the $[H_3O^+]$?

A. 8.5×10^{-9}
B. 5.5
C. 3.16×10^{-9}
D. 3.2×10^{-6}

13. If the $K_a = 1.8 \times 10^{-5}$, what is the pK_a?

A. 1.8
B. 4.7
C. 10.3
D. 5.5×10^{-10}

14. 25.0 mL of 0.200 M NaOH solution are required to titrate 30.0 mL of an HCl solution of unknown concentration. Calculate the initial HCl concentration.

A. 0.200 M
B. 0.100 M
C. 0.333 M
D. 0.167 M

15. Four different acids have the pK$_a$ values listed below. Which acid would be the strongest acid?

A. 3.5
B. 5.7
C. 7.8
D. 8.9

16. A student was conducting a experiment with the reaction

$$H_2S(aq) + NH_3(g) \rightleftharpoons NH_4^+(aq) + HS^-(aq)$$

in an open container. The black indicator for H$_2$S became lighter as the reaction started but the solution began to turn darker upon standing. Explain why this could be happening.

17. What is occurring if a flower is red at a pH of 4, but turns orange-yellow at a pH of 10?

18. Why aren't bases like NaOH or KOH used as antacids?

19. HBr is an example of a binary acid, which means it _____.

20. Write the formula for hydrofluoric acid.

21. In the equation,
$H_2S(aq) + NH_3(g) \rightleftharpoons NH_4^+(aq) + HS^-(aq)$, what is the conjugate base of H$_2$S?

22. In the equation,
$H_2S(aq) + NH_3(g) \rightleftharpoons NH_4^+(aq) + HS^-(aq)$, what is the conjugate acid of NH$_3$?

23. True or False: Bronsted-Lowry acids give off OH$^-$ ions in water.

24. Write the equilibrium expression for this reversible reaction.

$$H_2S(aq) + NH_3(g) \rightleftharpoons NH_4^+(aq) + HS^-(aq)$$

25. Molecules that can either gain or give off H$^+$ ions are called _____.

PRACTICE TEST Answers

1. C
2. A
3. D
4. B
5. B
6. A
7. A
8. C
9. B
10. D
11. A
12. C
13. B
14. D
15. A
16. Le Châtelier's principle states that once a reaction reaches equilibrium, if the equilibrium is disrupted, the reaction will shift to regain equilibrium. Since NH_3 is a gas, it will leave the open container over time, causing the reaction to shift to produce more NH_3. Therefore more H_2S causes the indicator to darken.
17. The organic molecule responsible for the color is probably a weak acid. At low pH (high H^+ ion concentration) the weak acid retains its color but in a high pH (low H^+ ion concentration) the H^+ will leave the organic molecule causing its color to change.
18. NaOH and KOH are strong bases and would be caustic to the skin and damage cell structures which have a high lipid content.
19. The acid is made up of *two* elements.
20. HF
21. HS^-
22. NH_4^+
23. F
24. $$\frac{[NH_4^+][HS^-]}{[H_2S][NH_3]}$$

25. amphoteric

181

ANSWER PAGE FOR BUILDING KNOWLEDGE

Match the terms or phrases to the correct meaning

___ j ___ 1.Catalyst

a. dissolve some metals, turn litmus red, release H^+, and those safe to ingest have a sour taste

___ q ___ 2. pH

b. feel slippery or soapy, turn litmus blue, accept H^+, and have a bitter taste

___ g ___ 3. Equilibrium constant

c. definition of acids and bases that states that acids release H^+ and bases accept H^+

___ v ___ 4. Buffer

d. the corresponding acid or base that is produced when a H^+ is gained or lost

___ u ___ 5. Titration

e. can act as either an acid or a base

___ f ___ 6. Equilibrium

f. the rate of the forward reaction and the rate of the reverse reaction are equal

___ c ___ 7. Bronsted-Lowery

g. a value that expresses the relationship between the forward rate and the reverse rate of a reaction

___ t ___ 8. Neutralization

h. the constant expressed for an acid in water

___ r ___ 9. Dissociates

i. when a reversible reaction is pushed out of equilibrium, the reaction responds to reestablish equilibrium

___ a ___ 10. Acids

j. increase reaction rates by lowering activation energy

___ s ___ 11. pK_a

l. equilibrium constant for water equal to 1.0×10^{-14}

___ o ___ 12. Basic

m. the formation of ions from molecules

___ p ___ 13. Neutral

n. $[H_3O^+]$ is greater than 1×10^{-7} and the pH is less that 7

___ m ___ 14. Ionization

o. $[H_3O^+]$ is less than 1×10^{-7} and the pH is greater that 7

___ n ___ 15. Acidic

p. $[H_3O^+]$ is equal to 1×10^{-7} and the pH is equal to 7

___ b ___ 16. Bases

q. $= -\log[H_3O^+]$

___ r ___ 17. Le Châtelier's principle

r. ions falling apart

___ l ___ 18. K_w

s. $= -\log K_a$

___ h ___ 19. Acidity constant

t. an acid and a base react to form water and a salt

___ e ___ 20. Amphoteric

u. technique used to determine the concentration of an acid or base

___ d ___ 21. Conjugate

v. a solution prepared from weak acids and their conjugate bases that resists changes in pH

Chapter 10
Carboxylic Acids, Phenols, and Amines

Chapter Overview

As you learned in chapter 9, using a strong base such as NaOH or KOH as an antacid isn't a great idea! Those bases are "caustic." You also learned that there was a group of acids and bases that did not ionize or dissociate completely, called weak acids and bases. In chapter 10 you will investigate a group of *organic* weak acids and bases.

The first part of the chapter examines how to distinguish organic weak acids and bases by identifying the characteristic functional group for the carboxylic acids, phenols, and amines. You'll learn about the intermolecular forces that provide the attraction between these molecules. This will be very important in later chapters when explaining how larger bimolecular substances are held together. Next, just as you did in Chapter 9, you will investigate the different ways in which these weak acids and bases react with strong acids and bases as well as other weak acids and bases. Carboxylic acid reactions that result in esters and amides are reviewed along with how to hydrolyze esters and amides to revert back to the carboxylic product. The final section introduces a characteristic of organic molecules; *chiral molecules* are composed of a carbon and four different atoms in a special arrangement.

Knowledge/Comprehension

Typically, there are new terms or phrases that you must memorize in order for the lesson to make sense to you. Look up this information and memorize their explanations and definitions:

phenol	quaternary	carboxylate ions
phenoxide ions	amide	chiral
chiral molecule	enantiomers	diastereomers
plane polarized light	dextrorotary	levorotatory

The following activity is designed to help you master the objectives outlined in the beginning of Chapter 10 and then summarized at the end of the chapter. To make the best use of your study time, *read* Chapter 10 before continuing. Next, work the matching activity provided as though you are taking a test *without first* referring to the answer page. Put your answers on a separate sheet of paper so that you can make more than one attempt without seeing your previous answers. Once you have completed the activity, check your answers. Make *flash cards* for those you missed. Have a study buddy drill you on them if you missed more than half of the terms. Once you have reviewed the ones you missed for at least 15 minutes, try the activity again. If you miss fewer than four, write in the answers so that you will have them to review just before taking the test.

Match the terms or phrases to the correct meaning.

_____1. Chiral

a. alcohols with a hydroxyl group on the first carbon of a benzene ring

_____2. Levorotatory

b. a nitrogen with four carbons bonded to it

_____3. Plane polarized light

c. formed when the H^+ ion comes of a carboxylic acid group

_____4. Dextrorotary

d. formed when the H^+ ion comes of a phenol carboxylic acid group

_____5. Enantiomers

e. molecule with an amine group attached to the single bonded oxygen of the carboxyl group.

_____6. Phenol

f. has a carbon with four different atoms or groups of atoms

_____7. Diastereomers

g. has one or more chiral carbons and are not superimposable on their mirror image

_____8. Amide

h. stereoisomers of a molecule that are nonsuperimposable mirror images

_____9. Chiral molecule

i. stereoisomers of a molecule that are not mirror images

_____10. Carboxylate ions

j. used to distinguish enantiomers

_____11. Phenoxide ions

k. (+) rotates plane polarized light clockwise

_____12. Quaternary

l. (-) rotates plane polarized light counterclockwise

In addition to knowing the terms or phrases, you should be able to compare, contrast, or list examples that show that you understand the following concepts:

- List the intermolecular forces that attract carboxylic acids, phenols, or amine molecules to one another.
- Describe the effect that intermolecular forces have on boiling point.
- Describe what takes place when carboxylic acids and phenols react with water or strong bases.
- Describe what takes place when amines react with water or with strong acids.

Once you know the appropriate tools for Chapter 10, you are ready to evaluate your understanding of those tools and their relationships. In this type of concept building, it is necessary to compare or contrast terms or concepts. In some cases you may be required to give examples or recognize a term or concept when it is depicted in a picture or sample scenario.

In naming organic compounds, the IUPAC system names the molecules according to the number and placement of carbons within the family classification based on the most dominant functional group. The table below summarizes the functional groups introduced in Chapter 10 that you will need to know.

Family	Functional Group	IUPAC "ending"
Carboxylic acid		"oic"
Phenols		"phenol"
Amines		"amine"
Amides		"amide"

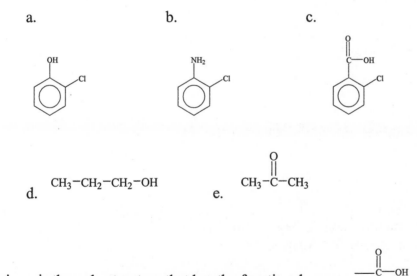

Sample Problems

1. Which of the following molecules would be classified as a carboxylic acid?

a.   b.   c.

d. $CH_3-CH_2-CH_2-OH$ e.

Choice c is the only structure that has the functional group,   ,
and is therefore the correct answer. **Answer is c.**

2.   Which of the following molecules would be classified as an amine?

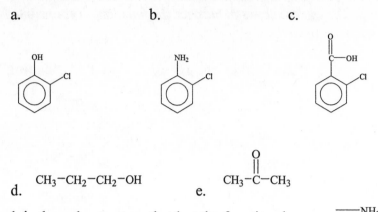

a.   b.   c.

CH₃−CH₂−CH₂−OH   CH₃−C−CH₃
d.   e.

Choice b is the only structure that has the functional group, $-NH_2$, and is therefore the correct answer. **Answer is b.**

3. Which of the following molecules would be classified as a phenol?

a.   b.   c.

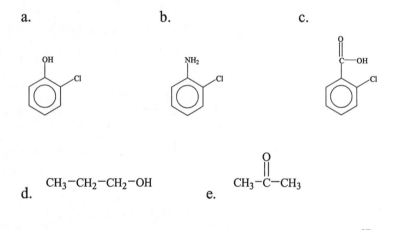

CH₃−CH₂−CH₂−OH   CH₃−C−CH₃
d.   e.

Choice a is the only structure that has the functional group, , and is therefore the correct answer. **Answer is a.**

4. The molecules of the carboxylic acid, phenol, and amine families share the characteristic of relatively high boiling points compared to other hydrocarbons of similar mass. Examine the molecules below and explain this phenomenon in terms of intermolecular forces.

Phenol   Amine   Carboxylic Acid

All of the molecules of these families have a hydrogen atom that is bonded to either an oxygen atom or a nitrogen atom. This characteristic makes it possible for these molecules to have intermolecular forces called hydrogen bonding. Increased hydrogen bonding leads to increased boiling points.

Carboxylic acids and phenols are classified as weak acids. Recall from Chapter 9 that a weak acid is a molecule that only partially ionizes in water to give off H^+ ions. To understand many of the reactions studied in Chapter 10 it is also important to remember that when H^+ ions are added (pH lowered) to a weak acid equilibrium, the reaction causes the H^+ ions to react with the conjugate base of the weak acid and creates molecules of the weak acid. Finally, acids donate H^+ ions to bases (higher pH).

5. Under which of the following conditions is the carboxylate ion, , most likely to be found?

     a. pH = 1

     b. when H^+ ions are added

     c. pH = 7

     d. pH = 13

Choices a and b are essentially the same since a pH of 1 means high H^+ ion concentration and adding H^+ ions will increase the H^+ ion concentration. In choice c, at a pH = 7, the H_2O acts as a base and the H^+ is removed from the carboxylic acid, creating the carboxylate ion. In choice d, a pH = 13 indicates a basic condition; the H^+ ions react off of the carboxylic acid, producing the carboxylate ions. Answer is c and d.

6. Under which of the following conditions is the phenoxide ion, , most likely to be found?

     a. pH = 1

     b. when H^+ ions are added

     c. pH = 7

     d. pH = 13

Choices a and b are essentially the same since a pH of 1 means high H^+ ion concentration and adding H^+ ions increases the H^+ ion concentration. In choice c, at a pH = 7, the H_2O acts as a base and the H^+ will be removed from the phenol molecule, creating the phenoxide ion. In choice d, a pH = 13 indicates a basic condition; the H^+ ions react of off the phenol, producing the phenoxide ions. **Answer is c and d.**

Note that the behavior of carboxylic acids and phenols in solution are the same. In acidic conditions the molecule is seen and in neutral water and basic solutions the carboxylate ion or phenoxide ions would be present.

7. Under which of the following conditions is the carboxylate ion, , most likely to be found?

      a. pH =1

      b. when H^+ ions are added

      c. pH = 7

      d. pH = 13

Choices a and b are essentially the same since a pH of 1 means high H^+ ion concentration and adding H^+ ions increases the H^+ ion concentration. Since bases accept H^+ ions, the amine molecule accepts these H^+ ions and becomes the ammonium ion. In choice c, at a pH = 7, the H_2O will act as an acid and give H^+ ions to the amine, creating the ammonium ion. In choice d, a pH = 13 indicates a basic condition; H^+ would not contribute to the amine molecule and the ammonium ion would not be formed. **Answer is a, b, and c.**

Amines only form ammonium ions at pH levels greater than 11.

Application

In order to apply what you've already learned and master the concepts, you must utilize all of the knowledge, understanding, facts, and/or techniques that you have gathered up to this point. Simply knowing an equation or process is not enough to master the concept. These you develop by *practice*. Solve as many different types of problems with as many different variations as you *strive to master the skill*. Your textbook provides a very good assortment of problems at the end of the chapter. Work as many as needed for skill mastery, not just those assigned.

- Explain how carboxylic acids can be converted into esters and amides.
- Explain the difference in the products obtained when an ester is hydrolyzed under basic conditions and under acidic conditions.
- Describe the products formed when an amide is hydrolyzed under acidic conditions.
- Distinguish between enantiomers and diastereomers.

In order to truly master any skill you must *practice*. In this section, problems related to the application objectives are discussed. Remember, the more you practice, the better your chances are of being able to successfully use these objectives to solve problems. Additional practice problems are at the end of the chapter in your textbook and in the online interactive lessons.

Sample Problems

1. Draw a reaction showing a carboxylic acid being converted to an ester.

Recall that an ester is a carboxylic acid with the hydrogen removed from the –OH group and replaced with another organic group. Ester formation is achieved by reacting carboxylic acid with an alcohol in the presence of an acid catalyst. In addition to the ester being formed, a molecule of water is also produced.

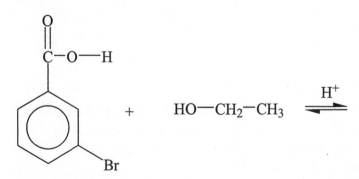

In order to get the products, remove both the hydrogen from the carboxylic acid and the hydroxy group from the alcohol to form water. Place the organic chain of the alcohol in place of the hydrogen on the carboxylic acid.

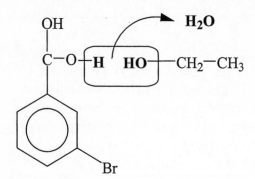

Place the products formed on the right side of the equation.

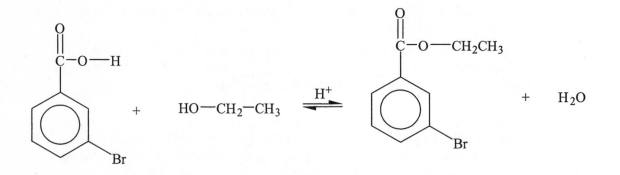

2. Draw a reaction showing a carboxylic acid being converted to an amide.

Recall that an amide is a carboxylic acid with the hydrogen removed from the –OH group and replaced with an amine group. Amide formation is achieved by first reacting carboxylic acid with the conjugate acid ammonia or an amine; this creates the carboxylate salt. Select the carboxylic acid and amine to be reacted.

$$CH_3(CH_2)_4 \overset{\overset{\displaystyle O}{\|}}{C}-OH \quad + \quad H_2N—CH_2—CH_3 \quad \rightleftharpoons$$

Next, remove the hydrogen from the carboxylic acid and place it on the amine.

$$CH_3(CH_2)_4 \overset{\overset{\displaystyle O}{\|}}{C}-O^- \overset{+}{H_3N}—CH_2—CH_3$$
When heat is applied, a water molecule is removed and the nitrogen is bonded to the carboxylate carbon. Because this is an intermediate step when the reaction is being heated, it is not shown as part of the equation. Therefore the final equation would be as shown below.

$$CH_3(CH_2)_4 \overset{\overset{\displaystyle O}{\|}}{C}-OH \quad + \quad H_2N—CH_2—CH_3 \quad \overset{heat}{\rightleftharpoons} \quad CH_3(CH_2)_4 \overset{\overset{\displaystyle O}{\|}}{C}-NH—CH_2—CH_3 \quad + \quad H_2O$$

190

3. Show the products of an ester that is hydrolyzed.

Recall that as the ester was formed, a water molecule was removed. The hydrolysis of an ester is essentially just putting that water molecule back; the resulting product is the carboxylic acid and alcohol used to create the ester. To illustrate this, the ester created in problem 1 will be used here. Since either the carboxylic acid or the carboxylate ion product may form, depending on the pH, both outcomes will be shown below.
In a neutral or basic solution:

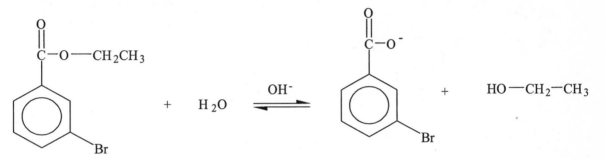

In an acidic solution:

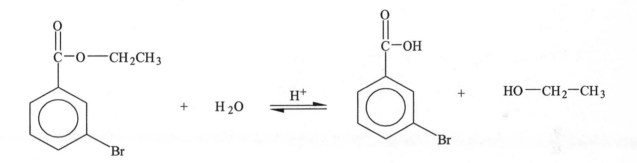

4. Draw the equation for an amide that is hydrolyzed under acidic conditions.

Had the question not specified the pH as in problem 3 above, the amide would also have two reaction possibilities. Under neutral or basic conditions, the products are the carboxylate ion and the amine used to produce the amide. In acidic conditions, the products are the carboxylic acid molecule and the ammonium ion of the amine used to create the amide. The amide from problem 2 will be used to illustrate this.

$$CH_3(CH_2)_4-\overset{\overset{\displaystyle O}{\|}}{C}-NH-CH_2-CH_3 + H_2O \underset{heat}{\overset{H^+}{\rightleftharpoons}} CH_3(CH_2)_4-\overset{\overset{\displaystyle O}{\|}}{C}-OH + \overset{+}{H_3N}-CH_2-CH_3$$

ammonium ion

5. Label the following molecule pairs as enantiomers, diastereomers, or neither. Justify your answer.

a.

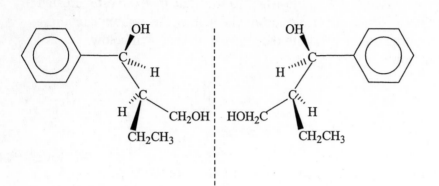

b.

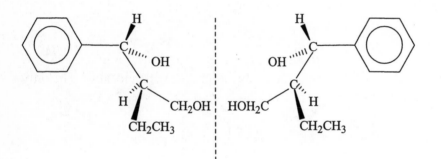

c.

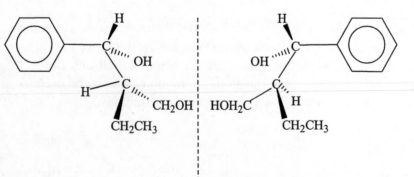

d.

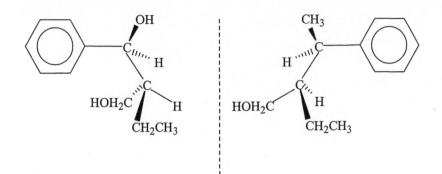

Choice a are enantiomers. In choice a, the two molecules have the same molecular formula (having the same number and kinds of atoms is the first requirement of isomers). The atoms or groups of atoms are arranged around the chiral carbons so that the molecules appear to be reflections of each other (like in a mirror), yet if you tried to turn them around and place them on top of each other, they are not superimposable.

Choice b are enantiomers. In choice b the two molecules have the same molecular formula (having the same number and kinds of atoms is the first requirement of isomers). The atoms or groups of atoms are arranged around the chiral carbons so that the molecules appear to be reflections of each other (like in a mirror), yet if you tried to turn them around and place them on top of each other, they are not superimposable. The only difference between choices a and b is that the –OH and –H of the center chiral carbon have exchanged places in choice b.

Choice c are diastereomers. In choice c the two molecules have the same molecular formula (having the same number and kinds of atoms is the first requirement of isomers). The atoms or groups of atoms are arranged around the chiral carbons so that the molecules appear to be reflections of each other (like in a mirror), yet if you look closely you will see that the H atom on the left is shown on the plane while the H on the right is shown going into the plane. Therefore, the two molecules are isomers but not enantiomers.

Choice d are not isomers. In choice d the two molecules do not have the same molecular formula (having the same number and kinds of atoms is the first requirement of isomers). The molecule on the right has a CH_3 group where the molecule on the left has an OH group. These molecules are not isomers.

Chapter 10 Practice Test

Now that you have had the chance to study and review Chapter 10's concepts and objectives, and work the practice problems provided in your textbook, you are ready to practice problems as you might see them in a testing situation. The questions and problems below are randomly selected in much the same way your instructor will select them; that is not necessarily ordered in the same way as they appear in the textbook. To make best use of this practice, have your paper, pencil and calculator ready as you would for a class exam. Use a clock and give yourself 45 minutes to complete this activity. Note that taking this test under similar conditions as a class exam provides two benefits:

1. Testing your knowledge of the chapter's contents and your mastery of the required skills.
2. Practicing working under the stress of a timed exam.

Most instructors have a type of testing format that they prefer and will most likely tell you about it before the test. For practice, this self-test uses a variety of test question formats such as matching, fill in the blank, listing, multiple choice, and essay questions. *Be sure to have a periodic table!*

Note the time and begin.

1. Which of the following is *not* found as part of a carboxylic acid?

A. –OH
B. organic chain
C. –NH$_2$
D. =O

2. Which of the following is the part of the carboxylic acid that gives off H$^+$?

A. –OH
B. organic chain
C. –NH$_2$
D. =O

3. Which of the following is the functional group of an amine?

A. –OH
B. organic chain
C. –NH$_2$
D. =O

4. Which of the following is the functional group of a phenol?

A. –OH
B. organic chain
C. –NH$_2$
D. =O

5. The characteristic organic chain of the phenols is:

A.   B.   C.   D.

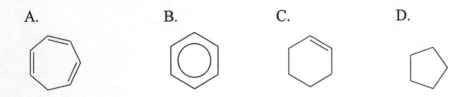

6. Which of the following is the correct formula for a carboxylic acid molecule in a basic solution?

A.
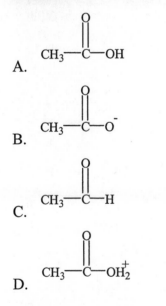

B.

C.

D.

7. Which of the following is the correct formula for a phenol molecule in an acidic solution?

A.

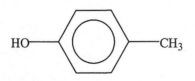

B.

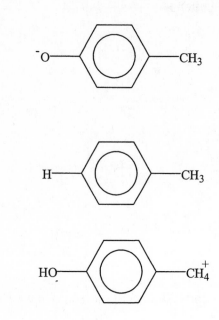

C.

D.

8. Carboxylic acids, phenols, and amines generally have higher boiling points than other organic compounds of the same size due to:

A. being acids and bases
B. London dispersion forces
C. more branching in the molecules
D. hydrogen bonding

9. When a carboxylic acid gives off an H^+ ion, the remaining ion is called a:

A. carbolic facilitator
B. carboxylate ion
C. ester
D. carbon-halite ion

10. When a phenol gives off an H^+ ion, the remaining ion is called a:

A. phenolic acetate ion
B. phenolate ion
C. phenoxide ion
D. phenotype

11. When an amine gains a H^+ ion, the ion created is called a:

A. ammonium ion
B. aminite ion
C. amino acid
D. aminylate ion

12. Carboxylic acids are usually found as their conjugate base form in the body because they have:

A. high pH values
B. pK_a values greater than 9
C. pK_a values less than 5
D. the blood is basic

13. When a carboxylic acid and alcohol react to create an ester, what other product is produced?

A. an amine
B. a phenol
C. carbon dioxide
D. water

14. The reaction of a carboxylic acid with an amine to produce an amide has an intermediate product called a(n):

A. ester
B. ammonium salt
C. amino acid
D. carboxylate ion

15. The nitrogen in amines has the ability to combine with four carbons. When this occurs the nitrogen is called:

A. primary
B. secondary
C. tertiary
D. quaternary

In questions 15 through 20, match the following terms or phrases to the correct meanings

_____16. Enantiomers   A. molecule with an amine group attached to the single bonded oxygen of the carboxyl group

_____17. Phenol

B. alcohols with a hydroxy group on the first carbon of a benzene ring

_____18. Diastereomers

C. has one or more chiral carbons and are not superimposable on their mirror image

_____19. Amide

D. stereoisomers of a molecule that are nonsuperimposable mirror images

_____20. Chiral molecule

E. stereoisomers of a molecule that are not mirror images

21. Which carbon is a chiral carbon?

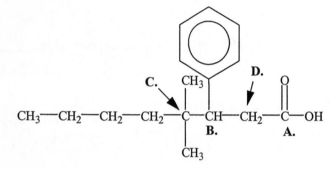

22. Draw the enantiomer of the molecule below.

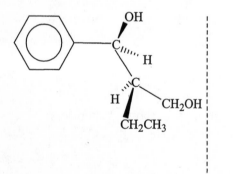

23. As an enantiomer, the molecule above will rotate _____-_____ light.

24. If the molecule of question 22 is dextrorotary its mirror image will be

_____.

25. If a molecule is named (+) esterone you know it will rotate light in a _____ direction.

PRACTICE TEST Answers

1. C
2. A
3. C
4. A
5. B
6. B
7. A
8. D
9. B
10. C
11. A
12. C
13. D
14. B
15. D
16. D
17. B
18. E
19. A.
20. C
21. C
22.

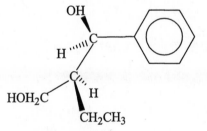

23. plane-polarized
24. levorotary
25. clockwise

ANSWER PAGE FOR BUILDING KNOWLEDGE

Match the terms or phrases to the correct meaning

___f___ 1. Chiral

a. alcohols with a hydroxy group on the first carbon of a benzene ring

___l___ 2. Levorotatory

b. a nitrogen with four carbons bonded to it

___j___ 3. Plane polarized light

c. formed when the H^+ ion comes of a carboxylic acid group

___k___ 4. Dextrorotary

d. formed when the H^+ ion comes of a phenol carboxylic acid group

___h___ 5. Enantiomers

e. molecule with an amine group attached to the single bonded oxygen of the carboxyl group.

___a___ 6. Phenol

f. has a carbon with four different atoms or groups of atoms

___i___ 7. Diastereomers

g. has one or more chiral carbons and are not superimposable on their mirror image

___e___ 8. Amide

h. stereoisomers of a molecule that are nonsuperimposable mirror images

___g___ 9. Chiral molecule

i. stereoisomers of a molecule that are not mirror images

___c___ 10. Carboxylate ions

j. used to distinguish enantiomers

___d___ 11. Phenoxide ions

k. (+) rotates plane polarized light clockwise

___b___ 12. Quaternary

l. (-) rotates plane polarized light counterclockwise

Chapter 11
Alcohols, Ethers, Aldehydes, and Ketones

Chapter Overview

In the previous chapters that introduced organic functional groups you became familiar with different bond types, carboxylic acids, amines, esters, and amides. There was also some discussion about alcohols. In Chapter 11, the remaining organic functional groups that make up the majority of organic families are introduced.

Alcohols are covered in more detail as you learn how they are classified within that family and some of their most common uses and reactions. The discussion of alcohols will include a lesser known group called thiols and sulfides. Thiols are very similar in structure to alcohols except that instead of the –OH functional group, thiols include the functional group–SH. Sulfides are created when sulfur atoms replace oxygen atoms in other functional groups. Next, the chapter covers the remaining types of functional groups that can be formed by placing oxygen in different bond locations. These groups include aldehydes, ketones (*these will play a critical role in carbohydrate identification in Chapter 12*), and ethers. In addition to identifying these groups, the chapter looks at some of the more common oxidation reactions involving alcohols, thiols, sulfides, and aldehydes. The chapter finishes with a discussion of reduction reactions involving aldehydes, ketones, and alcohols.

Knowledge/Comprehension

Typically, there are new terms or phrases that you must memorize in order for the lesson to make sense to you. Look up this information and memorize their explanations and definitions:

alcohols	ethers	thiols
sulfides	disulfides	carbonyl
1° alcohol	2° alcohol	3° alcohol
aldehydes	ketones	Benedict's solution
hemiacetals	acetals	Markovnikov's rule
nucleophilic substitution reactions		

The following activity is designed to help you master the objectives outlined in the beginning of Chapter 11 and then summarized at the end of the chapter. To make the best use of your study time, *read* Chapter 11 before continuing. Next, work the matching activity provided as though you are taking a test *without first* referring to the answer page. Put your answers on a separate sheet of paper so that you can make more than one attempt without seeing your previous answers. Once you have completed the activity, check your answers. Make *flash cards* for those you missed. Have a study buddy drill you on them if you missed more than half of the terms. Once you have reviewed the ones

you missed for at least 15 minutes try the activity again. If you miss fewer than four, write in the answers so that you will have them to review just before taking the test.

Match the terms or phrases to the correct meaning

_____1. Markovnikov's rule   a. –OH functional group

_____2. Hemiacetals   b. C–O–C functional group

_____3. 3° alcohol   c. –SH functional group

_____4. Benedict's solution   d. C–S–C functional group

_____5. 2° alcohol   e. C–S–S–C functional group

_____6. Ketones   f. C=O functional group

_____7. Aldehydes   g. the carbon atom carrying the –OH is attached to one other carbon

_____8. Disulfides   h. the carbon atom carrying the –OH is attached to two other carbons

_____9. Thiols   i. the carbon atom carrying the –OH is attached to three other carbons

_____10. Acetals   k. used to prepare alcohols, ethers, thiols, and sulfides

_____11. Alcohols   l. the alcohol formed by adding a hydrogen atom to the double-bonded carbon atom that carries the most hydrogen atoms, while the –OH goes on the other carbon

_____12. 1° alcohol   m. =O functional group on a primary carbon

_____13. Carbonyl   n. =O functional group on a secondary carbon

_____14. Sulfides   o. oxidizes aldehydes to carboxylic acids (used to test for the presence of aldehydes)

_____15. Ethers   p. the –OH and the –OC are attached to a single carbon atom

_____16. Nucleophilic   q. a single carbon atom attached to two –OC groups
substitution reactions

In addition to knowing the terms or phrases, you should be able to compare, contrast, or list examples that show that you understand the following concepts:

- Describe the structure of the molecules that belong to alcohol, ether, thiol, sulfide, disulfide, aldehyde, and ketone families.
- Describe the nucleophilic substitution reactions that can be used to prepare alcohols, ethers, thiols, and sulfides
- Explain what happens when an aldehyde or ketone is reacted with H_2 and Pt and when one of these compounds is reacted with one or two alcohol molecules, in the presence of H^+.

Once you know the appropriate tools for Chapter 11, you are ready to evaluate your understanding of those tools and their relationships. In this type of concept building, it is necessary to compare or contrast terms or concepts. In some cases you may be required to give examples or recognize a term or concept when it is depicted in a picture or sample scenario.

In naming organic compounds, the IUPAC system names the molecules according to the number and placement of carbons with the family classification based on the most dominant functional group. The table below summarizes the functional groups introduced in Chapter 11 that you will need to know.

Family	Functional Group R = carbon group	IUPAC "ending"
Alcohol	R——OH	"ol"
Thiol	R——SH	"thiol"
Ether	R——O——R	"ether"
Sulfide	R——S——R	"sulfide"
Disulfide	R——S——S——R	"disulfide"
Aldehyde	$\overset{\overset{\displaystyle H}{\mid}}{R——C}=O$	"al"
Ketone	$R——\overset{\overset{\displaystyle O}{\parallel}}{C}——R$	"one"

Sample Problems

1. Which of the following molecules would be classified as an alcohol?

a.

$$CH_3-CH_2-CH_2-\overset{\overset{\displaystyle CH_3}{\mid}}{CH}-CH_2-SH$$

b.

$$CH_3-CH_2-\overset{\overset{\displaystyle OH}{\mid}}{CH}-CH_3$$

c.

$CH_3-CH_2-O-CH_2-CH_3$

d.

$CH_3-CH_2-S-CH_2-CH_3$

e.

$$
\begin{array}{c}
O \\
\parallel \\
CH_3-CH_2-CH_2-CH
\end{array}
$$

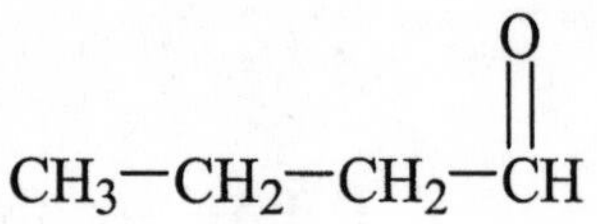

Choice b is the only structure that has the functional group, $R\!-\!\!-\!OH$, and is therefore the correct answer. **Answer is b.**

2. Which of the following molecules would be classified as a sulfide?

a.

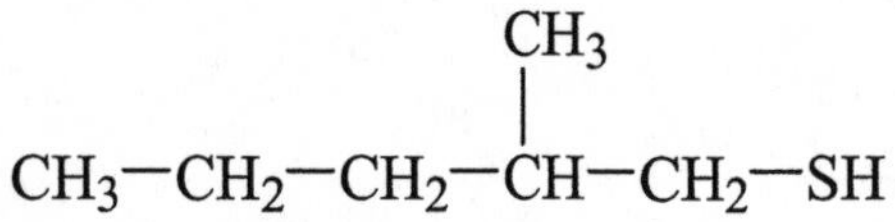

$$
\begin{array}{c}
CH_3 \\
\mid \\
CH_3-CH_2-CH_2-CH-CH_2-SH
\end{array}
$$

b.

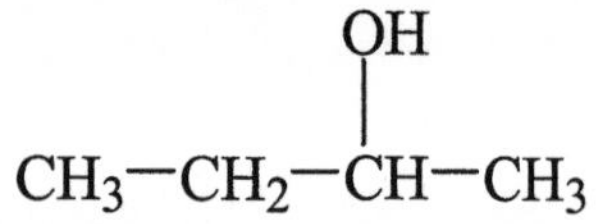

$$
\begin{array}{c}
OH \\
\mid \\
CH_3-CH_2-CH-CH_3
\end{array}
$$

c.

$CH_3-CH_2-O-CH_2-CH_3$

d.

$CH_3-CH_2-S-CH_2-CH_3$

e.

$$
\begin{array}{c}
O \\
\parallel \\
CH_3-CH_2-CH_2-CH
\end{array}
$$

204

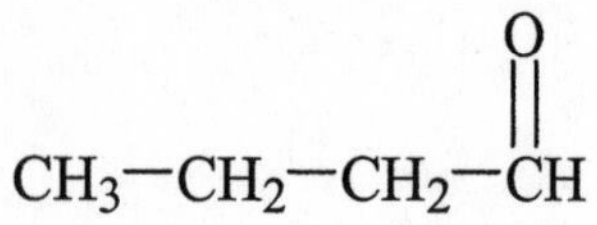

Here you are looking for the functional group, R—S—R , a sulfur atom with a carbon group on each side. **Answer is d.**

3. Which of the following molecules would be classified as an aldehyde?

a.

$$CH_3-CH_2-\overset{\overset{\displaystyle O}{\|}}{C}-CH_3$$

b.

$$CH_3-CH_2-\overset{\overset{\displaystyle OH}{|}}{CH}-CH_3$$

c.

$$CH_3-CH_2-O-CH_2-CH_3$$

d.

$$CH_3-CH_2-S-CH_2-CH_3$$

e.

$$CH_3-CH_2-CH_2-\overset{\overset{\displaystyle O}{\|}}{CH}$$

An aldehyde must have a C=O bond. Choices a and e are the only molecules that include a C=O bond, however, an aldehyde must have a C=O bond on an end (primary) carbon. **Answer is e.**

The clue to being successful when answering questions similar to questions 1 through 3 above is knowing the functional groups.

4. Which of the following reactions represents the nucleophilic substitution reaction used to prepare an alcohol?

a.

$$CH_3-CH_2-CH_2-\underset{\underset{CH_3}{|}}{CH}-CH_2-\ddot{B}r: \quad + \quad ^{-}\ddot{S}H \quad \longrightarrow \quad CH_3-CH_2-CH_2-\underset{\underset{CH_3}{|}}{CH}-CH_2-\ddot{S}H \quad + \quad :\ddot{B}r:^{-}$$

b.

$$CH_3-CH_2-CH_2-\underset{\underset{CH_3}{|}}{CH}-CH_2-\ddot{B}r: \quad + \quad ^{-}\ddot{O}H \quad \longrightarrow \quad CH_3-CH_2-CH_2-\underset{\underset{CH_3}{|}}{CH}-CH_2-\ddot{O}H \quad + \quad :\ddot{B}r:^{-}$$

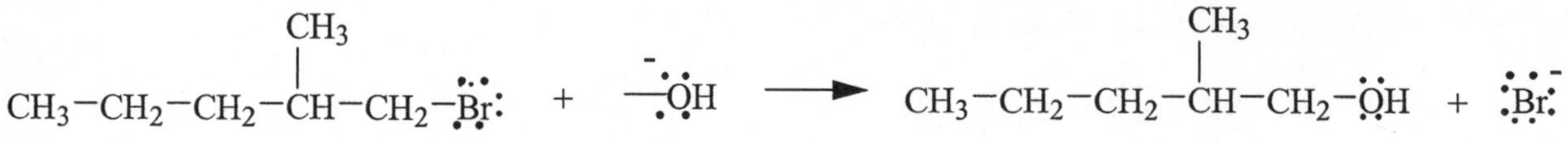

c.

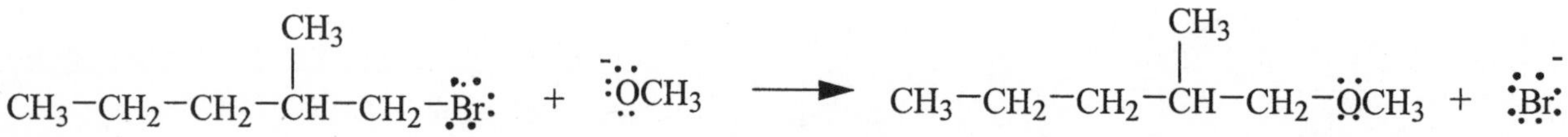

$$CH_3-CH_2-CH_2-\underset{\underset{CH_3}{|}}{CH}-CH_2-\ddot{B}r: \quad + \quad ^{-}\ddot{O}CH_3 \quad \longrightarrow \quad CH_3-CH_2-CH_2-\underset{\underset{CH_3}{|}}{CH}-CH_2-\ddot{O}CH_3 \quad + \quad :\ddot{B}r:^{-}$$

d.

$$CH_3-CH_2-CH_2-\underset{\underset{CH_3}{|}}{CH}-CH_2-\ddot{B}r: \quad + \quad ^{-}\ddot{S}CH_3 \quad \longrightarrow \quad CH_3-CH_2-CH_2-\underset{\underset{CH_3}{|}}{CH}-CH_2-\ddot{S}CH_3 \quad + \quad :\ddot{B}r:^{-}$$

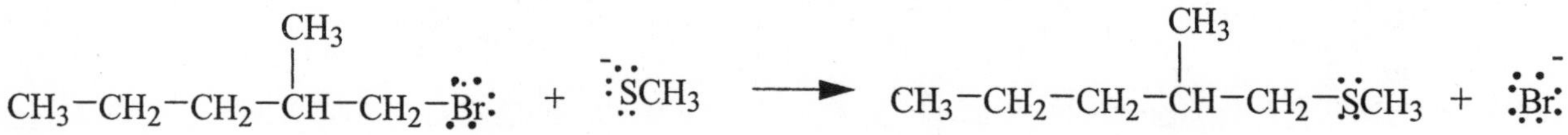

A nucleophilic reaction takes place when an organic group with a leaving group such as the chloride or bromide ion reacts with an electron rich nucleophile that includes the functional group to be added. Note that the reactions look very similar except for the group to be added. In this problem you are looking of the –OH as the nucleophile to be added. **Answer is b.**

5. In problem 4, two sets of the nucleophilic substitution reactions are identical except for one atom. Identify the products of those sets.

Alcohols and thiols are identical except that alcohols have an –OH functional group, while thiols have a –SH functional group: **reactions a and b are a set.** Ethers and sulfides are identical except that ethers have C–O–C, while sulfides have C–S–C: **reactions c and d are a set.**

6. When an aldehyde or ketone is reacted with H_2 and Pt, which of the following is *not* true?

a. the =O bond is broken

b. an –OH group replaces the =O

c. –H bond is added to the same carbon as the –OH

d. the –OH group always goes on the end (primary) carbon

Typical aldehyde and ketone reactions are shown below.

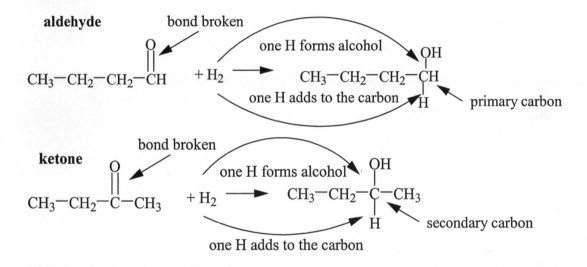

The =O bond is broken in both so choice a is not correct. An alcohol group replaces the =O so choice b is also not correct. An –H bond must be added to the carbon in order for the carbon to have four bonds so choice c is not correct. Aldehydes form the –OH on a primary carbon, but ketones do not. **Answer is d.**

Application

In order to apply what you've already learned and master the concepts, you must utilize all of the knowledge, understanding, facts, and/or techniques that you have gathered up to this point. Simply knowing an equation or process is not enough to master the concept. These you develop by *practice*. Solve as many different types of problems with as many different variations as you *strive to master the skill*. Your textbook provides a very good assortment of problems at the end of the chapter. Work as many as needed for skill mastery, not just those assigned.

- Distinguish 1°, 2°, 3° alcohols.
- Predict the major product of the addition reaction between an alkene and H_2O/H^+ and the major product for the elimination reaction between alcohol, H^+, and heat.

- Describe the oxidation reactions of alcohols, thiols, sulfides, and aldehydes.

In order to truly master any skill you must *practice*. In this section example problems related to the application objectives are discussed. Remember, the more you practice the better your chances are of being able to successfully use these objectives to solve problems. Additional practice problems are at the end of the chapter in your textbook and in the online interactive lessons.

Sample Problems

1. Name the following molecules.

a.

$$CH_3-CH_2-\overset{\overset{\displaystyle O}{\|}}{C}-CH_3$$

b.

$$CH_3-CH_2-\overset{\overset{\displaystyle OH}{|}}{CH}-CH_3$$

c.

$$CH_3-CH_2-O-CH_2-CH_3$$

d.

$$CH_3-CH_2-S-CH_2-CH_3$$

e.

$$CH_3-CH_2-CH_2-\overset{\overset{\displaystyle O}{\|}}{CH}$$

Before attempting to answer this type of question, review the basic naming rules for each family.

Family	Naming rule
Alcohol	Number the carbon chain starting from the end closest to the –OH . Follow IUPAC rules for naming an organic chain except use a number in front of the parent chain to indicate the location of the –OH. Drop "e" on parent chain name and add "ol".
Thiol	Follow same naming process as for alcohols except do

not drop the "e" and add "thiol".

Ether	Name the organic group on either side of the oxygen using IUPAC rules from Chapter 4, but end with "yl" as you would name a side chain. List the names in alphabetical order and then end the name with ether. If both groups are the same, use prefix "di" instead of using the name twice.
Sulfide	Same naming process as ethers except name will end with sulfide.
Disulfide	Same naming process as ethers except name will end with disulfide.
Aldehyde	Name the parent chain following IUPAC rules. The carbon with the =O is carbon 1. Drop the "e" and add "al" to the parent chain name.
Ketone	Name the parent chain following IUPAC rules. The end carbon closest to the carbon with the =O is carbon 1. Drop the "e" and add "al" to the parent chain name.

a. The =O is on a secondary carbon which means this is a ketone. Four carbons make it butane and the =O is on carbon 2. **Answer is 2-butanone.**

b. The –OH indicates this molecule is an alcohol. Four carbon make it butane and the –OH is on the carbon 2. **Answer is 2-butanol.**

c. The C–O–C means this molecule is an ether. The carbon group is the same on both sides, –CH$_2$CH$_3$, which is ethyl. **Since there are two groups, diethyl is used. Answer is diethyl ether.**

d. The C–S–C means this molecule is a sulfide. The carbon group is the same on both sides, –CH$_2$CH$_3$, which is ethyl. Since there are two groups, diethyl is used. **Answer is diethyl sulfide.**

e. The =O on the end carbon means this molecule is an aldehyde. Four carbons make it butane. Since the =O must be on the primary carbon in order to be an aldehyde, no preceding number is required. **Answer is butanal.**

2. Classify each of the following alcohols as 1°, 2°, or 3° alcohols.

a.   b.   c.

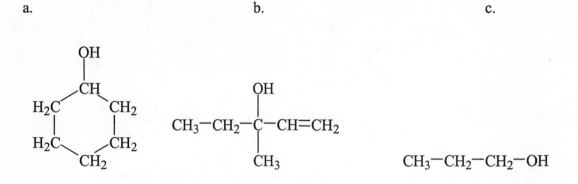

In choice a, the alcohol group, –OH, is bonded to a carbon that has two carbons bonded to it. **Molecule a is a 2° alcohol.**
In choice b, the alcohol group, –OH is bonded to a carbon that has three carbons bonded to it. **Molecule b is a 3° alcohol**.
In choice c, the alcohol group, –OH in choice c is bonded to a carbon that has one other carbon bonded to it. **Molecule c is a 1° alcohol.**

3. Complete the equation below showing the major product.

$$CH_3-CH_2-\overset{\overset{\displaystyle CH_3}{|}}{C}=CH-CH_3 \quad + \quad H_2O \quad \overset{H^+}{\longrightarrow}$$

This is called an addition reaction because it will add an –OH and a –H to the alkene. Markovnikov's rule states that in an addition reaction, the major product is the one in which a hydrogen atom has been added to the double-bonded carbon that originally carried the most hydrogen atoms. The hydroxy group goes on the carbon on the other side of the double bond.

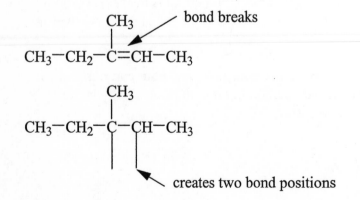

bond breaks

creates two bond positions

210

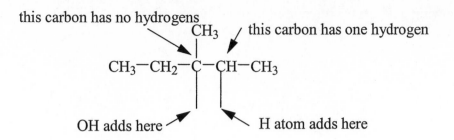

this carbon has no hydrogens

this carbon has one hydrogen

$CH_3-CH_2-C-CH-CH_3$

OH adds here

H atom adds here

The completed equation:

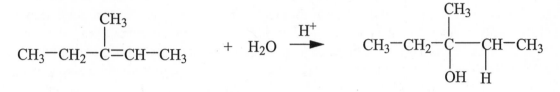

$$CH_3-CH_2-C=CH-CH_3 \quad + \quad H_2O \xrightarrow{H^+} CH_3-CH_2-C-CH-CH_3$$

4. Complete the equation below showing the major organic product.

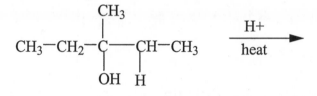

$$CH_3-CH_2-\underset{\underset{OH}{|}}{C}-\underset{\underset{H}{|}}{CH}-CH_3 \xrightarrow[heat]{H+}$$

When alcohols are heated in the presence of an acid catalyst, this is called an elimination reaction. Water is being eliminated from the molecule and a double bond is formed in its place.

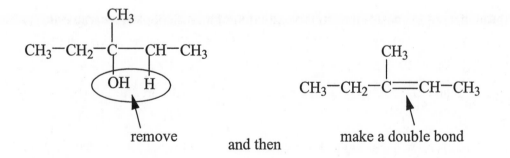

remove

and then

make a double bond

The completed equation will not show the water molecule since the question asked for the major organic product.

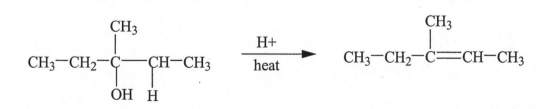

$$CH_3-CH_2-\underset{\underset{OH}{|}}{C}-\underset{\underset{H}{|}}{CH}-CH_3 \xrightarrow[heat]{H+} CH_3-CH_2-C=CH-CH_3$$

5. Which of the following oxidizers is often used to oxidize alcohols?

a. I_2

b. $K_2Cr_2O_7$

c. Cu^+

d. O_2

Since the oxidizer competes with oxygen for electrons, halogens such as I_2 are not used; choice a is not correct. In choice b, the Cr^{+6} in $K_2Cr_2O_7$ is very good at pulling electrons and creating Cr^{+3}; it is a good oxidizing agent and typically used to oxidize alcohols. In choice c, the ion Cu^+ is also a good oxidizer but not strong enough to oxidize alcohols; it can only oxidize aldehydes, therefore choice c. is not correct. Pure O_2 is a very good oxidizer, in fact too good. Using O_2 is likely to oxidize much more than just the alcohol and would not be used; choice d is not correct. **Answer is choice b.**

6. In order for a 1° alcohol to be oxidized, all of the following will occur except:

a. the –OH will be converted to =O

b. two hydrogen atoms are removed from the C that had the –OH attached to it

c. an –OH is attached to the carbon with the =O

d. the end carbon is removed and replaced by a =O

Choices a through c are the basic steps in the oxidation of a 1° alcohol to a carboxylic acid and would occur. Oxidation reactions of alcohols do not remove carbons. **Answer is d.**

7. If choices a through c of problem 6 above occur for 1° alcohols, are the same steps followed for 2° alcohols? Explain.

No. Carboxylic acids only occur on primary carbons. A secondary carbon would only have one hydrogen atom available to remove so the reaction would stop with the formation of a ketone.

8. Which of the following is *not* true for the oxidation of aldehydes?

a. oxidized with Benedict's solution

b. oxidized with a Cu^+ ion solution

c. produces ketones

d. produces carboxylic acids

Choices a and b are virtually the same solution since Benedict's solution is a source of Cu^+ ions and Cu^+ ions are used to oxidize aldehydes. The oxidation process will not shift the oxygen to a different carbon and therefore a ketone will not be produced. The product of choice d., carboxylic acids, is produced. **Answer is c.**

Chapter 11 Practice Test

Now that you have had the chance to study and review Chapter 11's concepts and objectives, and work the practice problems provided in your textbook, you are ready to practice problems as you might see them in a testing situation. The questions and problems below are randomly selected in much the same way your instructor will select them; that is not necessarily ordered in the same way as they appear in the textbook. To make best use of this practice, have your paper, pencil and calculator ready as you would for a class exam. Use a clock and give yourself 45 minutes to complete this activity. Note that taking this test under similar conditions as a class exam provides two benefits:

- Testing your knowledge of the chapter's contents and your mastery of the required skills.
- Practicing working under the stress of a timed exam.

Most instructors have a type of testing format that they prefer and will most likely tell you about it before the test. For practice, this self-test uses a variety of test question formats such as matching, fill in the blank, listing, multiple choice, and essay questions. *Be sure to have a periodic table!*

Note the time and begin.

1. Which of the following molecules is an ether?

    a.

$$\underset{\text{CH}_3-\text{CH}_2-\overset{\displaystyle\overset{\textstyle O}{\|}}{\text{C}}-\text{CH}_3}{}$$

    b.

$$\underset{\text{CH}_3-\text{CH}_2-\overset{\displaystyle\overset{\textstyle OH}{|}}{\text{CH}}-\text{CH}_3}{}$$

    c.

$$\text{CH}_3-\text{CH}_2-\text{O}-\text{CH}_2-\text{CH}_3$$

    d.

$$\text{CH}_3-\text{CH}_2-\text{S}-\text{CH}_2-\text{CH}_3$$

2. Which of the following molecules is a 2° alcohol?

a.

$$CH_3-CH_2-\overset{\overset{\displaystyle O}{\|}}{C}-CH_3$$

b.

$$CH_3-CH_2-\overset{\overset{\displaystyle OH}{|}}{CH}-CH_3$$

c.

$$CH_3-CH_2-CH_2-CH_2-OH$$

d.

$$CH_3-\overset{\overset{\displaystyle CH_3}{|}}{\underset{\underset{\displaystyle OH}{|}}{C}}-CH_3$$

3. Which of the following alcohols *cannot* be oxidized?

a.

b.

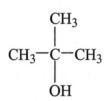

c.

$$CH_3-CH_2-CH_2-CH_2-OH$$

d.

$$CH_3-\overset{\overset{\displaystyle CH_3}{|}}{\underset{\underset{\displaystyle OH}{|}}{C}}-CH_3$$

4. Which of the following molecules is a 1° alcohol?

a.

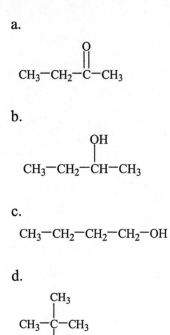

$CH_3-CH_2-\overset{\overset{\displaystyle O}{\|}}{C}-CH_3$

b.

$CH_3-CH_2-\overset{\overset{\displaystyle OH}{|}}{CH}-CH_3$

c.

$CH_3-CH_2-CH_2-CH_2-OH$

d.

$CH_3-\overset{\overset{\displaystyle CH_3}{|}}{\underset{\underset{\displaystyle OH}{|}}{C}}-CH_3$

5. Which of the following molecules is a possible product from the oxidation of 2° alcohol?

a.

$CH_3-CH_2-O-CH_2-CH_3$

b.

$CH_3-CH_2-\overset{\overset{\displaystyle OH}{|}}{CH}-CH_3$

c.

$CH_3-CH_2-\overset{\overset{\displaystyle O}{\|}}{C}-CH_3$

d.

$CH_3-CH_2-S-CH_2-CH_3$

In Questions 6 through 10 complete the table below.

Family	Functional Group R = carbon group	IUPAC "ending"
9	R——OH	"ol"
Thiol	**6**	"thiol"
Ether	R——O——R	"ether"
Sulfide	R——S——R	"**7**."
Disulfide	R——S——S——R	"disulfide"
Aldehyde	**8**	"al"
Ketone	R——C——R (with O double bonded to C)	**10**

6. _____

7. _____

8. _____

9. _____

10. _____

11. Complete the reaction below.

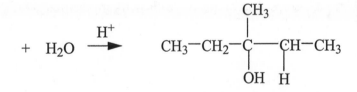

$$+ \quad H_2O \xrightarrow{H^+} CH_3{-}CH_2{-}\underset{\underset{OH}{|}}{\overset{\overset{CH_3}{|}}{C}}{-}\underset{\underset{H}{|}}{CH}{-}CH_3$$

In questions 12 through 19, match the following terms and phrases to the correct meaning

_____12. Aldehyde

A. used to prepare alcohols, ethers, thiols, and sulfides

_____13. Ketone

B. C-S-C functional group

_____14. Acetals

C. =O functional group on a primary carbon

_____15. Hemiacetal

D. =O functional group on a secondary carbon

217

_____16. Sulfides

E. oxidizes aldehydes to carboxylic acids (used to test of the presence of aldehydes)

_____17. Benedict's solution

F. the –OH and the –OC are attached to a single carbon atom

_____18. Nucleophilic substitution reactions

G. a single carbon atom attached to two –OC groups

19. Write the correct name for the following molecule. _____.

OH
|
CH₃—CH₂—CH—CH₃

20. Write the correct name for the following molecule. _____.

O
||
CH₃—CH₂—C—CH₃

21. Write the correct name for the following molecule. _____.

CH₃—CH₂—S—CH₂—CH₃

22. Write the correct name for the following molecule. _____.

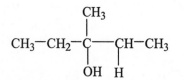

23. Draw the condensed molecular structure for propanal.

24. Draw the condensed molecular structure for 3,3-dimethyl-2-pentanone.

25. Draw the condensed molecular structure for ethyl propyl sulfide.

PRACTICE TEST Answers

1. C
2. A
3. B
4. D
5. C
6. R-SH
7. sulfide

8.

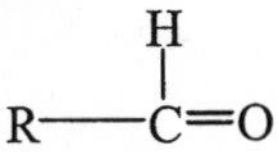

$$R-\overset{\overset{\displaystyle H}{|}}{C}=O$$

9. alcohol
10. "one"

11.

$$CH_3-CH_2-\overset{\overset{\displaystyle CH_3}{|}}{C}=CH-CH_3$$

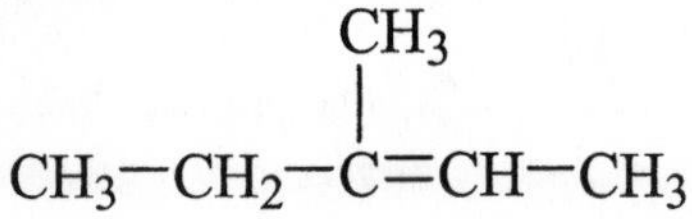

12. C
13. D
14. G
15. H
16. B
17. E
18. A.
19. 2-butanol
20. 2-butanone
21. diethyl sulfide
22. 3-methyl-3-pentanol
23.

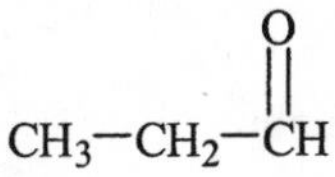

$$CH_3-CH_2-\overset{\overset{\displaystyle O}{\|}}{CH}$$

24.

$$CH_3-CH_2-\overset{\overset{\displaystyle CH_3}{|}}{\underset{\underset{\displaystyle CH_3}{|}}{C}}-\overset{\overset{\displaystyle O}{\|}}{C}-CH_3$$

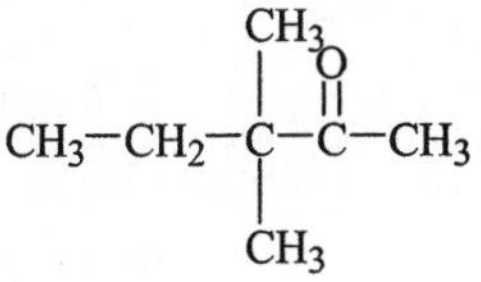

25.

$$CH_3-CH_2-S-CH_2-CH_2-CH_3$$

ANSWER PAGE FOR BUILDING KNOWLEDGE

Match the terms or phrases to the correct meaning

___l___ 1. Markovnikov's rule   a. –OH functional group

___p___ 2. Hemiacetals   b. C–O–C functional group

___i___ 3. 3° alcohol   c. –SH functional group

___o___ 4. Benedict's solution   d. C–S–C functional group

___h___ 5. 2° alcohol   e. C– S–S–C functional group

___n___ 6. Ketones   f. C=O functional group

___m___ 7. Aldehydes   g. the carbon atom carrying the –OH is attached to one other carbon

___e___ 8. Disulfides   h. the carbon atom carrying the –OH is attached to two other carbons

___c___ 9. Thiols   i. the carbon atom carrying the –OH is attached to three other carbons

___q___ 10. Acetals   k. used to prepare alcohols, ethers, thiols, and sulfides

___a___ 11. Alcohols   l. the alcohol formed by adding a hydrogen atom to the double-bonded carbon atom that carries the most hydrogen atoms, while the –OH goes on the other carbon

___g___ 12. 1° alcohol   m. =O functional group on a primary carbon

___f___ 13. Carbonyl   n. =O functional group on a secondary carbon

___d___ 14. Sulfides   o. oxidizes aldehydes to carboxylic acids (used to test for the presence of aldehydes)

___b___ 15. Ethers   p. the –OH and the –OC are attached to a single carbon atom

___k___ 16. Nucleophilic   q. a single carbon atom attached to two –OC groups
substitution reactions

Chapter 12
Carbohydrates

Chapter Overview

In chapters 4, 10 and 11, the basic building blocks of organic chemistry were discussed. The functional groups and reactions have been introduced and studied for each family. In living organisms, the functional groups are often combined to create various types of macromolecules, such as the lipids discussed in Chapter 8. Chapter 12 explains how the alcohol, aldehyde, and ketone functional groups serve as building blocks to construct *carbohydrates*.

The chapter begins by describing the structure of the monosaccharides, oligosaccharides, and polysaccharides. The four types of monosaccharide derivatives (compounds made from monosaccharides) that have biological importance are investigated. After discussing the monosaccharide derivatives, a few of the important reactions of monosaccharides are shown, including their conversion into cyclic forms. The oligosaccharides are discussed in detail, and some of the more biologically important structures, such as sucrose, are drawn and named. Finally, the structure of polysaccharides is presented and related to important living structural components such as chitin and cellulose.

Knowledge/Comprehension

Typically, there are new terms or phrases that you must memorize in order for the lesson to make sense to you. Look up this information and memorize their explanations and definitions:

monosaccharides	oligosaccharides	polysaccharides
residue	saccharide	aldoses
ketoses	suffix "ose"	Fisher-projection
glucose	galactose	D-glucose
D-galactose	D-fructose	alcohol sugar
deoxy sugar	amino sugars	chitin
Benedict's reagent	reducing sugars	hemiacetal carbon
anomer	α- anomer	β- anomer
furanose	pyranose	mutarotation
acetal carbon	glycosidic	glycosidic bond
oligosaccharides	disaccharides	maltose
cellobiose	sucrose	glycolipids
glycogen		

The following activity is designed to help you master the objectives outlined in the beginning of Chapter 12 and then summarized at the end of the chapter. To make the best use of your study time, *read* Chapter 12 before continuing. Next, work the matching activity provided as though you are taking a test *without first* referring to the answer page. Put your answers on a separate sheet of paper so that you can make more than one attempt without seeing your previous answers. Once you have completed the activity, check your answers. Make *flash cards* for those you missed. Have a study buddy drill you on them if you missed more than half of the terms. Once you have reviewed the ones you missed for at least 15 minutes try the activity again. If you miss fewer than four, write in the answers so that you will have them to review just before taking the test.

Match the terms or phrases to the correct meaning

_____1. Benedict's reagent

a. are the building blocks that produce oligosaccharides and are aldoses or ketoses

_____2. Fisher projection

b. are formed when between 2 and 10 monosaccharide residues are joined to one another by glycosidic bonds

_____3. Reducing sugars

c. contain more than 10 monosaccharide residues

_____4. Pyranose

d. the part of a reactant molecule that remains when it has been incorporated into a product

_____5. Alcohol sugar

e. comes from the Greek word sakcharon, or "sugar"

_____6. α- anomer

f. are those monosaccharides that contain an aldehyde group

_____7. β- anomer

g. are those monosaccharides that contain a ketone group

_____8. Maltose

h. indicates that a molecule is named a carbohydrate

_____9. Oligosaccharides

i. chiral carbon atoms sit at the intersection of a vertical and horizontal line

_____10. D-Fructose

j. has the enantiomers known as both D-glucose and L-glucose

_____11. Sucrose

k. diastereomer or non-mirror image stereoisomer of D-glucose and L-glucose

_____12. Glycosidic

l. also known as dextrose or blood sugar, and the most important monosaccharide in human biology

_____13. Disaccharides

m. combined with glucose produces lactose, an oligosaccharide, which gives milk its sweetness.

_____14. Glucose

n. the fruit sugar most commonly found in nature

_____15. Mutarotation

o. hydrogen atom replaces one or more of the –OH group in a monosaccharide

_____16. Glycosidic bond

p. an –OH group of monosaccharides that has been replaced by an amino (NH_2) group

_____17. Saccharide

q. the carbonyl group of a monosaccharide has been reduced to an alcohol group

_____18. Monosaccharides

r. can oxidize aldehydes but not alcohols

_____19. Polysaccharides

s. is when a given sugar is positive to the Benedict's test by becoming oxidized and reduced to the Cu^{2+}, which becomes present in the reagent

_____20. Glycolipids

t. attached to –OH and –OC

_____21. Chitin

u. cyclic form of a monosaccharide

_____22. Aldoses

v. a hemiacetal with the –OH pointing down

_____23. Ketoses

w. a hemiacetal with the –OH pointing up

_____24. Glycogen

x. has a 5-member ring

_____25. Residue

y. has a 6-member ring

_____26. Deoxy sugar

z. is the process of converting back and forth from an (alpha) anomer to the (beta) anomer

_____27. Cellobiose

aa. attached to two –OC groups

_____28. D-Glucose

bb. is the resulting acetal when the hemiacetal in question is the alpha or beta anomer of a monosaccharide

_____29. Amino sugars

cc. is the bond that connects the acetal carbon to the newly added –OC group

_____30. D-Galactose

ee. contain two monosaccharide residues and are the most common oligosaccharides found in nature

_____31. Suffix "ose"

dd. is produced as the product of the digestion of starch and glycogen and is categorized as a reducing sugar

_____32. Acetal carbon

ff. the product of broken down cellulose and is categorized as a reducing sugar

_____33. Galactose

gg. also known as table sugar

_____34. Hemiacetal carbon

hh. sugar-containing lipids present in nerve cell membranes, consisting of a mono- or oligosaccharide attached to an alcohol group of a lipid by a glycosidic bond

_____35. Furanose

ii. is the energy storage molecule for animals

_____36. Anomer

jj. is the material that makes up the exoskeleton of crustaceans and insects

In addition to knowing the terms or phrases, you should be able to compare, contrast, or list examples that show you understand the following concepts:

- Describe the differences between mono-, oligo, and polysaccharides.
- Explain the classification system used to categorize monosaccharides.

- Identify the four common types of monosaccharides.

Once you know the appropriate tools for Chapter 12, you are ready to evaluate your understanding of those tools and their relationships. In this type of concept building, it is necessary to compare or contrast terms or concepts. In some cases you may be required to give examples or recognize a term or concept when it is depicted in a picture or sample scenario.

Sample Problems

1. Which of the following molecules represent a monosaccharide?

a.

$$HO—CH_2—\overset{\overset{\displaystyle OH}{|}}{CH}—\overset{\overset{\displaystyle OH}{|}}{CH}—\overset{\overset{\displaystyle OH}{|}}{CH}—\overset{\overset{\displaystyle OH}{|}}{CH}—\overset{\overset{\displaystyle OH}{|}}{CH}—\overset{\overset{\displaystyle O}{||}}{CH}$$

b.

$$HO—CH_2—\overset{\overset{\displaystyle OH}{|}}{CH}—\overset{\overset{\displaystyle OH}{|}}{CH}—\overset{\overset{\displaystyle OH}{|}}{CH}—\overset{\overset{\displaystyle OH}{|}}{CH}—\overset{\overset{\displaystyle OH}{|}}{CH}—\overset{\overset{\displaystyle OH}{|}}{CH}—\overset{\overset{\displaystyle O}{||}}{C}—CH_2—OH$$

c.

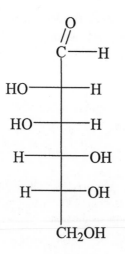

d.

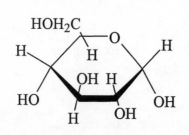

Saccharides are molecules that have both =O (or –O–, in the case of ring structures) and –OH groups attached to all carbons other than the one with the =O or –O–). Monosaccharides are the simplest of the saccharide structures and are the building blocks for the other saccharide families. Choice a meets the requirements of a monosaccharide and since the =O is on the primary carbon, it is an aldose. Choice b meets the requirements of a monosaccharide and because the =O is on a secondary carbon, it is a ketose. Choice c meets the requirements of monosaccharide and is also an aldose. When the molecule is drawn in this fashion it is called a Fischer projection. Choice d meets the requirements of a monosaccharide. In this representation the double bond of =O is broken and a new bond created with the next to the last carbon on the other end, forming a ring structure. **Answer is: all structures are monosaccharides.**

2. Which of the following molecules represent a oligosaccharide?

a.

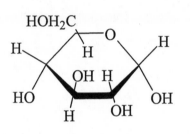

b.

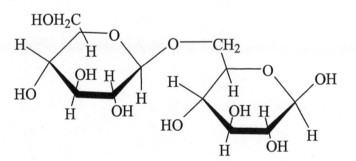

c.

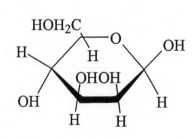

d.

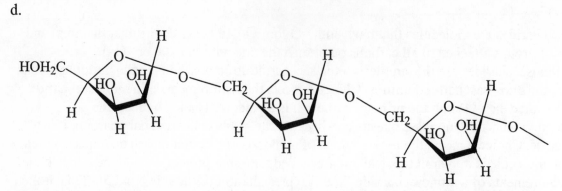

Choice a is composed of only one ring of aldose and is a monosaccharide. Choice b has two rings joined together, which makes it an oligosaccharide. Choice c has only one aldose ring and is a monosaccharide. Choice d. has three rings and at first appears to be an oligosaccharide, which can have between 2 and 10 monosaccharide residues. However, notice how the O bond on the right is open. Since it is not feasible to draw more than 10 monosaccharide residues, this open style typically denotes a polysaccharide. **Answer is b.**

3. Classify the following monosaccharides:

a.

$$HO-CH_2-\overset{\overset{\displaystyle OH}{|}}{CH}-\overset{\overset{\displaystyle OH}{|}}{CH}-\overset{\overset{\displaystyle OH}{|}}{CH}-\overset{\overset{\displaystyle OH}{|}}{CH}-\overset{\overset{\displaystyle OH}{|}}{CH}-\overset{\overset{\displaystyle O}{||}}{CH}$$

b.

$$HO-CH_2-\overset{\overset{\displaystyle OH}{|}}{CH}-\overset{\overset{\displaystyle OH}{|}}{CH}-\overset{\overset{\displaystyle OH}{|}}{CH}-\overset{\overset{\displaystyle OH}{|}}{CH}-\overset{\overset{\displaystyle OH}{|}}{CH}-\overset{\overset{\displaystyle OH}{|}}{CH}-\overset{\overset{\displaystyle O}{||}}{C}-CH_2-OH$$

c.

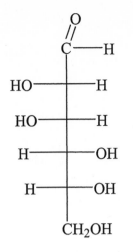

d.

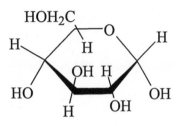

Monosaccharides are divided into two groups. Those with the =O on a primary carbon are classified as *aldoses*. Aldoses with the =O on a secondary carbon are *ketoses*. The two most common ring structures have six carbons and are called *pyranoses;* those with five carbons are called *furanoses*.

In choice a the =O is on a primary carbon making it an aldose. There are seven carbons in the chain so it is classified as an **aldoheptose**.

In choice b the =O is on a secondary carbon making it a ketose. There are nine carbons in the chain so it is classified as a **ketononose**.

In choice c the =O is on a primary carbon making it an aldose. There are six carbons in the chain so it is classified as an **aldohexose**.

In choice d there is a ring structure of six carbons making it a **pyranose**.

4. Which of the following *important* monosaccharides is incorporated into a larger biomolecule called RNA?

a.

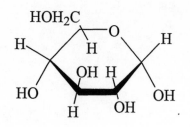

b.

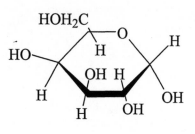

c.

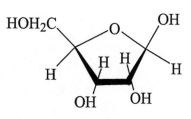

d.

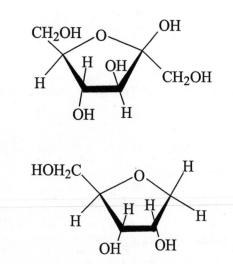

e.

If you are familiar with the term RNA (ribonucleic acid) you can eliminate choices a, b, and d since they all have six carbons; ribose is a pentose and only has five carbons in its structure. Choice c is a ribose and might be found incorporated into the structure of RNA. Choice e is also a ribose but upon closer inspection you will notice that it has a   –OH group that has been replaced by a hydrogen atom.

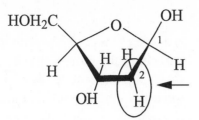

When this happens, this is called a deoxyribose. This molecule, 2-deoxyribose, is the building block for DNA. **Answer is c.**

5. What is the key structural difference of D-fructose when compared to the hexoses in problem 4?

You can eliminate choices c and e from consideration because you established in the previous problem that they are pentoses. If you have not familiarized yourself with the structures of these important monosaccharides you can find them in Section 12.3 of the textbook. Choice e is D-fructose. A close inspection of the structure reveals that the hemiacetal bond is formed on carbon 2 instead of carbon 1 as in choices a and b.

hemiacetal bond

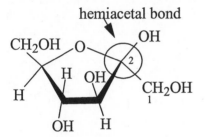

This is what happens when a ketose forms a ring structure.

Answer: The key structural difference between –D-fructose and the hexoses in problem 4 is that it is a ketone, not an aldehyde.

Application

In order to apply what you've already learned and master the concepts, you must utilize all of the knowledge, understanding, facts, and/or techniques that you have gathered up to this point. Simply knowing an equation or process is not enough to master the concept. These you develop by *practice*. Solve as many different types of problems with as many different variations as you *strive to master the skill*. Your textbook provides a very good assortment of problems at the end of the chapter. Work as many as needed for skill mastery, not just those assigned.

- Identify four important disaccharides and describe how their monosaccharide residues join together.
- Explain what happens when a monosaccharide is reacted with H_2 and Pt or with Benedict's reagent.

- Distinguish homopolysaccharides from heteropolysaccharides and give examples of each.

In order to truly master any skill you must *practice*. In this section example problems related to the application objectives are discussed. Remember, the more you practice, the better your chances of being able to successfully use these objectives to solve problems. Additional practice problems are at the end of the chapter in your textbook and in the online interactive lessons.

Sample Problems

1. Which of the following disaccharides is sucrose?

a.

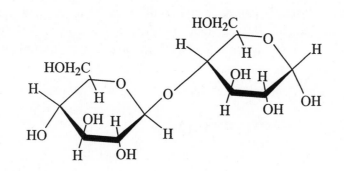

b.

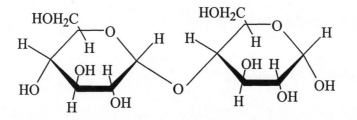

c.

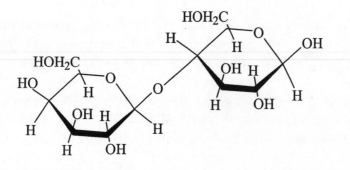

d.

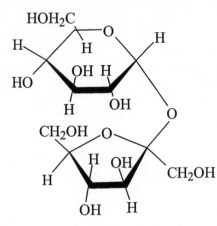

These four molecules represent the *four main disaccharides*. Choice a is *cellobiose* and is composed of two glucoses. Choice b is *maltose* and is composed of two glucoses; it differs from cellobiose by having an α-glycosidic bond instead of a β-glycosidic bond. Choice c is *lactose* and is composed of one galactose and one glucose. Choice d is *sucrose* and is composed of one glucose and one fructose (a ketose). **Answer is d.**

2. Which of the following molecules in problem 3 exhibits a α-(1→ 4) glycosidic bond?

The glycosidic bond type is determined by examining the glycosidic bond formed between the saccharide rings as shown below.

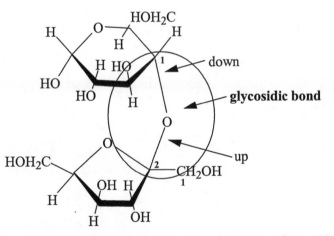

If both bonds are in the downward facing position it is an α-glycosidic bond. If both are in the upward facing position, it is a β-glycosidic bond. If one bond is facing down and the other is facing up, a combined name is used: (α-), (α, β-), (β, α-), and (β-) are used with the corresponding carbon numbers given from smallest to largest order. Recall that on a ring structure carbon that originally had the =O, or in the case of a ketone, the end carbon nearest the =O is numbered 1 as in the figure above. The figure above is choice d, sucrose, from problem 3 and has an α, β-(1→ 2) glycosidic bond. In figure a, cellobiose,

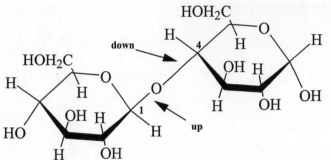

the 1 position is facing upward and the position 4 is facing downward so the bond is a β, α- (1→ 4).
In choice b, maltose,

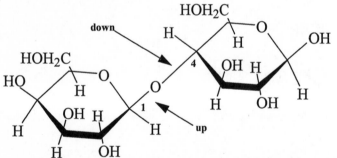

both bonds are in the downward facing position so the bond is a α- (1→ 4) glycosidic bond. Notice that when both bonds are the same, it is not necessary to list them twice.
In choice c, lactose,

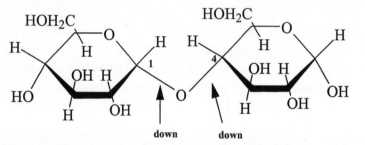

one bond is facing upward and one is in the downward facing position so the bond is a β,α- (1→ 4) glycosidic bond.
Answer is b.

3. Which of the following would be a product of the aldehyde shown below reacted with Benedict's reagent?

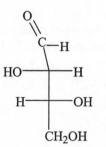

a.

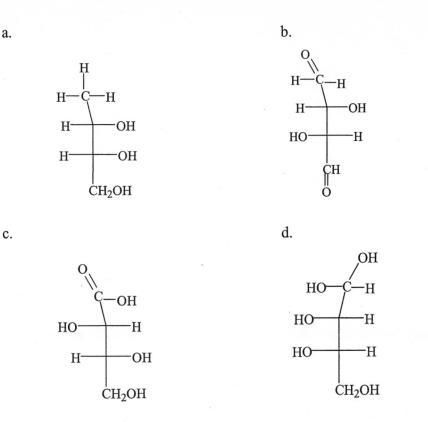

b.

c.

d.

Remember in Section 11.5 you studied that Benedict's reagent oxidizes aldehydes, but not alcohols. You've also studied that when organic compounds are oxidized, they typically remove hydrogens or add oxygens.

In choice a the =O was removed, which is a reduction, not an oxidation and it would not be a predicted product. In choice b, a hydrogen atom was removed from the –OH on the end carbon and replaced by a =O, which is an oxidation. However, it's the oxidation of an alcohol and Benedict's reagent does not oxidize alcohols. In choice c, the –H is replaced with a –OH, meaning that the aldehyde group was oxidized to a carboxylic acid group, which is a predicted product. In choice d, the =O was replaced by a –OH, adding a hydrogen atom, which is a reduction not an oxidation. **Answer is c**

4. Which of the following would be a product of the aldehyde shown below reacted with H_2 and Pt?

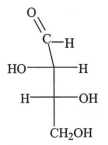

a.

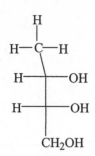

b.

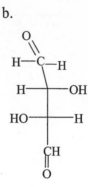

c.

d.

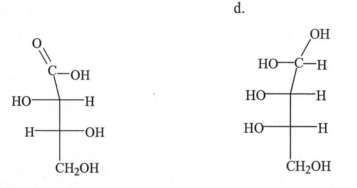

If the molecule reacts with H$_2$ in the presence of the Pt catalyst, reduction of the molecule will occur and at least one H atom will be added to the molecule.

Choices b and c can be eliminated because they represent oxidation products not reduction products. Choice a is not correct because it shows the complete removal of the =O, not just the addition of a hydrogen atom. Choice d is a product with a hydrogen atom added to the =O to create a –OH group. **Answer is d. Note: aldehydes and ketones are reduced to alcohols.**

5. Identify the following molecules as either homopolysaccharides or heteropolysaccharides and explain your answer.

a.

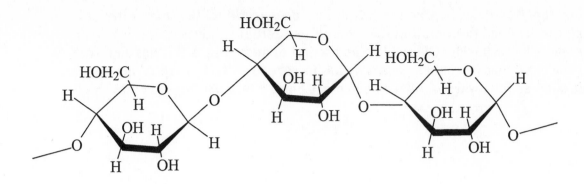

b.

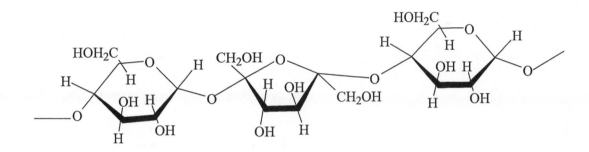

c.

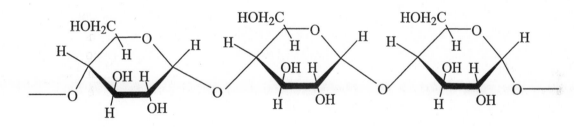

d.

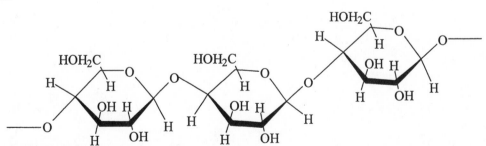

Homopolysaccharides are polysaccharide residues that are composed of only one type of monosaccharide. Heteropolysaccharides are polysaccharide residues that are composed of more than one type of monosaccharide.

Answer: Choice a is composed of three glucose residues and is therefore a homopolysaccharide. Choice b has a glucose-fructose-glucose residue and is a

235

heteropolysaccharide. Choice c has three glucose residues that have different glycosidic bond positions but does not have different residues so it is a homopolysaccharide. Choice d is the hardest to qualify in that it has a glucose-galactose-glucose residue. Galactose has its carbon 4 –OH group facing upward instead of downward, similar to glucose. Choice d. is a heteropolysaccharide.

Chapter 12 Practice Test

Now that you have had the chance to study and review Chapter 12's concepts and objectives, and work the practice problems provided in your textbook, you are ready to practice problems as you might see them in a testing situation. The questions and problems below are randomly selected in much the same way your instructor will select them; that is not necessarily ordered in the same way as they appear in the textbook. To make best use of this practice, have your paper, pencil and calculator ready as you would for a class exam. Use a clock and give yourself 45 minutes to complete this activity. Note that taking this test under similar conditions as a class exam provides two benefits:

1. Testing your knowledge of the chapter's contents and your mastery of the required skills.
2. Practicing working under the stress of a timed exam.

Most instructors have a type of testing format that they prefer and will most likely tell you about it before the test. For practice, this self-test uses a variety of test question formats such as matching, fill in the blank, listing, multiple choice, and essay questions. ***Be sure to have a periodic table!***

Note the time and begin

1. If a person suffers from lactose intolerance they:

A. flagellate violently after meals
B. become ill after consuming milk products
C. have a bias against persons from Lactosia
D. lack the enzyme needed to break down glucose

2. Oligosaccharides differ from polysaccharides in that they:

A. contain only glucose saccharide residues
B. always have the same saccharide residues
C. contain between 2 and 10 monosaccharide residues
D. they do not contain monosaccharide residues

3. The monosaccharide molecule below is an example of a(n)_____.

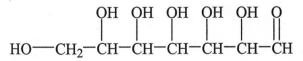

A. aldoheptose
B. ketononose

C. pyranose
D. furanose

4. If a molecule has 3 chiral carbons, how many stereoisomers are possible?

A. 1
B. 2
C. 3
D. 4

5. How many chiral carbon atoms does the molecule below contain?

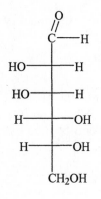

A. 1
B. 2
C. 3
D. 4

6. The molecules below are:

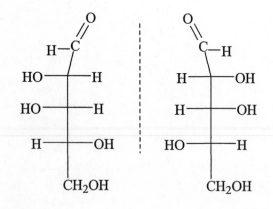

A. enantiomers
B. diastereomers
C. ketoses
D. not isomers

7. The molecules below are:

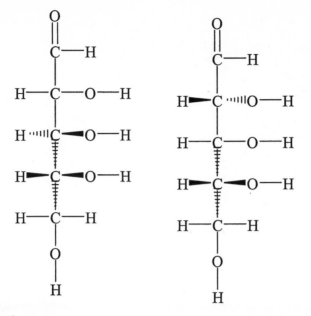

A. enantiomers
B. diastereomers
C. ketoses
D. not isomers

8. A monosaccharide is considered to be a D-residue when:

A. it is part of DNA
B. when the –OH on the chiral carbon next to the CH_2OH is on the right side of a Fischer projection
C. when the –OH on the chiral carbon next to the CH_2OH is on the left side of a Fischer projection
D. the monosaccharide is doubled in every other residue group of a heteropolysaccharide

9. The molecule below would be classified as a:

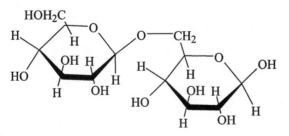

A. monosaccharide
B. oligosaccharide
C. polysaccharide
D. double saccharide

10. The molecule below would be classified as a:

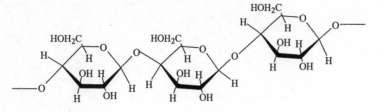

A. monosaccharide
B. oligosaccharide
C. polysaccharide
D. double saccharide

11. When an aldehyde reacts with H_2 in the presence of Pt the product is:

A. an alcohol
B. a carboxylic acid
C. a alkene
D. a saturated hydrocarbon

12. When an aldehyde reacts with Benedict's reagent the product is:

A. an alcohol
B. a carboxylic acid
C. a alkene
D. a saturated hydrocarbon

13. True or False: Only aldehydes will react with Benedict's reagent.

14. Draw the ring structure of a glucose anomer.

15. Describe the glycosidic bond for the molecule below and explain your answer.

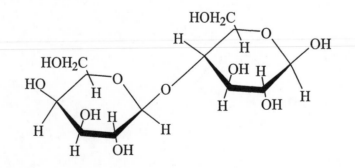

240

In 16 through 25, match the following terms and phrases to the correct meaning

_____16. Anomer

A. attached to –OH and –OC.

_____17. Acetal carbon

B. cyclic form of a monosaccharide

_____18. Mutarotation

C. a hemiacetal with the –OH pointing down

_____19. Glycosidic

D. a hemiacetal with the –OH pointing up

_____20. Alpha anomer

E. has a 5-member ring

_____21. Glycosidic bond

F. has a 6-member ring

_____22. Beta anomer

G. is the process of converting back and forth from an (alpha) anomer to the (beta) anomer.

_____23. pyranose

H. attached to two –OC groups

_____24.Hemiacetal carbon

I. is the resulting acetal when the hemiacetal in question is the alpha or beta anomer of a monosaccharide.

_____25. furanose

J. is the bond that connects the acetal carbon to the newly added –OC group.

1. B
2. C
3. A
4. D
5. C
6. A
7. B
8. B
9. B
10. C
11. A
12. B
13. F (FALSE)
14.

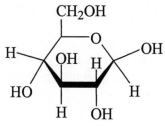

15. The glycosidic bond is β, α- (1→ 4). The bond is facing upward on carbon 1 making it β and the bond is facing downward on the carbon 4 of the other ring, making it α.
16. B
17. H
18. G
19. I
20. C.
21. J
22. D
23. E
24. B
25. F

ANSWER PAGE FOR BUILDING KNOWLEDGE

Match the terms or phrases to the correct meaning.

___r___ 1. Benedict's reagent

a. are the building blocks that produce oligosaccharides and are aldoses or ketoses

___i___ 2. Fisher-projection

b. are formed when between 2 and 10 monosaccharide residues are joined to one another by glycosidic bonds.

___s___ 3. Reducing sugars

c. contain more than 10 monosaccharide residues

___y___ 4. Pyranose

d. the part of a reactant molecule that remains when it has been incorporated into a product

___q___ 5. Alcohol sugar

e. comes from the Greek word sakcharon, or "sugar"

___v___ 6. α- anomer

f. are those monosaccharides that contain an aldehyde group

___w___ 7. β- anomer

g. are those monosaccharides that contain a ketone group

___ff___ 8. Maltose

h. indicates that a molecule is named a carbohydrate

___b___ 9. Oligosaccharides

i. chiral carbon atoms sit at the intersection of a vertical and horizontal line

___n___ 10. D-Fructose

j. has the enantiomers known as both D-glucose and L-glucose

___hh___ 11. Sucrose

k. diastereomer or non-mirror image stereoisomer of D-glucose and L-glucose

___bb___ 12. Glycosidic

l. also known as dextrose or blood sugar and the most important monosaccharides in human biology

___ee___ 13. Disaccharides

m. combined with glucose produces lactose, an oligosaccharide, that gives milk its sweetness.

___j___ 14. Glucose

n. the fruit sugar most commonly found in nature

___z___ 15. Mutarotation

o. hydrogen atom replaces one or more of the –OH group in a monosaccharide

___cc___ 16. Glycosidic bond replaced

p. are an –OH group of monosaccharides that has been   by an amino (NH_2) group

___e___ 17. Saccharide

q. the carbonyl group of a monosaccharide has been reduced to an alcohol group

___a___ 18. Monosaccharides

r. can oxidize aldehydes but not alcohols

___c___ 19. Polysaccharides

s. is when a given sugar is positive to the Benedict's test by becoming oxidized and reduced to the Cu^{2+}, which becomes present in the reagent

___ii___ 20. Glycolipids

t. attached to –OH and –OC

243

___kk___21. Chitin

___f___22. Aldoses

___g___23. Ketoses

___jj___24. Glycogen

___d___25. Residue

___o___26. Deoxy sugar

___gg___27. Cellobiose

___l___28. D-Glucose

___p___29. Amino sugars

___k___30. D-Galactose

___h___31. Suffix "ose"

___aa___32. Acetal carbon

___k___33. Galactose

___t___34. Hemiacetal carbon

___x___35. Furanose

___u___36. Anomer

u. cyclic form of a monosaccharide

v. a hemiacetal with the –OH pointing down

w. a hemiacetal with the –OH pointing up

x. has a 5-member ring

y. has a 6-member ring

z. is the process of converting back and forth from an (alpha) anomer to the (beta) anomer

aa. attached to two –OC groups

bb. is the resulting acetal when the hemiacetal in question is the alpha or beta anomer of a monosaccharide.

cc. is the bond that connects the acetal carbon to the newly added –OC group.

ee. contain two monosaccharide residues and are the most common oligosaccharides found in nature.

dd. is produced as the product of the digestion of starch and glycogen and categorized as a reducing sugar

ff. is produced as the product of cellulose is broken down and is categorized as a reducing sugar.

gg. also known as table sugar

hh. sugar-containing lipids present in nerve cell membranes, consisting of a mono- or oligosaccharide attached to an alcohol group of a lipid by a glycosidic bond

ii. is the energy storage molecule for animals

jj. is the material that makes up the exoskeleton of crustaceans and insects

Chapter 13
Peptides, Proteins, and Enzymes

Chapter Overview

In Chapter 8 we developed the structure of lipids; in Chapter 12 we explored carbohydrates. The last major group of macromolecules to investigate is proteins. Proteins are nitrogen-based macromolecules that play a critical role in the biological processes discussed in Chapters 14 and 15. You've almost completed building your knowledge of organic concepts, and at this point you should be able to recognize connections among the concepts. After learning about hydrocarbon and functional groups, we learned how saturated and unsaturated hydrocarbons along with carboxylic acids and alcohols come together to form lipids. In addition, the role of aldehydes, ketones, and alcohols as the building blocks of carbohydrates was discussed. Now, it is time to put the last few pieces of the organic puzzle together and see how ammonia, amides, and amides complete the building blocks of life.

In this chapter you'll learn how ammonia, amides or amines combine with carboxylic acids to create amino acids. The links that allow amino acids to be used as building blocks are peptide bonds. We'll discuss peptide bonds and the macromolecules they produce, known as proteins. In addition, the pH and structure of proteins are investigated. Once the nature of proteins are known, various types of protein reactions are discussed, beginning with how they become denatured (altered) from their original structures.. Finally, the chapter ends with a discussion of the role proteins play in biological processes as enzymes and how those enzymes can control reactions of other macromolecules.

Knowledge/Comprehension

Typically, there are new terms or phrases that you must memorize in order for the lesson to make sense to you. Look up this information and memorize their explanations and definitions:

a-amino	nonpolar	polar-acidic
polar-basic	polar-neutral	oligopeptides
polypeptides	protein	peptide bond
primary structure	secondary structure	tertiary structure
globular proteins	fibrous proteins	quaternary structure
prosthetic groups	denaturation	absolute specificity
competitive inhibitors	relative specificity	stereospecificity
noncompetitive inhibitors	irreversible inhibitors	allosteric enzymes
positive effectors	negative effectors	feedback inhibition
covalent modification	zymogens	
Michaelis-Menten enzyme catalysis		

The following activity is designed to help you master the objectives outlined in the beginning of Chapter 13 and then summarized at the end of the chapter. To make the best use of your study time, *read* Chapter 13 before continuing. Next, work the matching activity provided as though you are taking a test *without first* referring to the answer page. Put your answers on a separate sheet of paper so that you can make more than one attempt without seeing your previous answers. Once you have completed the activity, check your answers. Make *flash cards* for those you missed. Have a study buddy drill you on them if you missed more than half of the terms. Once you have reviewed the ones you missed for at least 15 minutes try the activity again. If you miss fewer than four, write in the answers so that you will have them to review just before taking the test.

Match the terms or phrases to the correct meaning

_____1. Polypeptides

a. contain an amino acid and a carboxyl group (pH dependent)

_____*2.* Quaternary structure

b. the amino group is attached to the carbon atom α to the carboxyl group

_____3. Stereospecificity

c. amino acid side chain composed of an alkyl group, an aromatic ring, or a nonpolar collection of atoms

_____4. Absolute specificity

d. amino acid side chain composed of a carboxyl group

_____5. Primary structure

e. amino acid side chain composed of an amine group

_____6. Allosteric enzymes

f. amino acid side chain composed of an alcohol, a phenol, or an amide

_____7. Denaturation

g. 2 to 10 amino acid residues

_____8. Amino acids

h. more than 10 amino acid residues

_____9. Zymogens

i. polypeptide with more than 50 amino acid residues

_____10. Relative specificity

j. amino acid residues in which either an oligo- or polypeptide are joined between the carboxyl group of one residue and the *a*-amino group of the other

_____11. Polar-basic

k. sequence of amino acid residues in a peptide or protein

_____12. Prosthetic groups

l. common structural features (*a*-helix and β-sheet); held together by hydrogen bonds between atoms in the polypeptide

_____13. Covalent modification

m. overall three-dimensional shape of a protein, including secondary structural features; held together by interactions between side chains

_____14. Positive effectors

n. spherical and water soluble

_____15. Michaelis-Menten enzyme catalysis

o. long fibers that are tough and water insoluble

_____16. α-amino

p. combination of more than one polypeptide chain that may be required to produce an active protein; held together by the same forces as tertiary structure (with the exception of disulfide bridges)

_____17. Secondary structure

q. tertiary structure can also include nonpeptide components attached to the polypeptide

_____18. Oligopeptides

r. any disruption on the noncovalent forces responsible for maintaining the native (biologically active) form of a protein

_____19. Polar-acidic

s. enzymes are limited to one particular substrate

_____20. Peptide bond

t. enzymes can use substrates containing the same functional group or similar structures

_____21. Nonpolar

u. when enzymes react with or produce a particular stereoisomer

_____22. Fibrous proteins

v. substrate (S) must bind to enzyme (E) and produce an enzyme substrate complex. Once the reaction has taken place, the product (P) dissociates and the enzyme is restored to its original form.

_____23. Feedback inhibition

w. Michaelis-Menten enzymes resemble the substrate, bind to the same site on the enzyme surface, change the value of $\underline{K}_M$ and have no effect on V_{max}

_____24. Irreversible inhibitors

x. do not resemble the substrate, bind to a different site, change V_{max} and leave K_M unchanged

_____25. Negative effectors

y. enzyme function is permanently destroyed

_____26. Polar-neutral

z. usually consist of more than one polypeptide chain, and the binding of the substrate to the active site on one subunit affects the binding of substrate at other sites

_____27. Globular proteins

aa. increase reaction velocity

_____28. Noncompetitive inhibitors

bb. decrease reaction velocity

_____29. Competitive inhibitors

cc. the product of a metabolic pathway prevents its own overproduction by inactivating one or more of the enzymes responsible for its synthesis

_____30. Tertiary structure

dd. an enzyme is switched "on" or "off" by making or breaking covalent bonds to the enzyme

_____31. Protein

ee. inactive enzyme precursors that are activated by the hydrolysis of peptide bonds in the protein

In addition to knowing the terms or phrases, you should be able to compare, contrast, or list examples that show that you understand the following concepts:

- Describe the structure of amino acids, and the system used to classify amino acids.

- List some ways to denature a protein.
- Describe allosteric enzymes and the role that effectors play in the ability of these enzymes to catalyze reactions.
- Describe K_M and V_{max}.

Once you know the appropriate tools for Chapter 13, you are ready to evaluate your understanding of those tools and their relationships. In this type of concept building, it is necessary to compare or contrast terms or concepts. In some cases you may be required to give examples or recognize a term or concept when it is depicted in a picture or sample scenario.

Sample Problems

1. Which of the following molecular structures is *not* an amino acid?

a.

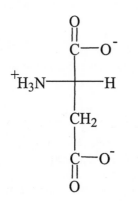

b.

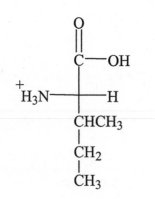

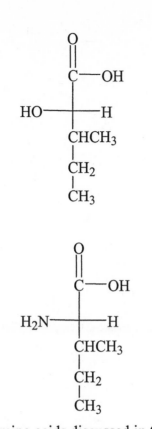

d.

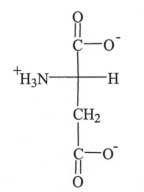

The amino acids discussed in Chapter 13 are called α-amino acids because they have an amino group attached to the carbon atom α to the carboxyl group (in other words, attached to the carbon that is attached to the carboxyl group). This tells you two things:

1. In order to be an amino acid, there must be an attached amino–NH_2 or ammonia $–NH_3^+$ group
2. The number 1 carbon is a carboxyl carbon.

Choices a, b, and d meet this criteria and are amino acids. Choice c has the carboxyl group but has a –OH (not an amino group –NH_2.) attached to the carbon α to the carboxyl group. **Answer is c.**

2. Which of the amino acids below would be classified polar-acidic?
a.

b.

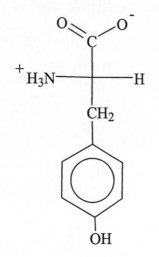

c.

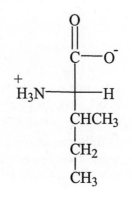

d.

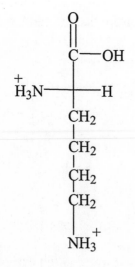

The amino acid classification scheme is based on the "side chain" since all of the 20 amino acids being considered in Chapter 13 have the same carboxyl carbon and amino positions. The classification is based on the side chain reaction when the amino acid is in a water solution with a pH of 7. Choice a has a carboxylic acid side chain which gives off a hydrogen at a pH of 7 and is therefore acidic. This molecule is classified as *polar-acidic*. Choice b has a phenol group as a side chain that is polar but at pH 7 the hydrogen is not removed, making the side chain neutral. This molecule is classified as *polar-neutral*. In choice c, the side chain is a hydrocarbon group that is nonpolar; acidic/basic is not a consideration. This molecule is classified simply as *nonpolar*. Choice d has an amine as a side chain that gains a hydrogen atom at pH 7, creating an ammonium ion. Since it gains a hydrogen atom, it is basic and this molecule is classified as *polar-basic*. **Answer is a.**

3. Which of the following is not typically used to denature a protein?

a. change in temperature

b. change in pH

c. change in light frequency

d. addition of soaps or detergents

Recall that proteins are strands of amino acids held together by peptide bonds and intermolecular forces. Therefore anything that causes agitation to the protein strand such as choice a, changing the *temperature* (measure of kinetic energy), can denature the protein. In choice b a change in *pH* can denature a protein because as you learned in problem 3 above, amino acids are so sensitive to pH that it is part of their classification scheme. In choice c changing the frequency (color) of light may, under certain circumstances, cause a protein to be denatured, but as a general rule taking a protein from one color of light to another does not denature it. Choice d, the addition of *soaps or detergents*, can cause the protein to be denatured. Recall from the classification scheme that some amino acids have nonpolar (hydrophobic) side chains that could be affected by soaps or detergents. **Answer is c.**

4. Which of the following is *not* true about allosteric enzymes?

a. they are regulated by positive effectors that increase the speed of the enzyme catalyzed reaction

b. they are regulated by negative effectors that reduce binding and slow down the reaction

c. they can be involved in feedback inhibition which prevents their own overproduction

d. they display a lack of cooperativity in which the binding of substrate to the active site on one subunit affects the binding of substrate at other subunits

Allosteric enzymes are regulated by ions or molecules called allosteric effectors so choices a and b are true. Some allosteric enzymes can provide feedback that regulates their own production so answer c is true. The distinguishing characteristic of allosteric enzymes is their display of cooperation in which the binding of substrate to the active site on one subunit affects the binding of substrate at other subunits. Choice d is not true. **Answer is d.**

5. Which of the following would *not* be true of K_M and V_{max} as used in describing Michaelis-Menten enzyme reaction?

a. they are used to characterize enzymes in much the same way melting points and boiling points are used to characterize simpler organic compounds

b. V_{max} is used to describe the maximum variation within mutations caused by enzyme activity

c. K_M is a measure of the strength of attraction between an enzyme and a substrate

d. each enzyme has a unique K_M and V_{max} under a particular set of reaction conditions (pH, temperature, etc.)

      Choice a is true; K_M and V_{max} were established to help characterize enzyme activity and are often used to identify an enzyme in much the same way as simpler organic compounds have a characteristic melting or boiling point. Choice b is not true; V_{max} actually describes the maximum velocity that a given concentration of enzyme can produce. Choice c is the definition of K_M and is therefore true. Choice d in fact is the reason why choice a is true; each enzyme has a characteristic K_M and V_{max} value under a particular set of conditions. **The answer is b.**

Application

In order to apply what you've already learned and master the concepts, you must utilize all of the knowledge, understanding, facts, and/or techniques that you have gathered up to this point. Simply knowing an equation or process is not enough to master the concept. These you develop by *practice*. Solve as many different types of problems with as many different variations as you *strive to master the skill*. Your textbook provides a very good assortment of problems at the end of the chapter. Work as many as needed for skill mastery, not just those assigned.

- Distinguish between oligopeptides, polypeptides, and proteins and describe the bond that joins amino residues in these compounds.
- Name the steps required for a typical Michaelis-Menten enzyme to convert a substrate into a product.
- Explain how competitive and noncompetitive inhibitors affect K_M and V_{max}.

- Distinguish between reversible and irreversible inhibitors.

In order to truly master any skill you must *practice*. In this section example problems related to the application objectives are discussed. Remember, the more you practice, the better your chances of being able to successfully use these objectives to solve problems. Additional practice problems are at the end of the chapter in your textbook and in the online interactive lessons.

Sample Problems

1. Which of the following is true for proteins?

a. they have an amino acid residue

b. they have 50 or more amino acid residues

c. they have between 11 and 50 amino acid residues

d. they have between 2 and 10 amino acid residues

Choice a is not correct because amino acids are the building blocks of peptide structures. Choice b is correct because proteins are composed of very long chains of 50 or more amino acid residues. Choice c is the characterization for polypeptides and therefore not the correct answer. Choice d is the characterization for oligopeptides and therefore is not the correct answer. **Answer is b.**

2. Which of the following is *not* true about peptides?

a. the end of the peptide chain with the unreacted amino group is called N-terminus

b. the end of the peptide chain with the free carboxyl group is called C-terminus

c. the peptide bond is another name for ammonia complex

d. the amino residue with the N-terminus is drawn on the left while the C-terminus is on the right

In order to have a point of reference when discussing peptide chains, the amino residue with the unreacted amine group ($-NH_2$) is called the N-terminus (note N for nitrogen). The amino residue with the unreacted carboxylic acid ($-COOH$) is called the C-terminus (note C for carbon); therefore choices a and b are true. Choice c is not true. A peptide bond is another name for an amide bond. Recall from Section 10.7 that an amide is formed when the NH_2 replaces the $-OH$ on the carboxyl carbon (see the figure below). Choice d is the preferred format for writing peptide chains and is true. **Answer is c.**

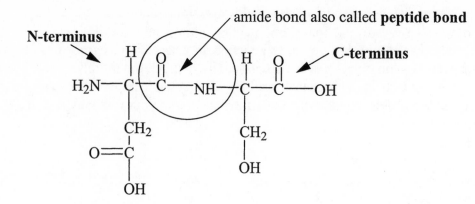

N-terminus

amide bond also called **peptide bond**

C-terminus

3. Which of the following is *not true* for a typical Michaelis-Menten enzyme process to convert a substrate into a new product?

a. the measure of the strength of the attraction between an enzyme and a substrate is K_M

b. enzyme (E) binds substrate (S) to form an enzyme-substrate complex (ES)

c. substrate (S) is transformed into product (P) which is released (ES → E + P)

d. the enzyme (E) is denatured to form the substrate (S) and then to product (P)

Choice a is true; the Michaelis constant, K_M, is a measure of the strength of attraction between an enzyme and a substrate during an enzyme catalyzed reaction. Choices b and c are the two defined steps of the Michaelis-Menten enzyme reaction and are therefore true. Choice d has no basis in fact, and only includes (E) as in choices c and d as a distracter! **Answer is d.**

4. Which of the following is *not* true about competitive inhibitors?

a. they compete with the enzyme at the active site of a substrate

b. they cause the K_M value to be reduced

c. they do not affect the V_{max}

d. they have a structure similar to that of the substrate and are mistaken for a substrate by an enzyme

Choice a is not true; the active site is located on the enzyme, not the substrate. Because the competitive inhibitor competes with the substrate for position on the enzyme and therefore keeps the enzyme from holding the substrate completely, the strength of attraction (K_M) is reduced; therefore choice b is true. Once the substrate is held by the enzyme, the conversion to product occurs at the same rate so V_{max} is unaffected; choice c is true. Choice d is true because the similarity to the substrate allows the competitive inhibitor to compete. **Answer is a.**

5. Which of the following is *not* true about noncompetitive inhibitors?

a. they attach to different sites than the substrate on the enzyme

b. their K_M value can be reduced

c. they reduce the V_{max}

d. they have a different structure than that of the substrate which interferes with the conversion process by reducing the enzyme's ability to convert the substrate

Choice a is true; because they have a different structure, the noncompetitive inhibitor attaches to a different site than the substrate. Therefore, the noncompetitive inhibitor does not compete with the substrate for position on the enzyme. Since it binds up other sites on the enzyme and keeps the enzyme from converting the substrate as quickly, the strength of attraction (K_M) is unaffected; choice b is not true. Once the substrate is held by the enzyme, the conversion to product cannot occur at the same rate due to the inhibitor's interference, so V_{max} is reduced and choice c is true. The noncompetitive inhibitor reduces the enzyme's ability to convert the substrate so choice d is true as well.
The answer is b.

6. Which of the following is true for a reversible inhibitor but not an irreversible inhibitor?

a. they are loosely bound to the enzyme and can dissociate

b. they can be a competitive inhibitor

c. they can be a noncompetitive inhibitor

d. none of the above

Choices a through c are all characteristics of a reversible reaction. Since the inhibitor is only loosely attached, it can be released and the enzyme returns to its original state. An irreversible inhibitor does not have any of these characteristics; it interferes by reacting with the enzyme and creating a protein that is not a catalyst, which is a nonreversible reaction. **Answer is d.**

Chapter 13 Practice Test

Now that you have had the chance to study and review Chapter 13's concepts and objectives, and work the practice problems provided in your textbook, you are ready to practice problems as you might see them in a testing situation. The questions and problems below are randomly selected in much the same way your instructor will select them; that is not necessarily ordered in the same way as they appear in the textbook. To make best use of this practice, have your paper, pencil and calculator ready as you would for a class exam. Use a clock and give yourself 45 minutes to complete this activity. Note that taking this test under similar conditions as a class exam provides two benefits:

1. Testing your knowledge of the chapter's contents and your mastery of the required skills. .
2. Practicing working under the stress of a timed exam.

Most instructors have a type of testing format that they prefer and will most likely tell you about it before the test. For practice, this self-test uses a variety of test question formats such as matching, fill in the blank, listing, multiple choice, and essay questions. ***Be sure to have a periodic table!***

Note the time and begin.

1. Which of the following molecules is *not* an amino acid?

    A.

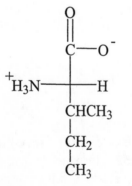

B.

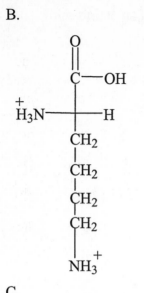

C.

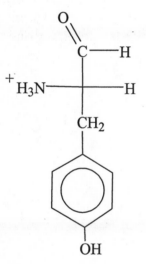

D.

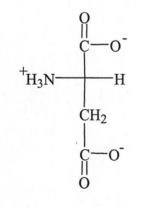

2. Which of the following drawings is a Fischer projection of an amino acid?

A.

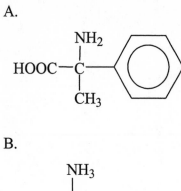

B.

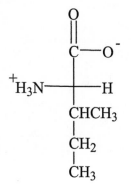

C.

D.

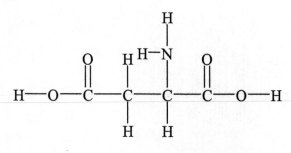

3. Based on the structure of the amino acid below, what is the most likely pH of the solution?

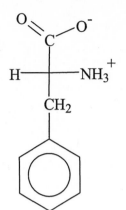

A. pH = 1
B. pH = 7
C. pH = 13
D. unable to tell with data provided

4. Based on the structure of the amino acid below, what is the most likely pH of the solution?

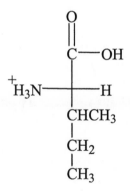

A. pH = 1
B. pH = 7
C. pH = 13
D. unable to tell with data provided

5. What is the net charge on the molecule below?

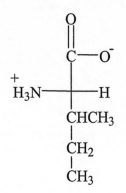

A. 0
B. +1
C. -1
D. 2

6. How many chiral carbons are in the molecule below?

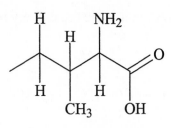

A. 0
B. 1
C. 2
D. 3

7. The molecule below is a(n):

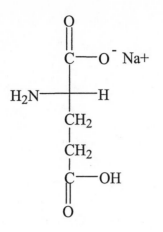

A. peptide
B. salt
C. amino acid
D. amide

8. Peptide bonds are also known as:

A. amide bonds
B. pepto-bonds
C. amino bonds
D. carboxylate bonds

9. The protein Ala-Ala-Val-Gly-Lys is an example of a(n):

A. amino acid residue
B. oligopeptide
C. polypeptide
D. nucleotide

10. The protein Ala-Ala-Val-Gly-Lys-Ala-Ala-Val-Gly-Lys-Ala-Ala-Val-Gly-Lys is an example of a:

A. amino acid residue
B. oligopeptide
C. polypeptide
D. nucleotide

11. Label the salt bridge in the peptide below.

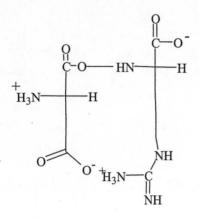

12. Label the N-terminus and the C-terminus for the following protein.

Ala-Val-Gly-Lys

13. How many peptide bonds are in the structure below?

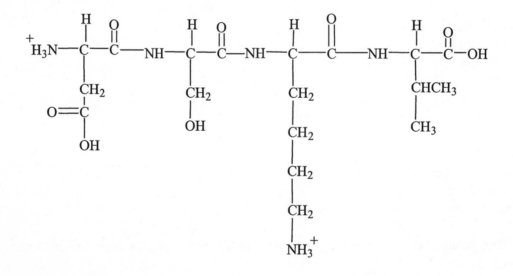

14. List the tripeptides that can be produced containing one copy each of methionine, arginine, and tyrosine.

15. What chemical difference distinguishes irreversible inhibitors from reversible inhibitors?

In questions 16 through 25, match the following terms or phrases with the correct meaning

_____16. Feedback inhibition

_____17. Irreversible inhibitors

_____18. Negative effectors

_____19. Covalent modification

_____20. Zymogens

_____21. Noncompetitive inhibitors

_____22. Competitive Inhibitors

_____23. Positive effectors

_____24. Protein

_____25. Allosteric enzymes

A. Michaelis-Menten enzymes resemble the substrate, bind to the same site on the enzyme surface, change the value of K_M and have no effect on V_{max}

B. do not resemble the substrate, bind to a different site, change V_{max} and leave K_M unchanged

C. enzyme function is permanently destroyed

D. usually consist of more than one polypeptide chain, and the binding of the substrate to the active site on one subunit affects the binding of substrate at other sites

E. increase reaction velocity

F. decrease reaction velocity

G. the product of a metabolic pathway prevents its own overproduction by inactivating one or more of the enzymes responsible for its synthesis

H. an enzyme is switched "on" or "off" by the making or breaking of covalent bonds to the enzyme

I. inactive enzyme precursors that are activated by the hydrolysis of peptide bonds in the protein

J. polypeptide with more than 50 amino acid residues

1. C
2. C
3. B
4. C
5. A
6. C
7. B
8. A
9. B
10. C
11.

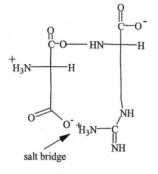

salt bridge

12. **N-terminus** → Ala-Val-Gly-Lys ← **C-terminus**
13. Answer is 3.
14.

Arg-Met-Tyr
Arg-Tyr-Met
Met-Arg-Tyr
Met-Tyr-Arg
Tyr-Met-Arg
Tyr-Arg-Met

15. The irreversible inhibitor chemically reacts with the enzyme to change it to a protein that is not a catalyst. The reversible inhibitor does not alter the actual structure of the enzyme.
16. G
17. C
18. F
19. H
20. I
21. B
22. A
23. E
24. J
25. D

ANSWER PAGE FOR BUILDING KNOWLEDGE

Match the terms or phrases to the correct meaning.

__h___1. Polypeptides

__p___2. Quaternary structure

__u___3. Stereospecificity

__s___4. Absolute specificity

__k___5. Primary structure

__z___6. Allosteric enzymes

__r__7. Denaturation

__a___8. Amino acids

_ee__ 9.Zymogens

__t_ 10. Relative specificity

__e__11. Polar-basic

__q__12. Prosthetic groups

_dd__13. Covalent modification

_aa_ 14. Positive effectors

__v__15. Michaelis-Menten
          enzyme catalysis

__b__16.. *a*-amino

__l._17. Secondary structure

__g__18. Oligopeptides

__d__19. Polar-acidic

a. contain an amino acid and a carboxyl group (pH dependent)

b. the amino group is attached to the carbon atom α to the carboxyl group

c. amino acid side chain composed of an alkyl group, an aromatic ring, or a nonpolar collection of atoms

d. amino acid side chain composed of a carboxyl group

e. amino acid side chain composed of an amine group

f. amino acid side chain composed of an alcohol, a phenol, or an amide

g. 2 to 10 amino acid residues

h. more than 10 amino acid residues

i. polypeptide with more than 50 amino acid residues

j. amino acid residues in which either an oligo- or polypeptide are joined between the carboxyl group of one residue and the *a*-amino group of the other

k. sequence of amino acid residues in a peptide or protein

l. common structural features (*a*-helix and *β*-sheet); held together by hydrogen bonds between atoms in the polypeptide

m. overall three-dimensional shape of a protein, including secondary structural features; held together by interactions between side chains

n. spherical and water soluble

o. long fibers that are tough and water insoluble

p. combination of more than one polypeptide chain that may be required to produce an active protein; held together by the same forces for tertiary structure (with the exception of disulfide bridges)

q. tertiary structure can also include nonpeptide components attached to the polypeptide

r. any disruption on the noncovalent forces responsible for maintaining the native (biologically active) form of a protein

s. enzymes are limited to one particular substrate

__j__ 20. Peptide bond

__c__ 21. Nonpolar

__o__ 22. Fibrous proteins

__cc__ 23. Feedback inhibition

__y__ 24. Irreversible inhibitors

__bb__ 25. Negative effectors

__f__ 26. Polar-neutral

__n__ 27. Globular proteins

__x__ 28. Noncompetitive inhibitors

__w__ 29. Competitive inhibitors

__m__ 30. Tertiary structure

__i__ 31. Protein

t. enzymes can use substrates containing the same functional group or similar structures

u. when enzymes react with or produce a particular stereoisomer

v. substrate (S) must bind to enzyme (E) and produce an enzyme substrate complex. Once the reaction has taken place, the product (P) dissociates and the enzyme is restored to its original form.

w. Michaelis-Menten enzymes resemble the substrate, bind to the same site on the enzyme surface, change the value of K_M and have no effect on V_{max}

x. do not resemble the substrate, bind to a different site, change V_{max} and leave K_M unchanged

y. enzyme function is permanently destroyed

z. usually consist of more than one polypeptide chain, and the binding of the substrate to the active site on one subunit affects the binding of substrate at other sites

aa. increase reaction velocity

bb. decrease reaction velocity

cc. the product of a metabolic pathway prevents its own overproduction by inactivating one or more of the enzymes responsible for its synthesis

dd. an enzyme is switched "on" or "off" by making or breaking covalent bonds to the enzyme

ee. inactive enzyme precursors that are activated by the hydrolysis of peptide bonds in the protein

Chapter 14
Nucleic Acids

Chapter Overview

In addition to our discussion of the biological macromolecules lipids, carbohydrates, and proteins, there is one more critically important group of molecules that highlight how life intersects with chemistry. The building blocks for this group consist of a monosaccharide, phosphoric acid, and an organic base. These building blocks are used to construct the nucleic acids known as RNA and DNA, the genetic code material of all living organisms.

Chapter 14 begins by reviewing the structure of each of the building blocks, phosphoric acid, monosaccharides, and organic bases. After studying how these building blocks are arranged to construct the two major nucleosides, ribonucleosides and deoxyribonucleosides, the nucleosides are used to demonstrate how polynucleotides are constructed and named. The two polynucleotides discussed will be RNA and DNA. A closer look at the structure of RNA and DNA shows how their structure can be denatured. The latter part of the chapter demonstrates the role of DNA and RNA in genetic coding as well as RNA's role in the process of creating new DNA. Finally, factors such as viruses, mutation, and recombination will be presented as ways that can change DNA patterns.

Knowledge/Comprehension

Typically, there are new terms or phrases that you must memorize in order for the lesson to make sense to you. Look up this information and memorize their explanations and definitions:

nucleotides	phosphoric acid	major bases
DNA replication	translation	origin
template strand	transfer RNAs	messenger RNAs
ribosomal RNAs	ribosomes	anticodon
base pairing	double-Stranded DNA	mutation
genetic disease	BRCA1 and BRCA2	recombinant DNA
nucleotides	initiation	elongation
termination	post- translational	restriction enzymes
semiconservative replication		short tandem repeats
post- transcriptional modification		

The following activity is designed to help you master the objectives outlined in the beginning of Chapter 14 and then summarized at the end of the chapter. To make the best use of your study time, *read* Chapter 14 before continuing. Next, work the matching activity provided as though you are taking a test *without first* referring to the answer page. Put your answers on a separate sheet of paper so that you can make more than one attempt without seeing your previous answers. Once you have completed the activity, check your answers. Make *flash cards* for those you missed. Have a study buddy drill you on them if you missed more than half of the terms. Once you have reviewed the ones you missed for at least 15 minutes try the activity again. If you miss fewer than four, write in the answers so that you will have them to review just before taking the test.

Match the terms or phrases to the correct meaning

_____1. Restriction enzymes   a. phosphoric acid, a monosaccharide, and an organic base

_____2 Ribosomes   b. has the ability to form phosphate esters

_____3 . Nucleotides   c. found in the nucleotides are and the three pyrimidine bases

_____4. Messenger RNAs   d. to pass the information stored in DNA to a new generation of cells

_____5. Termination   e. the primary structure of a particular form of RNA

_____6. Post- transcriptional modification   f. in each of the daughter DNA strands, one strand from the parent DNA is present

_____7. Anticodon   g. separating two strands and forming a bubble where single DNA strands are exposed

_____8. Elongation   h. the DNA strand that RNA polymerase reads

_____9. Recombinant DNA   i. the smallest of the three types of RNA that carry the correct amino acid to the site of protein synthesis

_____10. DNA Replication   j. carry the information that specifies which protein should be made

_____11. Initiation   k. are relatively long RNA strands that combine with proteins

_____12. Phosphoric acid   l. the multi-subunit complexes in which protein synthesis takes place

_____13. Origin   m. a series of three bases that is complementary to a codon

_____14. Post- translational modification   n. two strands of DNA combine

_____15. BRCA1 and BRCA2   o. hydrogen bonding between the complementary bases A and G or C and T to form antiparallel strands

_____16 Short Tandem Repeats   p. In the primary structure of (sequence of nucleotide residues in) DNA

_____17.Genetic Disease   q. mutations that take place in egg or sperm cells can be inherited (diseases caused by inherited mutations)

_____18. Double-Stranded DNA

r. are genes that code for proteins indirectly involved in a form of DNA repair called recombinational repair

_____19. Transfer RNAs

s. the processes involved in DNA replication have given researchers the ability to create DNA which contains DNA from two or more sources

_____20. Mutation

t. relatively small stretch of DNA that contains short repeating sequences

_____21. Major Bases

u. are a combination of ribose (in RNA) or 2-deoxyribose (in DNA)

_____22. Semiconservative replication

v. RNA undergoes modification following transcription

_____23. Ribosomal RNAs

w. a ribosome mRNA, and tRNA come together to form a complex

_____24. Template Strand

x. amino acids are joined to the growing polypeptide chain

_____25. Translation

y. when the protein has been synthesized and the ribosome -mRNA- tRNA complex dissociates

_____26. Base pairing

z. removal of fMet at the N-terminus, trimming and cutting of the polypeptide chain

_____27. Nucleotides

aa. are used to cut repeats in each of the STR

In addition to knowing the terms or phrases, you should be able to compare, contrast, or list examples that show you understand the following concepts:

- Describe the make up of nucleosides, nucleotides, oligonucleotides, and polynucleotides.
- Describe the primary structure of DNA and RNA and secondary and tertiary structure of DNA.
- Name the three types of RNA and identify the role of each in translation.
- List the steps involved in producing recombinant DNA.

Once you know the appropriate tools for Chapter 14, you are ready to evaluate your understanding of those tools and their relationships. In this type of concept building, it is necessary to compare or contrast terms or concepts. In some cases you may be required to give examples or recognize a term or concept when it is depicted in a picture or sample scenario.

Sample Problems

1. Which of the following is *not* part of a nucleoside?

a. a ribose in its β furanose form

b. a 2-deoxyribose in its β furanose form

c. an organic base

d. a phosphoric acid as a phosphate bond

A nucleoside is composed of a monosaccharide with an organic base attached to it. The two nucleosides used to make RNA and DNA contain ribose and deoxyribose in their β furanose form so choices a and b are true. Choice c is also true. When the phosphate bond is added, the term nucleotide is used, so d is not true. **Answer is d.**

2. Which of the following is *not* a nucleotide?

a. nucleoside diphosphate

b. phosphoester bond

c. nucleoside triphosphate

d. cyclic nucleotide

Nucleotides are classified by the phosphoester bonds that are attached to the nucleoside, therefore answers a and c are true. In some cases, a side oxygen from the phosphoester bond can attach in place of another –OH group on the monosaccharide ring. When this occurs, it is a cyclic nucleotide so choice d is true. Choice b, the phosphoester bond, is only one component of the nucleotide. **Answer is b.**

3. Which of the molecules below is a nucleoside diphosphate?

a.

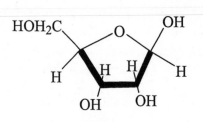

b.

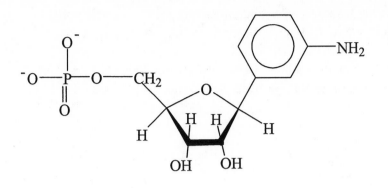

c.

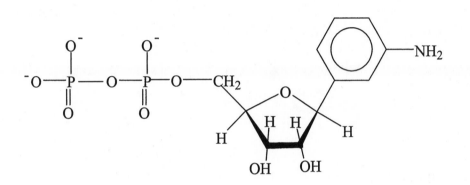

d.

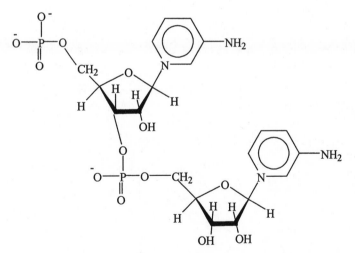

Choice a is easily eliminated since it is only ribose and not a nucleoside. In the other three choices, which are nucleosides, you need to count the phosphate

groups   in each molecule and recall that the prefix "di" means two. This eliminates choice b since it only has one phosphate. Choice c is a nucleoside diphosphate because it has two phosphates attached in sequence on a nucleoside. Choice d has two phosphates, but is not a nucleoside diphosphate. Instead, it is two nucleoside monophosphates bonded together by phosphoester bonds. **Answer is c.**

4. On the molecule below, locate and label the phosphoester bond(s), ribose(s), organic base(s), the N-glycosidic bond(s), the 5'-terminus, and the 3'-terminus.

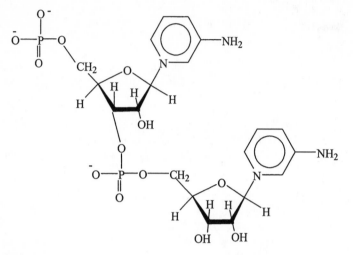

Answer

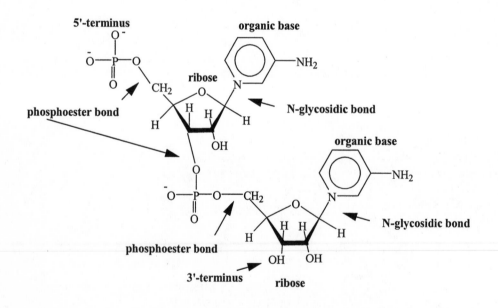

5. The sequence of nucleotide residues gives DNA its:

a. primary structure

b. secondary structure

c. tertiary structure

d. ability to resist denaturation

DNA molecules are described by the base sequence attached to the deoxyribose back-bone such as thymine, guanine, cytosine, and adenine abbreviated as TGCA. Therefore the sequence of the bases is the primary structure of DNA and choice a. is correct. Choice b is not correct because the secondary structure is one type of antiparallel strand arrangement, such as the α-helix, or the β-sheet. Likewise, choice c is not correct because the tertiary structure describes the overall three-dimensional shape of DNA, such as supercoiling. There is no particular sequence of nucleotide residues that is more resistant than another, so choice d is not applicable. **Answer is a.**

6. Which of the following is *not* a type of RNA?

a. transfer RNA (tRNA)

b. chromosomal RNA (cRNA)

c. messenger RNA (mRNA)

d. ribosomal RNA (rRNA)

There are only three types of RNA, transfer (tRNA), messenger (mRNA), and ribosomal (rRNA). Therefore choice b is not a type of RNA. **Answer is b.**

7. Which of the following is the role of messenger RNA (mRNA) in translation?

a. combines with proteins to form ribosomes, the multi-subunit complexes in which protein synthesis takes place

b. carries the correct amino acid to the site of protein synthesis

c. carries the information that specifies which protein should be made

d. removes sections of tRNA from the middle and the ends, whose code is not used in protein production

In choice a the use of the term ribosome is your clue that this is not a role of mRNA; in fact it is the role of rRNA. Choice b is the role of tRNA; note that amino acids are transferred, denoting smaller groups of nucleotide residues. Choice c is the role of mRNA; once all of the correct amino acids are in place, it is mRNA's role to carry the information that specifies which protein is made. Choice d has no basis in fact. **Answer is c.**

8. Which of the following is not a step involved in producing recombinant DNA?

a. identifying a gene of interest and its primary structure

b. cutting the gene of interest from the DNA strand in such a way that it can be combined with DNA from another source

c. combining the sections of DNA to one another and treating with the appropriate enzyme to connect the sugar-phosphate backbones

d. placing the recombinant DNA into a mRNA-rich solution and keeping it at body temperature to promote growth of more recombinant DNA

Choices a through c are the first steps in producing recombinant DNA. Choice d is not correct. To date, most recombinant DNA is made by combining desired gene strands with bacterial DNA. The recombinant DNA is placed back into the bacteria to allow it to be reproduced by the bacteria. **Answer is d.**

Application

In order to apply what you've already learned and master the concepts, you must utilize all of the knowledge, understanding, facts, and/or techniques that you have gathered up to this point. Simply knowing an equation or process is not enough to master the concept. These you develop by *practice*. Solve as many different types of problems with as many different variations as you *strive to master the skill.* Your textbook provides a very good assortment of problems at the end of the chapter. Work as many as needed for skill mastery, not just those assigned.

- Explain how replication takes places and describe the roles of DNA polymerase in this process.
- Explain how transcription takes places and describe the roles of RNA polymerase in this process.
- Explain how *E. coli* controls the expression of genes.
- Explain the term "mutation."

In order to truly master any skill you must *practice*. In this section example problems related to the application objectives are discussed. Remember, the more you practice, the better your chances of being able to successfully use these objectives to solve problems. Additional practice problems are at the end of the chapter in your textbook and in the online interactive lessons.

Sample Problems

1. Which of the following is *not* a part of the replication of DNA?

a. enzymes alter the secondary and tertiary structure of DNA by separating two strands and forming a bubble where single DNA strands are exposed

b. single-stranded binding proteins (SSBs) attach to the newly exposed single strands of DNA

c. DNA polymerases attach to each of the existing single strands of DNA

d. DNA polymerase, using dATP, dGTP, dCTP, and dTTP as substrates, match each base of the parent chain with a complimentary base all the way down the parent chain

e. RNA duplicates the sequence of the DNA

Choices a through d represent the basic steps in DNA replication. RNA does not play a role in replication but takes part in a process called transcription and is not part of the DNA replication process. **Answer is e.**

2. Which of the following is *not* part of the transcription process?

a. portions of DNA are copied, making additional RNA used in protein synthesis

b. RNA polymerase attaches to a strand of DNA at a promoter site

c. RNA polymerase uses ATP, GTP, CTP, and UTP as substrates and catalyzes the addition of the organic bases, guanine, cytosine, thymine, and uracil

d. when the RNA reaches the termination sequence, the enzyme detaches and releases the RNA

Choice b is the first step in transcription. Choice a is the next step that is accomplished by RNA polymerase using ATP, GTP, CTP, and UTP as substrates, catalyzing the addition of ribonucleic acids to the growing RNA strand. Since choice c states that the RNA polymerase adds bases, it is false and not part of transcription. The final step of transcription is choice d; when the RNA reaches the termination sequence the enzyme detaches and releases the RNA. **Answer is c.**

3. Gene expression is controlled in *E. coli* bacteria using all but:

a. an operon under the control of one promoter site

b. three genes, lacZ, lacY, and lacA

c. repressor proteins to bind to operator sites O_1 and O_2

d. an abundance of glucose to provide the necessary energy

The *E. coli* bacteria controls the gene expression for the production of β-galactosidase by utilizing choices a through c however, if choice d were true, the presence of glucose would inhibit the production of cAMP, preventing effective polymerase binding and making the production of β-galactosidase less effective. **Answer is d.**

4. Which of the following is *not* true about mutations?

a. they are always destructive and cause abnormalities in cell structures

b. genetic mutations occurring in an egg or sperm cell cause inherited diseases

c. diseases such as breast cancer, cystic fibrosis, and sickle cell anemia are linked to mutations

d. mutations permanently change the primary structure (the nucleotide residue sequence) of DNA

Choice a is not true; mutations generally occur over a relatively small section of the DNA and genes make up only about 5% of the DNA structure. Many mutations have no effect on the genetic function of the DNA. Choice b is true; genetic mutations that occur in an egg or sperm are passed to future generations and may cause inherited diseases. Choice c is an example of some diseases that have been linked to genetic mutations. Choice d is true since it is the definition of a mutation. **Answer is a.**

Chapter 14 Practice Test

Now that you have had the chance to study and review Chapter 14's concepts and objectives, and work the practice problems provided in your textbook, you are ready to practice problems as you might see them in a testing situation. The questions and problems below are randomly selected in much the same way your instructor will select them; that is not necessarily ordered in the same way as they appear in the textbook. To make best use of this practice, have your paper, pencil and calculator ready as you would for a class exam. Use a clock and give yourself 45 minutes to complete this activity. Note that taking this test under similar conditions as a class exam provides two benefits:

1. Testing your knowledge of the chapter's contents and your mastery of the required skills.
2. Practicing working under the stress of a timed exam. Most instructors have a type of testing format that they prefer and will most likely tell you about it before the test. For practice, this self-test uses a variety of test question formats such as matching, fill in the blank, listing, multiple choice, and essay.

1. The molecule shown below is a:

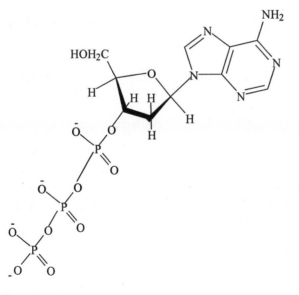

A. nucleoside monophosphate
B. nucleoside diphosphate
C. nucleoside triphosphate
D. cyclic nucleotide

2. Which of the following is *not* a part of a nucleotide?

A. phosphoric acid

B. phenylcarbolate
C. monosaccharide
D. organic base

3. The base components of the nucleotides that make up DNA do *not* include:

A. adenine
B. guanine
C. cytosine
D. uracil

4. The furanose ring structure that composes part of the DNA backbone is:

A. ribose
B. lactose
C. glucose
D. phosphoester

5. The bases on the sugar phosphate backbone of DNA are attached at:

A. random
B. alternating carbon from C-1
C. the C-1 carbon
D. the 3' terminus

6. The structure of DNA that consists of a helix formed by the interaction of two DNA strands is a:

A. primary structure
B. secondary structure
C. tertiary structure
D. quaternary structure

7. During replication the number of origins opened up is:

A. 1
B. 2
C. 3
D. too many to count

8. Polymerases move along an existing DNA strand in:

A. a 5' to a 3' direction
B. a 3' to a 5' direction
C. a 1' to a 5' direction

D. alternating directions

9. Once DNA has been denatured can it be renatured? Explain.

10. List the name of the bases indicated in the sequence dGGCAT.

In questions 11 through 15, identify the missing labels on the drawing below

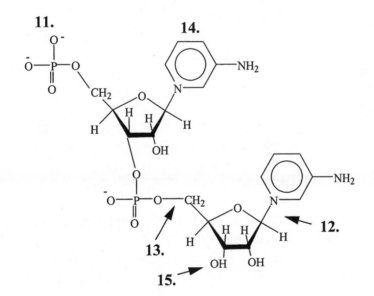

11. _____

12. _____

13. _____

14. _____

15. _____

In questions 16 through 25, match the following terms and phrases to the correcting meaning

_____16. Messenger RNAs

A. separating two strands and forming a bubble where single DNA strands are exposed

_____17. Mutation

B. the DNA strand that RNA polymerase reads

_____18. Ribosomes

C. the smallest of the three types and carries the correct amino acid to the site of protein synthesis

_____19. Anticodon

D. carries the information that specifies which protein should be made

_____20. Double –stranded DNA

E. are relatively long RNA strands that combine with proteins

_____21. Template strand

F. the multi-subunit complexes in which protein synthesis takes place

_____22. Origin

G. a series of three bases that is complementary to a codon

_____23. Ribosomal RNAs

H. combination of two strands of DNA

_____24. Transfer RNAs

I. hydrogen bonding between the complementary bases A and G or C and T to form antiparallel strands

_____25. Base pairing

J. In the primary structure of (sequence of nucleotide residues in) DNA

PRACTICE TEST Answers

1. C
2. B
3. D
4. A
5. C
6. B
7. D
8. B
9. Yes. The DNA can be renatured to its original secondary or tertiary forms as long as the primary structure was not destroyed.
10. guanine, guanine, cytosine, adenine, and thymine
11. 5'-terminus
12. N-glycosidic bond
13. phosphoester bond
14. organic base
15. 3'-terminus
16. D
17. J
18. F
19. G
20. I
21. B
22. A
23. E
24. C
25. H

ANSWER PAGE FOR BUILDING KNOWLEDGE

Match the terms or phrases to the correct meaning

__aa__ 1. Restriction enzymes

__l__ 2 Ribosomes

__u__ 3 . Nucleotides

__j__ 4. Messenger RNAs

__y__ 5. Termination

__v__ 6. Post- transcriptional modification

__m__ 7. Anticodon

__x__ 8. Elongation

__s__ 9. Recombinant DNA

__d__ 10. DNA Replication

__w__ 11. Initiation

__b__ 12. Phosphoric acid

__g__ 13. Origin

__z__ 14. Post- translational Modification

__r__ 15. BRCA1 and BRCA2

__t__ 16 Short Tandem Repeats

__q__ 17.Genetic Disease

__o__ 18. Double-Stranded DNA

a. phosphoric acid, a monosaccharide, and an organic base

b. has the ability to form phosphate esters

c. found in the nucleotides are and three pyrimidine bases

d. to pass the information stored in DNA to a new generation of cells

e. the primary structure of a particular form of RNA

f. in each of the daughter DNA strands, one strand from the parent DNA is present

g. separating two strands and forming a bubble where single DNA strands are exposed

h. the DNA strand that RNA polymerase reads

i. the smallest of the three types of RNA and carries the correct amino acid to the site of protein synthesis

j. carry the information that specifies which protein should be made

k. are relatively long RNA strands that combine with proteins

l. the multi-subunit complexes in which protein synthesis takes place

m. a series of three bases that is complementary to a codon

n. two strands of DNA combine

o. hydrogen bonding between the complementary bases A and G or C and T to form antiparallel

p. In the primary structure of (sequence of nucleotide residues in) DNA

q. mutations that take place in egg or sperm cells can be inherited (diseases caused by inherited mutations)

r. are genes that code for proteins indirectly involved in a form of DNA repair called recombinational repair

282

___i__ 19. Transfer RNAs

s. the processes involved in DNA replication have given researchers the ability to create DNA which contains DNA from two or more sources

__p___ 20. Mutation

t. relatively small stretch of DNA that contains short repeating sequences

__c___ 21. Major Bases

u. are a combination of ribose (in RNA) or 2-deoxyribose (in DNA)

__f___ 22. Semiconservative Replication

v. RNA undergoes modification following transcription

__k___ 23. Ribosomal RNAs

w. a ribosome mRNA, and tRNA come together to form a complex

__h___ 24. Template Strand

x. amino acids are joined to the growing polypeptide chain

__e___ 25. Translation

y. when the protein has been synthesized and the ribosome -mRNA- tRNA complex dissociates

__n___ 26. Base Pairing

z. removal of fMet at the N-terminus, trimming and cutting of the polypeptide chain

__a___ 27. Nucleotides

aa. are used to cut repeats in each of the STR

Chapter 15
Metabolism

Chapter Overview

Throughout chapters 8, 12, 13, and 14 we've discussed the assembling and disassembling of molecular building blocks. If you have ever built anything you understand that it does not just happen, it takes work and lots of it! What is the source of energy that drives these reactions? Energy in the body is derived from a process known as metabolism.

Chapter 15 first develops the two fundamental metabolic processes, called catabolism and anabolism. The sum of all of the reactions taking place is the process called metabolism. Metabolism begins with the intake of food; food is broken down into respective building blocks through digestion. Once glucose is available it undergoes catabolism by glycolysis. The energy involved in glycolysis is analyzed and studied. Then, gluconeogenesis is explained as the method used in the body to create glucose when it is not readily available from food. Next, pyruvates, created by glycolysis, are taken through the citric acid cycle to show how they help generate energy. Lipid metabolism is also discussed, followed by a final section exploring amino acid metabolism.

Knowledge/Comprehension

Typically, there are new terms or phrases that you must memorize in order for the lesson to make sense to you. Look up this information and memorize their explanations and definitions:

metabolism	catabolism	anabolism
linear	spiral	coupled Reaction
metabolic pathways	circular	transamination
glycolysis	anaerobic	aerobic
gluconeogenesis	citric acid cycle	electron transport chain
oxidation spiral	oxidative phosphorylation	

The following activity is designed to help you master the objectives outlined in the beginning of Chapter 15 and then summarized at the end of the chapter. To make the best use of your study time, *read* Chapter 15 before continuing. Next, work the matching activity provided as though you are taking a test *without first* referring to the answer page. Put your answers on a separate sheet of paper so that you can make more than one attempt without seeing your previous answers. Once you have completed the activity, check your answers. Make *flash cards* for those you missed. Have a study buddy drill you on them if you missed more than half of the terms. Once you have reviewed the ones you missed for at least 15 minutes try the activity again. If you miss fewer than four, write in the answers so that you will have them to review just before taking the test.
Match the terms or phrases to the correct meaning

_____1. Circular

a. the sum of all reactions that take place in a living organism that can be divided into two parts, catabolism and anabolism

_____2. Citric acid cycle

b. reactions that break large molecules into small ones, usually releasing energy

_____3. Aerobic

c. reactions that biosynthesize large molecules from small ones, usually consuming energy

_____4. Transamination

d. continuous series of reactions in which the product of one reaction is the reactant in the next

_____5. Oxidation spiral

e. series of repeated reactions used to break down or build up a molecule

_____6. Metabolic pathways

f. spontaneous reaction provides the energy needed by a nonspontaneous one

_____7. Oxidative phosphorylation

g. group of reactions that detail the manufacturing or breaking down of carbohydrates, lipids, or members of any other biochemical class of compounds.

_____8. Metabolism

h. series of reactions where the final product is an initial reactant

_____9. Catabolism

i. 10-step linear pathway that converts one monosaccharide molecule into 2 pyruvates, producing 2 ATP and 2 NADH in the process

_____10. Electron Transport Chain

j. condition in which pyruvates are converted to lactate or, in yeast, are converted to CO_2 and ethanol

_____11. Anabolism

k. pyruvate is converted into acetyl CoA

_____12. Glycolysis

l. pathway in which glucose is biosynthesized from lactate and other non-carbohydrate sources

_____13. Gluconeogenesis

m. an 8-step circular pathway that begins with a reaction between oxaloacetate and acetyl-CoA and ends with manufacture of oxaloacetate

_____14. Anaerobic

n. series of proteins and other molecules embedded in the inner mitochondrial membrane that removes two elections from NADH or $FADH_2$ and uses the energy released while transferring the electrons down the chain to pump H^+ into the intermembrane space

_____15. Coupled Reaction

o. the process in which H^+ moves back into the mitochondrial matrix though channels provided by ATP synthase; ATP is manufactured from ADP and P_i

_____16. Spiral	p. each cycle removes 2 carbon atoms from a fatty acid and produces 1 acetyl-CoA, 1 FADH$_2$, and 1 NADH
_____17. Linear	q. the process in which an α-keto acid is the –NH$_2$ acceptor

In addition to knowing the terms or phrases, you should be able to compare, contrast, or list examples that show you understand the following concepts:

- Describe three different types of metabolic pathways.
- Name the products formed during the digestion of polysaccharides, triglycerides, and proteins and state where the digestion takes place.
- Describe the catabolism of triglycerides, the β-oxidation spiral, and how the β-oxidation spiral differs from fatty acid biosynthesis.
- Explain the fate of the amino groups in amino acids.

Once you know the appropriate tools for Chapter 15, you are ready to evaluate your understanding of those tools and their relationships. In this type of concept building, it is necessary to compare or contrast terms or concepts. In some cases you may be required to give examples or recognize a term or concept when it is depicted in a picture or sample scenario.

Sample Problems

1. Which of the following is not a type of metabolic pathway?

a. catabolism

b. anabolism

c. cannibalism

d. glycolysis

Metabolism is the sum of all reactions that take place in a living thing. Therefore, the processes that lead to those reactions are considered metabolic pathways. The larger scope of the pathways is divided into two categories; choices a, catabolism and b, anabolism, are pathways and not correct choices for the question. While choice c, cannibalism could certainly be the beginning of the digestive process, it is not part of the metabolic process. Choice d, glycolysis, is a portion of the catabolic process and would be a type of metabolic pathway. **Answer is c.**

2. Which of the following metabolic processes is not considered a catabolic pathway?

a. glycolysis

b. gluconeogenesis

c. citric acid cycle

d. electron transport chain

Catabolic pathways are processes that break compounds down into smaller ones. Choice c, glycolysis, is the pathway that breaks glucose into smaller compounds so it is not the correct answer. Choice b, gluconeogenesis, uses molecules from other catabolic processes to build glucose molecules; since it is a building up process, it would be considered anabolic and therefore the correct answer. Choice c, citric acid cycle, breaks down pyruvates created during glycolysis into acetyl-CoA so it is a catabolic process. Choice d, electron transport chain, is the process that uses NADH and $FADH_2$ produced in both glycolysis and the citric cycle and breaks them down them into their oxidative states, so it is a catabolic process. **Answer is b.**

3. Metabolic pathways are used to derive energy from all of the following sources *except*:

a. nucleic acid

b. carbohydrates

c. lipids

d. amino acids

Choice a is found in RNA and DNA. These molecules are not used as an energy source and genetic material is not changing or breaking down. Choice b, carbohydrates, are broken down for energy as are choices c, lipids (fats) and d, amino acids (protein). **Answer is a.**

4. Complete the table below.

Food group	Digestion product	Body organ where digestion occurs
Polysaccharides	_____	Mouth and stomach
Triglycerides	Fatty acids and _____	Small intestines
Proteins	Polypeptides, oligopeptides Amino acids	_____ _____

Digestion of polysaccharides begins in the mouth with the breakdown of amylase; other starches are broken down in the stomach. The end result of these reactions is the production of monosaccharides. The digestion of triglycerides occurs in the small intestine where they are broken down into fatty acids, glycerol, and monoacylglycerides. Proteins are initially digested in the stomach into polypeptides, oligopeptides, and some amino acids. Further digestion occurs in the small intestine; the final products are amino acids.

Food group	Digestion product	Body organ where digestion occurs
Polysaccharides	**monosaccharides**	Mouth and stomach
Triglycerides	Fatty acids and **glycerol**	Small intestines
Proteins	Polypeptides, oligopeptides Amino acids	**stomach** **small intestine**

5. Which of the following characteristics is similar in both the β-oxidation spiral and fatty acid biosynthesis?

a. involves removing two carbons at a time

b. uses a spiral pathway

c. utilizes NAD^+ and FAD for oxidation

d. activates with Co-A

Choice a is not true for fatty acid biosynthesis because this process adds two carbons each time through the cycle. The addition of carbons in fatty acid biosynthesis is achieved in a spiral pathway just as the carbons are removed in β oxidation, so choice b is correct. Choice c is not correct because fatty acid biosynthesis uses NADPH for oxidation. Choice d is not correct because fatty acid biosynthesis uses an acyl carrier protein (ACP) to activate the growing fatty acid. **Answer is b.**

6. Complete the following flow chart for the catabolism of triglycerides.

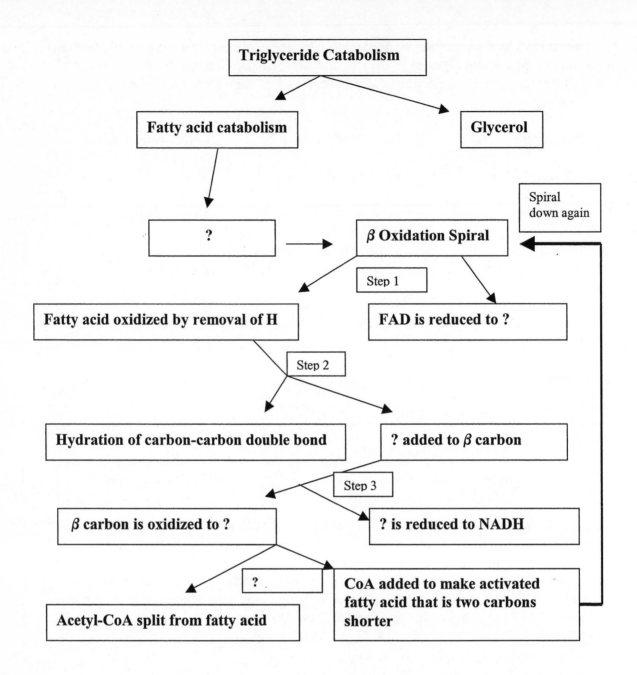

Answer:

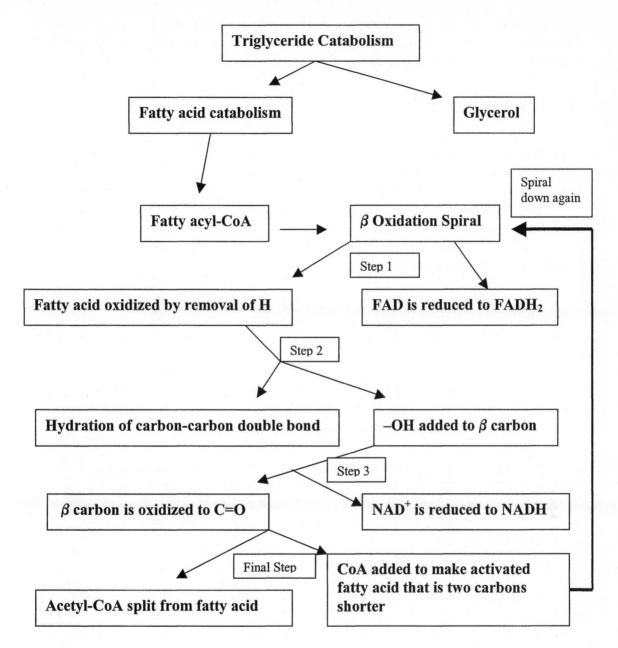

7. Which of the following is the final "fate" of amino acids in humans?

a. transferred to a α-keto acid

b. forms glutamate

c. NH^+ released into the blood stream

d. urea

Choices a through c are the intermediate catabolic steps in the breakdown of an amino acid and therefore not the final "fate." Choice d is an anabolic process in which NH^+ is turned into urea so that it can be excreted from the body. **Answer is d.**

Application

In order to apply what you've already learned and master the concepts, you must utilize all of the knowledge, understanding, facts, and/or techniques that you have gathered up to this point. Simply knowing an equation or process is not enough to master the concept. These you develop by *practice*. Solve as many different types of problems with as many different variations as you *strive to master the skill*. Your textbook provides a very good assortment of problems at the end of the chapter. Work as many as needed for skill mastery, not just those assigned.

- Identify the initial reactant and final products of glycolysis, describe how this pathway is controlled, and explain how gluconeogenesis differs from glycolysis. Describe how the manufacture and breakdown of glycogen are related to each of these pathways

- Provide an overview of the citric acid cycle, explain how it is controlled, and describe how the products of this circular pathway are used by the electron transport chain and oxidative phosphorylation

In order to truly master any skill you must *practice*. In this section example problems related to the application objectives are discussed. Remember, the more you practice, the better your chances of being able to successfully use these objectives to solve problems. Additional practice problems are at the end of the chapter in your textbook and in the online interactive lessons.

Sample Problems

1. Which of the following is the initial reactant of glycolsis?

a. pyruvate ions

b. glucose

c. 2 ATP and 2 NADH

d. glucose 6-phosphate

Glycolysis starts with glucose (choice b) and an attached phosphate, creating a glucose 6-phosphate (choice d). Glucose 6-phosphate is then converted into two pyruvate ions (choice a) and 2ATP and 2 NADH (choice c). **Answer is b.**

2. Which of the following is a final product of the glycolysis process?

a. glucose

b. glucose 6-phosphate

c. pyruvate

d. 2-phosphoglycerate

Glycolysis starts with glucose (choice a) and an attached phosphate, creating a glucose 6-phosphate (choice b). Glucose 6-phosphate is then converted to a 2-phosphoglycerate (choice d) and finally two pyruvate ions (choice c). **Answer is c.**

3. Which of the following does not play a role in controlling the rate of glycolysis?

a. citrate

b. NADH

c. AMP

d. ATP

The rate of glycolysis is controlled at steps 1,3, and 10. Step 3 is catalyzed by the enzyme phosphofructokinase, considered the main control point of the glycolysis rate. ATP (choice d) and citrate (choice a) are negative effectors and switch the enzyme off when energy levels are high. AMP (choice c) and ADP are positive effectors that turn the enzyme on when energy is low. While NADH (choice b) is a product of glycolysis, it does not serve to control it. **Answer is b.**

4. Which of the following statements is *not* true?

a. Gluconeogenesis is an anabolic process, while glycolysis is a catabolic process.

b. Gluconeogenesis is a process with glucose as a product, while glycolysis uses glucose as a reactant.

c. Gluconeogenesis is simply the reverse of glycolysis.

d. The synthesis of glycogen begins with glucose 6-phosphate, which is formed early in glycolysis and late in gluconeogenesis.

Gluconeogenesis is a process that creates glucose from non-carbohydrate sources and so it is an anabolic (building-up) process; glycolysis uses glucose as a reactant and is a catabolic (breaking down) process; therefore choices a and b are true. Gluconeogenesis is not simply the reverse of glycolysis; it utilizes several different enzymes not used in glycolysis so choice c is not true. The synthesis of glycogen begins with glucose 6-phosphate, which is the product of the first step (early) in glycolysis; glucose 6-phosphate is produced in gluconeogenesis during steps 6 and 7 (late) which means choice d is also true. **Answer is c.**

5. The manufacture and breakdown of glycogen is important to both glycolysis metabolism and the gluconeogenesis metabolism because:

a. it catalyzes the change from forward to reverse reactions

b. it is a glucose storage unit that can be quickly broken to release glucose

c. it can serve to stop glycogenesis from occurring

d. it is a source of stored lipids that can serve as an energy reserve

Glycogen is a highly branched homopolysaccharide found mainly in liver and muscle cells; it is not a catalyst so choice a is not true. Since glycogen is made from many glucose residues it serves as a glucose storage unit that can be quickly broken down as needed so choice b is true. Glycogenesis is the synthesis of glycogen; glycogen does not stop it from occurring, which means choice c is not true. Since glycogen is a homopolysaccharide, it is a carbohydrate and would not contain lipids, therefore choice d is not true. **Answer is b.**

6. In terms of energy, if a citric acid cycle produced 3 NADH and 2 $FADH_2$, what is the equivalent number of ATPs produced?

a. 8.5 ATPs

b. 5.0 ATPs

c. 10.5 ATPs

d. 7.0 ATPs

During the electron transport chain NADH is converted back to NAD^+ generating enough energy to produce 2.5 ATPs. $FADH_2$ is converted to FAD generating enough energy to produce 1.5 ATPs. Using these equivalents, calculate the total numbers of ATP that could be generated.

$$3 \text{ NADH} \quad \times \quad \frac{2.5 \text{ ATP}}{1 \text{ NADH}} \quad = \quad 7.5 \text{ ATP}$$

$$2 \text{ FADH}_2 \quad \times \quad \frac{1.5 \text{ ATP}}{1 \text{ FADH}_2} \quad = \quad 3.0 \text{ ATP}$$

Finally, total up the ATPs produced $7.5 \text{ ATP} + 3.0 \text{ ATP} = 10.5 \text{ ATP}$. **Answer is c.**

7. Which of the following is *not* part of the citric cycle?

a. 2 CO_2 molecules, 3 NADH, 1 FADH$_2$, and 1 GTP

b. step 1 is the combination of acetyl-CoA with oxaloacetate (the final pathway product)

c. the citrate formed in step one converted to isocitrate which is oxidized to α-ketoglutarate

d. α-ketoglutarate is decarboxylated and is converted into oxaloacetate

Choice a includes the final products of the citric acid cycle and is true. Choice b is also true; the process is called a "cycle." Choice c is the second and third steps of the citric cycle and is true. Choice d is not true. There are eight steps in the citric acid cycle, so there are four more steps after step 3 (isocitrate is oxidized to α-ketoglutarate) before the oxaloacetate is produced. **Answer is d.**

8. Which of the following does *not* play a role in controlling the citric acid cycle?

a. ATP

b. NADH

c. pyruvate

d. acetyl-CoA

Choices a, b, and d all serve as negative effectors to control the citric acid cycle. Pyruvate is the beginning reactant and although its concentration is a factor in the cycle, it is not a control agent. **Answer is c.**

9. Which of the following does *not* occur in the electron transport chain and oxidative phosphorylation of the citric acid cycle products?

a. the energy generated by the electron transport chain is used to produce ATP during oxidative phosphorylation

b. NADH is converted to NAD^+ during oxidative phosphorylation

c. $FADH_2$ is converted to FAD during the electron transport chain

d. the electron chain transport occurs in the mitochondria following facilitated diffusion

The energy generated during the electron transport chain produces 2.5 ATPs for each NADH and 1.5 ATPs for each $FADH_2$ during oxidative phosphorylation so choice a is true. NADH is converted to NAD^+ during the electron transport chain, not during oxidative phosphorylation, so choice b is not true. $FADH_2$ is converted to FAD during the electron transport chain so choice c is true. Pyruvates are moved by facilitated transport into the mitochondria, where the citric acid cycle begins; therefore the electron chain transport occurs in the mitochondria following facilitated diffusion; choice d is true. **Answer is b.**

Chapter 15 Practice Test

Now that you have had the chance to study and review Chapter 15's concepts and objectives, and work the practice problems provided in your textbook, you are ready to practice problems as you might see them in a testing situation. The questions and problems below are randomly selected in much the same way your instructor will select them; that is not necessarily ordered in the same way as they appear in the textbook. To make best use of this practice, have your paper, pencil and calculator ready as you would for a class exam. Use a clock and give yourself 45 minutes to complete this activity. Note that taking this test under similar conditions as a class exam provides two benefits:

1. Testing your knowledge of the chapter's contents and your mastery of the required skills.
2. Practicing working under the stress of a timed exam.

Most instructors have a type of testing format that they prefer and will most likely tell you about it before the test. For practice, this self-test uses a variety of test question formats such as matching, fill in the blank, listing, multiple choice, and essay questions. **Be sure to have a periodic table!**

Note the time and begin.

1. The H^+ ion moves from the intermembrane space, through ATP synthase, and into the mitochondrial matrix by:

A. electron chain transport
B. diffusion
C. facilitated diffusion
D. active transport

2. A group of reactions that show the progression of a biochemical process is:

A. metabolism
B. botulism
C. nitrolysis
D. carboxylation

3. For the following reaction, what is the ΔG for the overall reaction?

$$\text{glucose} + P_i \rightarrow \text{glucose 6-phophate} + H_2O \quad \Delta G = +3.3 \text{ kcal/mol}$$
$$\text{ATP} + H_2O \rightarrow \text{ADP} + P_i \quad \Delta G = -7.3 \text{ kcal/mol}$$

A. +7.3 kcal/mol
B. -3.3 kcal/mol
C. 10.6 kcal/mol

D. -4.0 kcal/mol

4. The metabolic process that provides the energy necessary to make ATP is:

A. catabolism
B. anabolism
C. gluconeogenesis
D. oxidative phosphorylation

5. In order for FAD to convert to FADH$_2$ it requires:

A. active transport
B. reduction
C. oxidation
D. acidosis

6. When acetyl-CoA is hydrolyzed the bond is broken between carbon and:

A. oxygen
B. hydrogen
C. sulfur
D. nitrogen

7. Which of the following is *not* a product obtained when triglycerides are hydrolyzed in the small intestine?

A. fatty acids
B. amino acids
C. glycerol
D. monoacylglycerides

8. Which of the following is *not* a product of glycolysis?

A. two pyruvic ions
B. 2 ATP
C. 2 NADH
D. glycerol

9. Which of the following is a positive effector in step 3 of glycolysis?

A. citrate
B. ADP
C. ATP
D. glucose 6-phosphate

10. The conversion of pyruvate to lactate is:

A. active transport
B. reduction
C. oxidation
D. acidosis

11. It is more efficient for monosaccharides other than glucose to use the same catabolic pathways because:

A. it reduces the number of different enzymes
B. glucose can create an active transport for the other monosaccharides
C. glucose catalyzes the other reactions
D. the other monosaccharides can use the energy glucose is producing to help them react

12. Which of the glycolysis steps *cannot* be directly reversed during gluconeogenesis?

A. 1
B. 3
C. 7
D. 10

13. Glycolsis and gluconeogenesis both produce the intermediate glucose 6-phosphate, which is used to produce:

A. pyruvate
B. amino acids
C. citric acid
D. glycogen

14. In terms of energy, if a citric acid cycle produced 2 NADH and 1 $FADH_2$ what is the equivalent number of ATP produced?

A. 6.5 ATPs
B. 5.0 ATPs
C. 10.5 ATPs
D. 7.0 ATPs

15. Which of the compounds below donates electrons during the electron transport chain?

A. NAD^+
B. NADH
C. FAD
D. ATP

In questions 16 through 25, match the following terms and phrases to the correct meaning

_____16. Oxidative Phosphorylation

A. group of reactions involved in how living things manufacture or break down carbohydrates, lipids, or members of any other biochemical class of compounds

_____17. Metabolic pathways

B. each cycle removes 2 carbon atoms from a fatty acid and produces 1 acetyl-CoA, 1 FADH2, and 1 NADH

_____18. Aerobic

C. a spontaneous process that supplies the energy required for a nonspontaneous one to take place

_____19. Citric acid cycle

D. 10-step linear pathway that converts one monosaccharide molecule into 2 pyruvates, producing 2 ATP and 2 NADH in the process

_____20. Electron Transport Chain

E. condition in which pyruvates are converted to lactate or, in yeast, is converted to CO_2 and ethanol

_____21. Oxidation spiral

F. pyruvate is converted into acetyl CoA

_____22. Glycolysis

G. pathway in which glucose is biosynthesized from lactate and other non carbohydrate sources

_____23. Gluconeogenesis

H. 8-step circular pathway that begins with a reaction between oxaloacetate and acetyl-CoA and ends with manufacture of oxaloacetate

_____24. Anaerobic

I. series of proteins and other molecules embedded in the inner mitochondrial membrane that removes two elections from NADH or $FADH_2$ and uses the energy released while transferring the electrons down the chain to pump H^+ into the intermembrane space

_____25. Coupled Reaction

J. the process in which H$^+$ moves back into the mitochondrial matrix though channels provided by ATP synthase, ATP is manufactured from ADP and P$_i$

PRACTICE TEST Answers

1. C
2. B
3. D
4. A
5. B
6. C
7. B
8. D
9. B
10. B
11. A
12. C
13. D
14. A
15. B
16. J
17. A
18. F
19. H
20. I
21. B
22. D
23. G
24. E
25. C

ANSWER PAGE FOR BUILDING KNOWLEDGE

Match the terms or phrases to the correct meaning

___h___ 1. Circular

___m___ 2. Citric Acid

___k___ 3. Aerobic

___q___ 4. Transamination

___p___ 5. Oxidation Spiral

___g___ 6. Metabolic Pathways

___o___ 7. Oxidative Phosphorylation

___a___ 8. Metabolism

___b___ 9. Catabolism

___n___ 10. Electron Transport Chain

___c___ 11. Anabolism

___i___ 12. Glycolysis

___l___ 13. Gluconeogenesis

a. the sum of all reactions that take place in a living organism that can be divided into two parts, catabolism and anabolism

b. reactions that break large molecules into small ones, usually releasing energy

c. reactions that biosynthesize large molecules from small ones, usually consuming energy

d. continuous series of reactions in which the product of one reaction is the reactant in the next

e. series of repeated reactions used to break down or build up a molecule

f. spontaneous reaction that provides the energy needed by a nonspontaneous one

g. Group of reactions that details the manufacturing or breaking down of carbohydrates, lipids, or members of any other biochemical class of compounds

h. series of reactions where the final product is an initial reactant

i. 10-step linear pathway that converts one monosaccharide molecule into 2 pyruvates, producing 2 ATP and 2 NADH in the process

j. condition in which pyruvates are converted to lactate or, in yeast, are converted to CO_2 and ethanol

k. pyruvate is converted into acetyl CoA

l. pathway in which glucose is biosynthesized from lactate and other non carbohydrate sources is the exact reverse of glycolysis, except at step 1, 3, and 10

m. 8-step circular pathway that begins with a reaction between oxaloacetate and acetyl-CoA and ends with manufacture of oxaloacetate

___j___ 14. Anaerobic

n. series of proteins and other molecules embedded in the inner mitochondrial membrane that removes two elections from NADH or $FADH_2$ and uses the energy released while transferring the electrons down the chain to pump H^+ into the intermembrane space

___f___ 15. Coupled Reaction

o. the process in which H^+ moves back into the mitochondrial matrix though channels provided by ATP synthase; ATP is manufactured from ADP and P_i

___e___ 16. Spiral

p. each cycle of which removes 2 carbon atoms from a fatty acid and produces 1 acetyl-CoA, 1 $FADH_2$, and 1 NADH

___d___ 17. Linear

q. the process in which an α-keto acid is the $-NH_2$ acceptor

Chapter 1
Science and Measurements

Solutions to Problems

1.1   *Is the statement "What goes up must come down" a scientific law or scientific theory? Explain.*

A law. It describes what is observed but does not explain why it happens.

1.3   *How is a theory different from a hypothesis?*

A hypothesis is a tentative explanation based on presently known facts while a theory is an experimentally tested explanation that is consistent with existing experimental evidence and accurately predicts the results of future experiments.

1.5   *What are the three states of matter?*

solid, liquid, gas

1.7   *A gymnast runs across a mat and does a series of flips.*

a. Describe the changes in kinetic energy that take place.

Kinetic energy increases as the gymnast accelerates across the mat. Some kinetic energy is converted to potential energy as the gymnast leaves the floor and starts upward into the flip. At the point where the upward motion stops and the downward motion begins, the potential energy is beginning to convert back to kinetic energy.

b. Describe the changes in potential energy that take place.

Potential energy increases as the gymnast leaves the floor until the maximum height is reached. At that point the potential energy decreases as the gymnast comes back down.

c. What work is done?

Work is done as the gymnast changes position.

1.9   a. *List some of the physical properties of a piece of copper wire.*

It is shiny, orange-brown in color, ductile, malleable, and conducts heat and electricity.

b. *Give examples of some physical changes that a piece of copper wire could undergo.*

It can be stretched, pounded flat, and melted.

1.11   *Which is larger?*
a. *1 pint of 1 L?*   One quart is a bit smaller than one liter, and one pint is smaller than one quart.

1L

b. *1°F or 1°C*   One degree Celsius is 1.8 times larger than one degree Fahrenheit.

1°C

c. *1 gal or 1 L?*   One gallon is four quarts and one quart is just a bit smaller than one liter.

1 gal

1.13   *Which is larger?*

a. 1 mg or 1 µg?   From Table 1.2, 1 mg equals 1000 µg.

1 mg

b. *1 gr or 1 mg?*   From Table 1.2, 1 gr equals 65 mg.

1 mg

c. *1 T or 1 t?*   From Table 1.2, 1 T equals 15 mL which equals 3 t (or 3 tsp)

1 T

d. *1 T or 1 oz?*   From Table 1.2, 1 oz equals 2T.

1oz

1.15   *Express each number using metric prefixes (example: one-tenth = deci or d). Give both the name and the abbreviation for each prefix.*

a. *one-thousandth* = milli or m

b. *one million* = mega or M

c. *one-hundredth* = centi or c

1.17. *Express each distance in scientific notation and ordinary (decimal) notation, without using metric prefixes (example: 6.2 cm = 6.2 x 10^{-2} m = 0.062 m)*

a. *1.5 km* = 1.5 x 10^3 m = 1,500 m

b. *5.67 mm* = 5.67 x 10^{-3} m = 0.00567 m

c. *5.67 nm* = 5.67 x 10^{-9} m = 0.00000000567 m

d. *0.3 cm* = 3 x 10^{-3} m = 0.003 m

1.19 *Which is the greater amount of energy?*

a. *1 kcal or 1 kJ?* One calorie is larger than one joule (1 cal = 4.184 J), so 1 kcal (1000 cal) is larger than 1 kJ (1000 J).

1 kcal

b. *4.184 cal or 1 J?* One calorie is larger than one joule (1 cal = 4.184 J), so 4.184 cal is larger than 1 J.

4.184 cal

1.21 a. *How many meters are in 1 km?*

The prefix kilo (k) stand for 10^3. Making this substitution, 1 km = 1 x 10^3 m.

b. *How many meters are in 5 km?*

The prefix kilo (k) stand for 10^3. Making this substitution, 5 km = 5 x 10^3 m.

c. *How many millimeters are in 1 m?*

The prefix milli (m) stands for 10^{-3} or one-thousandth. This means that one millimeter equals one-thousandth of a meter (1 mm = 1 x 10^{-3} m) and that there are one thousand (1 x 10^3) millimeters in one meter.

1.23 *How many significant figures does each number have? Assume that each is a measured value.*

a. *1000000.5* All the zeroes count because they are between non-zero digits.

307

8

b. *887.60*   The ending zero is counted because there is a decimal to its left.

5

c. *0.668*   Zeroes at the beginning of a number are not significant.

3

d. *45*   All non-zero digits are significant.

2

e. *0.00045*   Zeroes at the beginning of a number are not significant.

2

1.25   *Solve each calculation, reporting each answer with the correct number of significant figures. Assume that each value is a measured value.*

a. *14 x 3.6*

In multiplication and division, the answer is rounded to have the same number of significant figures as the measurement with the fewest significant figures. Since both measured values have two significant figures, the calculator answer (50.4) rounds to 50. or 5.0×10^1. The decimal point is necessary to indicate that the zero is significant.

5.0×10^1

b. *0.0027 ÷ 6.7784*

The same rule applies as in part a. Dividing 0.0027 (2 significant figures) by 6.7784 (5 significant figures) gives 0.000398324, which round to 0.00040 or 4.0×10^{-4} (2 significant figures).

4.0×10^{-4}

c. *12.567 + 34*

When adding or subtracting, the answer should have the same of decimal places as the quantity with the fewest significant figures.

12. 567   3 decimal places
34   0 decimal places
_____

46.567   rounds to 47 (0 decimal places)

47

d. $(1.2 \times 10^3 \times 0.66) + 1.0$

When more than one operation is involved, calculate the part in parentheses first and round it to the appropriate significant figures and then perform the next part of the calculation. In the first calculation, both numbers have two significant figures, so the answer ($1.2 \times 10^3 \times 0.66 = 792$) is rounded to two significant figures (790).

790   10s place is significant
1.0   10ths place is significant
_____

791.0   rounds to 790 with the 10s place significant

7.9×10^2

1.27   *A microbiologist wants to know the circumference of a cell being viewed through a microscope. Estimating the diameter of the cell to be 11 μm and knowing that circumference = π x diameter (we will assume that the cell is round, even though that is usually not the case), the microbiologist uses a calculator and gets the answer 34.55751919 μm. Taking significant figures into account, what answer should actually be reported? (π = 3.141592654......).*

In the calculation π x diameter, the diameter (11 μm) has the fewest number of significant figures (2). The answer (34.55751414... μm) is reported with two significant figures.

35 mm

1.29   *Give the two conversion factors that are based on each equality.*

a. 12 eggs = 1 dozen $\dfrac{12 \text{ eggs}}{1 \text{ dozen}}$ or $\dfrac{1 \text{ dozen}}{12 \text{ eggs}}$

b. 1×10^3 m = 1 km $\dfrac{1 \times 10^3 \text{ m}}{1 \text{km}}$ $\dfrac{1 \text{ km}}{1 \times 10^3 \text{ m}}$

c. 0.946 L = 1 qt $\dfrac{0.946 \text{ L}}{1 \text{ qt}}$ $\dfrac{1 \text{ qt}}{0.946 \text{ L}}$

1.31   a. *17 ft is how many yards?*

3 ft = 1 yd

$$17 \text{ ft} \quad \times \quad \frac{1 \text{ yd}}{3 \text{ ft}} \quad = 5.7 \text{ yd} \quad (2 \text{ significant figures})$$

b. *36.8 ft is how many inches?*

1 ft = 12 in

$$36.8 \text{ ft} \quad \times \quad \frac{12 \text{ in}}{1 \text{ ft}} \quad = \quad 442 \text{ in} \quad (3 \text{ significant figures})$$

1.33   *Convert*

a. *92 μg into grams.*

$1 \mu g = 1 \times 10^{-6} \text{ g}$

$$92 \mu g \quad \times \quad \frac{1 \times 10^{-6} \text{ g}}{1 \mu g} \quad = \quad 9.2 \times 10^{-5} \text{ g}$$

b. *27.2 ng into milligrams.*

$1 \text{ ng} = 1 \times 10^{-9} \text{ g};\ 1 \text{ mg} = 1 \times 10^{-3} \text{ g}.$

$$27.2 \text{ ng} \times \frac{1 \times 10^{-9} \text{ g}}{\text{ng}} \times \frac{1 \text{ mg}}{1 \times 10^{-3} \text{ g}} \quad = \quad 2.72 \times 10^{-5} \text{ mg}$$

c. *0.33 kg into milligrams.*

$1 \text{ kg} = 1 \times 10^{3} \text{ g};\ 1 \text{ mg} = 1 \times 10^{-3} \text{ g}$

$$0.33 \text{ kg} \times \frac{1 \times 10^{3} \text{ g}}{1 \text{ kg}} \times \frac{\text{mg}}{1 \times 10^{3} \text{ g}} \quad = \quad 3.3 \times 10^{5} \text{ mg}$$

d. *7.27 mg into micrograms.*

$1 \text{ mg} = 1 \times 10^{-3} \text{ g};\ 1 \mu g = 1 \times 10^{-6} \text{ g}$

$$7.27 \text{ mg} \quad \times \quad \frac{1 \times 10^{-3} \text{ g}}{1 \text{ mg}} \times \frac{\mu g}{1 \times 10^{-6} \text{ g}} \quad = \quad 7.27 \times 10^{3} \mu g$$

1.35   *Convert your weight from pounds to kilograms.*

Answers will vary depending on your weight. Below is a sample setup.

1 kg = 2.205 lb   If your weight is 175 lb then

$$175 \text{ lb} \times \frac{1 \text{ kg}}{2.205 \text{ lb}} = 79.4 \text{ kg}$$

Answer: Your pound weight $\times \dfrac{1 \text{ kg}}{2.205 \text{ lb}}$

1.37   *Convert*

a. *103° F into degrees Celsius.*

The conversion equation is $°C = \dfrac{°F - 32}{1.8}$, and substituting 103 for °F gives

$$°C = \frac{103 - 32}{(1.8)}$$

$$°C = \frac{71}{1.8}$$

$$°C = 39$$

39 °C

b. *25 °C to Fahrenheit.*

The conversion equation is °F = (1.8 x °C) + 32, and substituting 25 for °C gives

$$°F = (1.8 \times 25) + 32$$

$$°F = 45 + 32$$

$$°F = 77$$

77 °F

c. *35 °C into Kelvin.*

The conversion equation is $K = °C + 273.15$ and substituting 35 for °C gives

$$K = 35 + 273.15$$

$$K = 308$$

308 K

d. *405 K into degrees Fahrenheit.*

There is not a direct conversion equation for this one so first convert K to °C using $K = °C + 273.15$ and then convert to °F using $°F = (1.8 \times 25) + 32$.

$$°C = 405 - 273.15$$
$$°C = 132$$

$$°F = (1.8 \times 132) + 32$$
$$°F = 238 + 32$$
$$°F = 270$$

$$°F = 270$$

1.39   *It is estimated that an accordion player expends 9.2 kJ of energy per minute of playing time. Convert this value into Calories (1 food Calorie = 1000 cal).*

$1 kJ = 1 \times 10^3 J$, $1 cal = 4.184 J$

$$9.2 \; \cancel{kJ} \; \times \; \frac{1 \times 10^3 \cancel{J}}{\cancel{kJ}} \; \times \; \frac{cal}{4.184 \; J} \; \times \; \frac{Cal}{1000 \; \cancel{cal}} \; = \; 2.2 \; Cal$$

1.41   *As an alternative to ear tags and lip tattoos, tetracycline (an antibiotic) is used to mark polar bears. The advantages of using tetracycline in this fashion are that it leaves a detectable deposit on the bear's teeth, it can be administered remotely, and using it doesn't require that the animal be sedated. If 25 mg/kg is an effective dose, how much tetracycline is needed (in grams) to mark a 1000 kg polar bear?*

First determine the number of milligrams needed and then convert that to grams. Note that 25 mg/kg (25 mg of tetracycline per kilogram of body weight) is used as a conversion factor.

$$1000 \; \cancel{kg} \text{ polar bear} \; \times \; \frac{25 \; mg}{\cancel{kg} \text{ polar bear}} \; = \; 25000 \; mg \; drug$$

$$25000 \, \cancel{\text{mg}} \quad x \quad \frac{1 \times 10^{-3} \, \text{g}}{\cancel{\text{mg}}} \quad = \quad 25 \, \text{g}$$

1.43 *Chloroquine is used to treat malaria. Studies have shown that an effective dose for children is 3.5 mg per kilogram (3.5 mg/kg) of body weight, every 6 hours. If a child weighs 12 kg, how many milligrams of this drug should be given in a 24 hour period?*

This problem can be solved in three steps:

a. 1 dose every 6 hours

$$24 \, \cancel{\text{hr}} \quad x \quad \frac{\text{dose}}{6 \, \cancel{\text{hr}}} \quad = \quad 4 \, \text{doses}$$

b. 3.5 mg drug/kg body weight = 1 dose

$$4 \, \cancel{\text{doses}} \quad x \quad \frac{3.5 \, \text{mg drug/kg body weight}}{\cancel{\text{dose}}} \quad = \quad 14 \, \text{mg drug/kg body weight}$$

c. $12 \, \cancel{\text{kg body weight}} \quad x \quad \dfrac{14 \, \text{mg drug}}{\cancel{\text{kg body weight}}} \quad = \quad 1.7 \times 10^2 \, \text{mg drug}$

1.45 *The tranquilizer Valium is sold in 2.0 mL syringes that contain 50.0 mg of drug per 1.0 mL of liquid (50.0 mg/1.0 mL). If a physician prescribes 25 mg of this drug, how many milliliters should be administered?*

Note that the size of the syringes does not have anything to do with the solution of the problem, since it asks for milliliters needed and not the number of syringes. Use 50.0 mg = 1.0 mL as a conversion factor.

$$25 \, \cancel{\text{mg}} \quad x \quad \frac{1.0 \, \text{mL}}{50.0 \, \cancel{\text{mg}}} \quad = \quad 0.50 \, \text{mL}$$

1.47 *In the past 200 years, in what ways have scientific discoveries led to changes in the treatment of diabetes?*

Insulin became available, the purity of insulin was improved, genetically engineered human insulin was put on the market, and oral drugs were developed.

1.51 *Suppose that you take your temperature orally and see that it is 99.1°F. Does this necessarily mean that you are running a fever? Explain.*

No. A normal body temperature of approximately 98.6°F is based on the average of temperatures in healthy people. A difference of only 0.5 °F could easily be normal for you.

1.53   *If your temperature is taken orally, one measurement is usually sufficient. If it is taken at the eardrum, however, more than one measurement is recommended. Why?*

Tympanic temperature measurements can give false readings.

Chapter 2
Atoms and Elements

Solutions to Problems

2.1 *Why does the nucleus of an atom have a positive charge?*

It consists of protons (positive charge) and neutrons (no charge).

2.3 *Describe the structure of an atom.*

An atom is constructed from three subatomic particles: protons (1+ charge, 1 amu), neutrons (no charge, 1 amu), and electrons (1- charge, 1/2000 amu). Protons and neutrons make up the nucleus. Electrons are located outside of the nucleus.

2.5 *Give the atomic symbol for each element*

a. *lithium*   Li

b. *bromine*   Br

c. *boron*   B

d. *aluminum*   Al

e. *fluorine*   F

2.7 *Give the name of each element.*

a. *Be*   beryllium

b. *Ne*   neon

c. *Mg*   magnesium

d. *P*   phosphorus

2.9 *How many protons and neutrons are present in the nucleus of each?*

The number of protons is equal to the atomic number. The number of neutrons is equal to the mass number minus the atomic number.

a. $_{9}^{19}F$ 9 protons, 19 - 9 = 10 neutrons

b. $_{11}^{23}Na$ 11 protons, 23 - 11 = 12 neutrons

c. $^{238}_{92}U$ 92 protons, 238 - 92 = 146 neutrons

2.11   *Which of the following statements do not accurately describe isotopes of an element?*

a. *same number of protons*

b. *same mass number*

c. *same atomic number*

Isotopes are atoms of the same element that have the same number of protons but different atomic masses. Since the number of protons and atomic number are the same thing, then answers a and c are correct. The mass number would not be the same for isotopes of an element.

Choice b, the same mass number, would not accurately describe isotopes of an element.

2.13   *In nature, the element chlorine exists as two different isotopes $^{35}_{17}Cl$ and $^{37}_{17}Cl$. The atomic weight of chlorine is 35.5 amu. Which chlorine isotope predominates?*

Atomic weight is the average mass of the atoms of an element, as it is found in nature. For chlorine, the average mass is closer to 35 than 37

.

$^{35}_{17}Cl$ isotope predominates.

2.15   *Name and give the atomic notation for an isotope that has 65 protons and 83 neutrons.*

Element 65 on the periodic table is terbium, Tb. To get the mass number, add the number of protons and neutrons (65 + 83 = 148). The atomic notation symbol is $^{148}_{65}Tb$.

terbium-148, $^{148}_{65}Tb$

2.17   *Calculate the number of protons, neutrons, and electrons in each neutral atom.*

The number of protons is equal to the atomic number. The number of neutrons is equal to the mass number minus the atomic number. For a neutral atom, the number of electrons is equal to the number of protons.

a. $^{24}_{12}Mg$ 12 protons, 24-12 = 12 neutrons, 12 electrons

316

b. $^{55}_{25}Mn$ 25 protons, 55-25 = 30 neutrons, 25 electrons

c. $^{64}_{30}Zn$ 30 protons, 64-30 = 34 neutrons, 30 electrons

d. $^{74}_{34}Se$ 34 protons, 74-34 = 40 neutrons, 34 electrons

2.19   *Which of the two atoms is the most metallic?*

The metallic character of elements increases as you move to the left across a period and down a group on the periodic table.

a. *Al and Si*   Al (it is further to the left than Si)

b. *Ca and Mg*   Ca (it is further down the column than Mg)

2.21   *Arrange each set of three atoms in order of size (largest to smallest).*

Atomic size decreases as you move to the right across a period and up a group on the periodic table.

a. *I, F, and Br*   I, Br, F

b. *Ne, F, and O*   O, F, Ne

2.23   *List some of the physical properties of nonmetals.*

They are poor conductors of heat and electricity. As solids they are nonlusterous and brittle.

2.25   *What is the*

a. *atomic weight of helium (He)?*

The answer is found under the He symbol on the periodic table.

 4.00 amu (rounded to two decimal places)

b. *mass (in grams) of 5.00 mol of helium?*

One mole of He has a mass of 4.00g.

5.00 ~~mol~~   x $\dfrac{4.00 \text{ g}}{1 \text{ mol}}$ =   20.0 g

c. *mass (in grams) of 0.100 mol of helium?*

One mole of He has a mass of 4.00g.

0.100 ~~mol~~   x $\dfrac{4.00\ g}{1\ mol}$ =   0.400 g

d. *mass (in grams) of 6.02 x 10^{23} helium atoms?*

One mole of He has a mass of 4.00g and contains 6.02 x 10^{23} atoms.

6.02 x 10^{23} ~~atoms~~   x $\dfrac{1\ mol}{6.02\ x\ 10^{23}\ atoms}$ x $\dfrac{4.00\ g}{1\ mol}$ =   4.00 g

2.27   *How many atoms are present in 2.00 mol of aluminum?*

One mole of aluminum contains 6.02 x 10^{23} atoms.

2.00 ~~mol~~   x $\dfrac{6.02\ x\ 10^{23}\ atoms}{1\ mol}$ =   1.20 x 10^{24} atoms

2.29   a. *What is the atomic weight of sulfur (S)?*

Looking under the symbol S on the periodic table you can find the atomic weight.

32.1 amu (rounded to one decimal place)

b. *How many sulfur atoms are contained in 32.1 g of sulfur?*

One mole of sulfur (6.02 x 10^{23} atoms) has a mass of 32.1 g..

32.1 ~~g~~   x $\dfrac{1\ mol}{32.1\ g}$ x $\dfrac{6.02\ x\ 10^{23}\ atoms}{1\ mol}$ =   6.02 x 10^{23} atoms

2.31   *How many atoms are contained in the following?*

a. *1.0 mol of carbon*

One mole of carbon contains 6.02 x 10^{23} atoms

6.02 x 10^{23} atoms

b. *1.22 x 10^{-9} mol of carbon*

1.22 x 10^{-9} ~~mol~~   x $\dfrac{6.02\ x\ 10^{23}\ atoms}{1\ mol}$ =   7.34 x 10^{14} atoms

c. *12.0 g of carbon*

One mole of carbon atoms has a mass of 12.0 g and contains 6.02 x 10²³ atoms.

$$12.0 \; \cancel{g} \; \times \; \frac{1 \; \cancel{mol}}{12.0 \; g} \; \times \; \frac{6.02 \times 10^{23} \; atoms}{1 \; \cancel{mol}} \; = \; 6.02 \times 10^{23} \; atoms$$

d. *4.5 ng of carbon*

$1 \; ng = 1.0 \times 10^{-9} \; g$

$$4.5 \; \cancel{ng} \; C \; \times \; \frac{1 \times 10^{-9} \; \cancel{g}}{\cancel{ng}} \; \times \; \frac{1 \; \cancel{mol}}{12.0 \cancel{g}} \; \times \; \frac{6.02 \times 10^{23} \; atoms}{1 \; \cancel{mol}} \; = \; 2.26 \times 10^{14} \; atoms$$

2.33   *Write a balanced nuclear equation for each process.*

To find the atomic number of a particle produced in the nuclear reaction, the atomic number of the emitted particle is subtracted from the atomic number of the original radioisotope. Looking this number up on the periodic table also gives the atomic symbol of the newly produced isotope. The mass number of the emitted particle is subtracted from the mass number of the original radioisotope to give the new mass number.

a. $^{187}_{80}Hg$ *emits an alpha particle* $^{187}_{80}Hg \; \rightarrow \; ^{183}_{78}Pt \; + \; ^{4}_{2}\alpha$

b. $^{266}_{88}Ra$ *emits an alpha particle* $^{226}_{88}Ra \; \rightarrow \; ^{222}_{86}Rn \; + \; ^{4}_{2}\alpha$

c. $^{238}_{92}U$ *emits an alpha particle* $^{238}_{92}U \; \rightarrow \; ^{234}_{90}Th \; + \; ^{4}_{2}\alpha$

2.35   *Identify the missing product in each nuclear equation.*

To find the atomic number of the particle produced in the nuclear reaction, the atomic number of the emitted particle is subtracted from the atomic number of the original radioisotope. Looking this number up on the periodic table also gives the atomic symbol of the newly produced isotope. The mass number of the emitted particle is subtracted from the mass number of the original radioisotope to give the new mass number. Remember that for some types of nuclear radiation, including α, β, β^+ and γ, charge is treated as an atomic number.

a. $^{14}_{8}O \; \rightarrow \; ? \; + \; ^{0}_{1}\beta^+$

$^{14}_{8}O \; \rightarrow \; ^{14}_{7}N \; + \; ^{0}_{1}\beta^+$

b. $^{3}_{1}H \; \rightarrow \; ^{3}_{2}He \; + \; ?$

A β particle has a charge of 1- and a mass number of 0.

$$\,_1^3H \rightarrow \,_2^3He + \,_{-1}^0\beta$$

c. $\,_6^{14}C \rightarrow ? + \,_{-1}^0\beta$

$$\,_6^{14}C \rightarrow \,_7^{14}N + \,_{-1}^0\beta$$

2.37 a. *Write a balanced nuclear equation for the loss of an alpha particle from $\,_{16}^{35}S$.*

To find the atomic number of the particle produced in the nuclear reaction, the atomic number of the emitted particle is subtracted from the atomic number of the original radioisotope. Looking this number up on the periodic table also gives the atomic symbol of the newly produced isotope. The mass number of the emitted particle is subtracted from the mass number of the original radioisotope to give the new mass number. Remember that for a β particle, charge is treated as an atomic number.

$$\,_{16}^{35}S \rightarrow \,_{14}^{31}Si + \,_2^4\alpha$$

b. *Write a balanced nuclear equation for the loss of a beta particle from $\,_{12}^{27}Mg$.*

Remember for a beta particle, charge is treated as an atomic number.

$$\,_{12}^{27}Mg \rightarrow \,_{13}^{27}Al + \,_{-1}^0\beta$$

2.39 *Smoke detectors contain an alpha emitter. Considering the type of radiation released and the usual placement of a smoke detector, do these detectors pose a radiation risk? Explain.*

No. Since alpha particles travel only 4 -5 cm in air and smoke detectors are usually on the ceiling, the risk of exposure to alpha radiation is small.

2.41 *Radioisotopes used for diagnosis are beta, gamma, or positron emitters. Why are alpha emitters not used for diagnostic purposes?*

Alpha particles are relatively large and do not penetrate tissue very deeply. Radiation emissions must be able to pass through the body and reach a detector to be useful for diagnostic purposes.

2.43 $\,_{80}^{197}Hg$, *a radioisotope used in brain scans, has a half-life of 66 hours.*
a. Beginning with a 1.00 mg sample of this isotope, how many milligrams of the isotope will remain after 264 hours?

After one half-life (66 hours), 1.00 mg will be reduced to 0.500 mg. After a second half-life (66 + 66 = 132 hours), the 0.500 mg will have decayed to 0.250 mg. After a third half-life (66 + 66 + 66 = 198 hours), the 0.250 mg will be reduced to 0.125 mg and after a fourth half-life (66 + 66 + 66+ 66 =264 hours), the 0.125 mg will be reduced to 0.0625 mg.

0.0625 mg

b. *This isotope decays by emitting 1 neutron ($_0^1 n$) and 1 gamma ray ($_0^0 \gamma$). Write a balanced nuclear reaction for this decay process.*

To find the atomic number of the isotope produced in the nuclear process, the atomic numbers of the emitted radiation are subtracted from the atomic number of the original radioisotope. Looking this number up on the periodic table gives the atomic symbol of the newly produced radioisotope. The mass numbers of the emitted radiation are subtracted from the mass number of the original radioisotope

$$_{80}^{197}\text{Hg} \longrightarrow \ _{80}^{196}\text{Pt} + \ _0^1 \text{n} + \ _0^0 \gamma$$

2.45   *What is the recommended daily allowance of iron for a male aged 19-24 years old, in grams?*

From Table 2.3 on page 29, 36 the RDA for iron in males is 10 mg.

$1 \text{ mg} = 1 \times 10^{-3} \text{ g}$

$10 \ \cancel{\text{mg}} \ \times \ \dfrac{1 \times 10^{-3} \text{ g}}{\cancel{\text{mg}}} \ = \ 1 \times 10^{-2} \text{ g}$

$1.0 \times 10^{-2} \text{ g}$

2.49   *A nurse is assisting a patient who has just undergone cancer treatment that involved exposure to gamma radiation from cobalt-60. Should the nurse be concerned that he will be exposed to gamma radiation given off by the patient?*

No. The patient contains no cobalt-60 and will not give off gamma radiation.

Chapter 3
Compounds

Solutions to Problems

3.1   *Give the total number of protons and electrons in each ion.*

The number of protons is equal to the atomic number (refer to a periodic table). Neutral atoms contain the same number of protons and electrons. Cations contain fewer electrons than protons, one less electron for each positive charge. Anions contain more electrons than protons, one more for each negative charge.

a. K^+ 19 protons, $19 - 1 = 18$ electrons

b. Mg^{2+} 12 protons, $12 - 2 = 10$ electrons

c. P^{3-} 15 protons, $15 + 3 = 18$ electrons

3.3   *Give the total number of protons, neutrons, and electrons in each ion.*

The number of protons is equal to the atomic number. The number of neutrons is equal to the mass number minus the atomic number. The number of electrons in each ion may be calculated as outline in the solution to Problem 3.10.

a. $^{63}_{29}Cu^+$

29 protons $63 - 29 = 34$ neutrons, $29 - 1 = 28$ electrons

b. $^{19}_{9}F^-$

9 protons $19 - 9 = 10$ neutrons, $9 + 1 = 10$ electrons

c. $^{35}_{17}Cl^-$

17 protons $37 - 17 = 20$ neutrons, $17 + 1 = 18$ electrons

3.5   *Give the name of each ion.*

a. F^- fluoride ion

The element is fluorine. As an anion, "ine" is replaced with "ide"

b. O^{2-} oxide ion

The element is oxygen. As an anion, "ygen" is replaced with "ide"

c. Ca^{2+} calcium ion

The element is calcium. As a cation, the name stays the same as for the element.

d. Br^- bromide ion

The element is bromine. As an anion, "ine" is replaced with "ide".

3.7   *Give the name of each ion.*

The name of polyatomic ions cannot be predicted in the same way as monatomic ions. Refer to Table 3.2 in the text.

a. $CO_3{}^{2-}$ carbonate ion

b. $NO_3{}^-$ nitrate ion

c. $SO_3{}^{2-}$ sulfite ion

d. $CH_3CO_2{}^-$ acetate ion

3.9   *Write the formula of each ion.*

These are polyatomic ions. Formulas cannot be predicted in the same way as monatomic ions. Refer to Table 3.2 in the text.

a. *hydrogen carbonate ion* $HCO_3{}^-$

b. *nitrite ion* $NO_2{}^-$

c. *sulfate ion* $SO_4{}^{2-}$

3.11   *For a helium atom, the energy separation between the ground state and the excited state electron energy levels is different than that for a hydrogen atom. Does this cause helium to have a different emission spectrum than hydrogen? Explain.*

Yes. The emission spectrum is created when electrons move from excited state energy levels to more stable ones. A different energy separation between levels would produce different colors in an emission spectrum.

3.13   *Specify the number of electrons held in each energy level of each atom.*

Electrons are placed into atoms from the lowest energy level to the highest. The first energy level can hold only 2 electrons, the second energy level can hold 8, the third 18, and the fourth 32. The total number of electrons is equal to the atomic number.

a. *B*

The atomic number of boron is 5. Place 2 electrons in the first energy level and the remaining 3 in the second.

n = 1	n = 2	n = 3	n = 4
2	3		

b. *C*

The atomic number of carbon is 6. Place 2 electrons in the first energy level and the remaining 4 in the second.

n = 1	n = 2	n = 3	n = 4
2	4		

c. *Mg*

The atomic number of magnesium is 12. Place 2 electrons in the first energy level 8 electrons in the second energy level, and the remaining 2 in the third.

n = 1	n = 2	n = 3	n = 4
2	8	2	

3.15   *Specify the number of valence electrons for each atom.*

The number of valence electrons is the same as the group number to which the element belongs.

a. *Li*   Li is in group IA, 1 valence electron

b. *Si*   Si is in group IVA, 4 valence electrons

c. *Al*   Al is in group IIIA, 3 valence electrons

d. *Kr*   Kr is in group VIIIA, 8 valence electrons

e. *P*   P is in group VA, 5 valence electrons

3.17   *For each, give the total number electrons, the number of valence electrons,*
        and the number of the energy level that holds the valence electrons.

The total number of electrons is equal to the atomic number, the number of
valence electrons is the same as the group number, and the energy level that
holds the valence electrons is the same as the row of the periodic table
(period) that the element is in.

a. *Br*

     35 total electrons, 7 valence electrons, held in level 4

b. *Kr*

     36 total electrons, 8 valence electrons, held in level 4

c. *As*

     33 total electrons, 5 valence electrons, held in level 4

d. *I*

     53 total electrons, 7 valence electrons, held in level 5

3.19   *When a potassium atom is converted into an ion, it becomes isoelectronic with*
        argon.

a. *What is the name of this ion?*

Metal ions are given the name of the atom from which they are derived.

potassium ion

b. *How many electrons does potassium lose when it forms the ion?*

One. A potassium atom has one valence electron and loses it to achieve an octet.

c. *What is the charge on the ion?*

1+. Losing one electron produces an ion with one more proton than electrons.

3.21   *Draw the electron dot structure of each atom and of the ion that it is expected to*
        form.

Electron dot notation is a way to represent valence electrons. The number of
valence electrons and, therefore, the number of electron dots is the same as the
group number to which the element belongs. For monoatomic anions, all valence

electrons are shown. For monoatomic cations, no valence electrons are shown – the original valence electrons were removed when the ion formed.

a. *K*

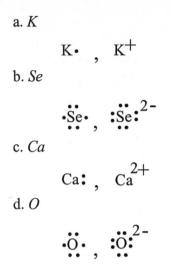

$K\cdot$, K^+

b. *Se*

$\cdot\ddot{Se}\cdot$, $:\ddot{Se}:^{2-}$

c. *Ca*

$Ca{:}$, Ca^{2+}

d. *O*

$\cdot\ddot{O}\cdot$, $:\ddot{O}:^{2-}$

3.23   *Name each ionic compound.*

a. *MgO*

MgO is combination of magnesium ions (Mg^{2+}) and oxide ions (O^{2-}).

magnesium oxide

b. *Na₂SO₄*

Na_2SO_4 is a combination of sodium ions (Na^+) and sulfate ions (SO_4^{2-}). In the names of ionic compounds of representative metals, the relative number of each ion is not specified.

sodium sulfate

c. *CaF₂*

CaF_2 is a combination of calcium ions (Ca^{2+}) and fluoride ions (F^-).

calcium fluoride

d. *FeCl₂*

$FeCl_2$ is a combination of iron ions (Fe^{2+}) and chloride ions (Cl^-). For cations formed from transition metals, the charge of the metal, represented by a Roman numeral, is part of the name.

iron(II) chloride

3.25   *Write the formula of each ionic compound.*

a. *calcium hydrogenphosphate*

Calcium ion is Ca^{2+} and the polyatomic ion hydrogenphosphate is HPO_4^{2-}. For the compound to be neutral, an equal number of Ca^{2+} and HPO_4^{2-} are required. $CaHPO_4$

b. *copper(II) bromide*

Copper(II) ion is Cu^{2+} and bromide ion is Br^-. For the compound to be neutral, twice as many Br^- ions as Cu^{2+} are required.

$CuBr_2$

c. *copper(II) sulfate*

Copper(II) ion is Cu^{2+} and sulfate ion is SO_4^{2-}. For the compound to be neutral, an equal number of Cu^{2+} and SO_4^{2-} are required.

$CuSO_4$

d. *sodium hydrogensulfate*

Sodium ion is Na^+ and the polyatomic ion hydrogensulfate is HSO_4^-. For the compound to be neutral, an equal number of Na^+ and HSO_4^- are required.

$NaHSO_4$

3.27   *Give the formula of each ionic compound.*

a. *lithium sulfate (an antidepressant)*

Lithium ion is Li^+ and the sulfate ion is SO_4^{2-}. For the compound to be neutral, twice as many Li^+ as SO_4^{2-} are required.

Li_2SO_4

b. *calcium dihydrogenphosphate (used in foods as a mineral supplement)*

Calcium ion is Ca^{2+} and the dihydrogenphosphate ion is $H_2PO_4^-$. For the compound to be neutral, twice as many $H_2PO_4^-$ as Ca^{2+} and are required.

$Ca(H_2PO_4)_2$

c. *barium carbonate (used as a rat poison)*

Barium ion is Ba^{2+} and the carbonate ion is CO_3^{2-}. For the compound to be neutral, an equal number of Ba^{2+} and CO_3^{2-} are required.

$BaCO_3$

3.29   *Write the formula of the ionic compound that forms between*

a. *magnesium ions and fluoride ions*

Twice as many (F^-) ions as magnesium ions (Mg^{2+}) are required to give a neutral compound.

MgF_2

b. *potassium ions and bromide ions*

An equal number of potassium ions (K^+) and bromide ions (Br^-) give a neutral compound.

KBr

c. *potassium ions and sulfide ions*

Twice as many potassium ions (K^+) as sulfide ions (S^{2-}) are required to form a neutral compound.

K_2S

d. *aluminum ions and sulfide ions*

To obtain a neutral compound, for every two aluminum ions (Al^{3+}), three sulfide ions (S^{2-}) are required.

Al_2S_3

3.31   *Predict the number of covalent bonds formed by each nonmetal atom.*

The number of covalent bonds that a nonmetal atom usually forms is the same as the number of electrons required to achieve an octet.

a. *N*   A nitrogen atom (group VA) has five valence electrons and needs 3 more for an octet.

3 covalent bonds

b. *Cl*   A chlorine atom (group VIIA) has seven valence electrons and needs 1 more for an octet.

1 covalent bond

c. *P*   A phosphorus atom (group VA) has five valence electrons and needs 3 more for an octet.

3 covalent bonds

3.33   *Draw the electron dot structure of each molecule.*

In the molecular drawings given, replace each single bond with two dots (:)and each double bond with four dots (::).

a.

$$
\begin{array}{ccc}
\text{H} & & \text{H} \\
\text{H}\!:\!\text{C}\!:\!\ddot{\text{O}}\!:\!\text{C}\!:\!\text{H} \\
\text{H} & & \text{H}
\end{array}
$$

b.

$$
\begin{array}{cccc}
\text{H} & \text{H} & \ddot{\text{O}}: \\
\text{H}\!:\!\text{C}\; :\; \text{C}\; :\; \text{C}\!:\!\text{H} \\
\text{H} & \text{H}\!:\!\text{N} & \text{H} \\
 & \text{H}
\end{array}
$$

3.35   *Draw the line-bond structure of pyruvic acid, a compound formed during the breakdown of sugars by the body.*

Substitute a line for each pair of dots that represent a bond.

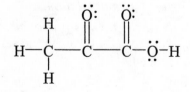

3.37   *Draw the electron dot structure of ethylene glycol (used as an antifreeze).*

Substitute a pair of dots for each line that represents a bond. In this answer an H atom has been moved to allow better spacing.

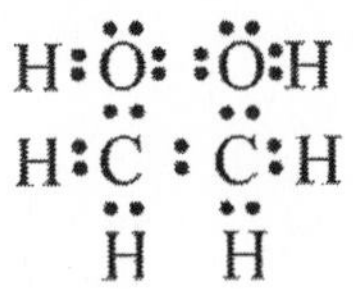

$$
\begin{array}{ccc}
H\!:\!\ddot{\underset{..}{O}}\!: & :\!\ddot{\underset{..}{O}}\!:\!H \\
H\!:\!\underset{..}{C}\!: & :\!\underset{..}{C}\!:\!H \\
H & H
\end{array}
$$

3.39   *Name each molecule.*

When naming binary molecules, the relative number of each atom is specified using a prefix. (mono = 1, di = 2, tri = 3, tetra =4, penta = 5, and hexa = 6) Note that the mono prefix is not used on the first element written when only one is present.

a. NCl_3 nitrogen trichloride

b. PCl_3 phosphorus trichloride

c. PCl_5 phosphorus pentachloride

3.41   *Phosphine (PH_3) is a poisonous gas that has the odor of decaying fish. Give another name for this binary molecule.*

Follow the same format as given in Problem 3.39.

phosphorus trihydride

3.43   *Dentists use nitrous oxide (N_2O) as an anesthetic. Give another name for this binary molecule.*

Follow the same format as given in Problem 3.39

dinitrogen monoxide

3.45   *Draw the electron dot structure of the molecule formed when sufficient H atoms are added to give each atom an octet of valence electrons.*

First draw the electron dot structure for the atom and then use H· to finish pairing up the dots in the structure.

a. *F*

$$\overset{\displaystyle\cdot\cdot}{\underset{\displaystyle\cdot\cdot}{\cdot\text{F}\!:}} \quad \text{becomes} \quad \overset{\displaystyle\cdot\cdot}{\underset{\displaystyle\cdot\cdot}{\text{H}\!:\!\text{F}\!:}}$$

b. *P*

$$\overset{\displaystyle\cdot\cdot}{\underset{\displaystyle\cdot}{\cdot\text{P}\cdot}} \quad \text{becomes} \quad \overset{\displaystyle\cdot\cdot}{\underset{\displaystyle\cdot\cdot}{\text{H}\!:\!\text{P}\!:\!\text{H}}}$$
$$\text{H}$$

c. *Br*

$$\overset{\displaystyle\cdot\cdot}{\underset{\displaystyle\cdot\cdot}{\cdot\text{Br}\!:}} \quad \text{becomes} \quad \overset{\displaystyle\cdot\cdot}{\underset{\displaystyle\cdot\cdot}{\text{H}\!:\!\text{Br}\!:}}$$

3.47   a. *What is the formula weight of Li_2CO_3 ?*

Count the number of atoms of each element in the formula, multiply that number by the atomic weight for the element as given on the periodic table (keeping one decimal place is sufficient), and then add the total weight contributed by each element to get the formula weight.

	# of atoms	atomic weight	total weight contributed by each element
Li	2	6.9	13.8
C	1	12.0	12.0
O	3	16.0	48.0
		Answer:	73.8 amu

b. *What is the mass of 1.33×10^{-4} mol of Li_2CO_3 ?*

Use the formula weight just calculated as a conversion factor (73.8 g /1 mol).

$$1.33 \times 10^{-4} \text{ mol} \quad \times \quad \frac{73.8 \text{ g}}{1 \text{ mol}} \quad = \quad 9.82 \times 10^{-3} \text{ g}$$

c. *How many moles of CO_3^{2-} ions are present in 73.5 g of Li_2CO_3?*

Use the formula weight just calculated as a conversion factor (73.8 g /1 mol) to convert grams of Li_2CO_3 to moles. Then use the fact that 1 mol Li_2CO_3 contains 1 mol CO_3^{2-} as a conversion factor to calculate moles of CO_3^{2-}.

$$73.5 \text{ g } \cancel{Li_2CO_3} \quad \text{x} \quad \frac{1 \text{ mol } \cancel{Li_2CO_3}}{73.8 \text{ g } \cancel{Li_2CO_3}} \quad \text{x} \quad \frac{1 \text{ mol } CO_3^{2-}}{1 \text{ mol } \cancel{Li_2CO_3}} \quad = 0.996 \text{ mol}$$

3.49   a. *What is the formula weight of magnesium iodide?*

First, write the correct formula, MgI_2. Next, count the number of atoms of each element in the formula, multiply that number by the atomic weight for the element as given on the periodic table (keeping one decimal place is sufficient), and then add the total weight contributed by each element together to get the formula weight.

	# of atoms	atomic mass	total mass contributed by each element
Mg	1	24.3	24.3
I	2	126.9	253.8
		Answer:	278.1 amu

b. *How many magnesium ions are present in 7.5 x 10^{-6} mol of magnesium iodide?*

One mole of MgI_2 contains one mole of Mg^{2+} ions (6.02 x 10^{23} ions).

$$7.5 \text{ x } 10^{-6} \cancel{\text{ mol } MgI_2} \text{ x } \frac{1 \cancel{\text{ mol } Mg^{2+}}}{1 \cancel{\text{ mol } MgI_2}} \text{ x } \frac{6.02 \text{ x } 10^{23} Mg^{2+}}{\cancel{\text{mole } Mg^{2+}}} = 4.5 \text{ x } 10^{18} Mg^{2+}$$

c. *How many iodide ions are present in 7.5 x 10^{-6} mol of magesium iodide?*

One of MgI_2 contains two moles of I^- ions. One mole of I^- ions is 6.02 x 10^{23} ions.

$$7.5 \text{ x } 10^{-6} \cancel{\text{ mol } MgI_2} \text{ x } \frac{2 \cancel{\text{ mol } I^-}}{1 \cancel{\text{ mol } MgI_2}} \text{ x } \frac{6.02 \text{ x } 10^{23} I^-}{1 \cancel{\text{mol } I^-}} = 9.0 \text{ x } 10^{18} I^-$$

d. *How many magnesium ions are present in 4.5 mg of magnesium iodide?*

Convert milligrams of MgI_2 to grams of MgI_2 and then use the formula weight just calculated as a conversion factor (278.1g /1 mol). Then use the fact that one mole of MgI_2 contains one mole of Mg^{2+} ions (6.02 x 10^{23} ions).

$$4.5 \cancel{\text{ mg } MgI_2} \text{ x } \frac{1 \text{ x } 10^{-3} \text{ g}}{\cancel{\text{mg}}} \text{ x } \frac{1 \cancel{\text{ mol } MgI_2}}{278.1 \cancel{\text{g } MgI_2}} \text{ x } \frac{1 \cancel{\text{ mol } Mg^{2+}}}{1 \cancel{\text{ mol } MgI_2}} \text{ x } \frac{6.02 \text{ x } 10^{23} Mg^{2+}}{1 \cancel{\text{ mol } Mg^{2+}}}$$

$$= 9.7 \text{ x } 10^{18} Mg^{2+}$$

e. How many iodide ions are present in 4.5 mg of magnesium iodide?

Follow the approach outlined in steps c and d.

$$4.5 \text{ mg MgI}_2 \ \times \frac{1 \times 10^{-3} \text{ g}}{\text{mg}} \ \times \ \frac{1 \text{ mol MgI}_2}{278.1 \text{g MgI}_2} \ \times \ \frac{2 \text{ mol I}^-}{1 \text{ mol MgI}_2} \ \times \ \frac{6.02 \times 10^{23} \text{ I}^-}{1 \text{ mol I}^-}$$

$$= 1.9 \times 10^{19} \text{ I}^-$$

3.51 *To control manic-depressive behavior, some patients are administered up to 2000 mg of lithium carbonate per day. Convert this dosage into millimoles.*

Lithium carbonate has the formula Li_2CO_3. Using the procedure outlined in the solution to problem 3.47, the formula weight of Li_2CO_3 is calculated to be 73.8 amu. The final answer is reported with one significant figure.

$$2000 \text{ mg } Li_2CO_3 \times \frac{1 \times 10^{-3} \text{ g}}{\text{mg}} \ \times \ \frac{1 \text{ mol}}{73.8 \text{ g}} \ \times \ \frac{\text{mmol}}{1 \times 10^{-3} \text{ mol}} \ = \ 30 \text{ mmol}$$

3.53 *a. What is the molecular weight of CCl_4?*

Count the number of atoms of each element in the formula, multiply that number by the atomic weight for the element as given on the periodic table (keeping one decimal place is sufficient), and then add the total weight contributed by each element to get the molecular weight.

	# of atoms	atomic mass	total mass contributed by each element
C	1	12.0	12.0
Cl	4	35.5	142.0
		Answer:	154.0 amu

b. What is the mass of 61.3 mol of CCl_4?

Use the molecular weight just calculated as a conversion factor (154.0 g/mol).

$$61.3 \text{ mol} \ \times \ \frac{154.0 \text{ g}}{1 \text{ mol}} \ = \ 9.44 \times 10^3 \text{ g}$$

c. How many moles of CCl_4 are present in 0.465 g of CCl_4?

Use the molecular weight just calculated as a conversion factor (154.0 g/mol).

$$0.465 \text{ g} \quad \text{x} \quad \frac{1 \text{ mol}}{154.0 \text{ g}} \quad = \quad 3.02 \text{ x } 10^{-3} \text{ mol}$$

d. *How many molecules of CCl_4 are present in $5.50 \text{ x } 10^{-3}$ g of CCl_4?*

Use the molecular weight just calculated (154.0 g /1 mol) as a conversion factor to convert grams of CCl_4 to moles. Then use the equivalency 1 mol = 6.02 x 10^{23} molecules to convert to molecules of CCl_4.

$$5.50 \text{ x } 10^{-3} \text{ g} \quad \text{x} \quad \frac{1 \text{ mol}}{154.0 \text{ g}} \quad \text{x} \quad \frac{6.02 \text{ x } 10^{23} \text{ molecules}}{1 \text{ mol}} \quad = 2.15 \text{ x } 10^{19} \text{ molecules}$$

3.55   One tablet of a particular analgesic contains 250 mg acetaminophen ($C_8H_9NO_2$). How many acetaminophen molecules are contained in the tablet?

First convert milligrams into grams. Then calculate the molecular weight of $C_8H_9NO_2$ to obtain a conversion factor (151.2 g /1 mol) for converting grams of $C_8H_9NO_2$ to moles. Then use the fact that 1 mol = 6.02 x 10^{23} molecules to calculate the number of molecules.

$$250 \text{ mg} \quad \text{x} \quad \frac{1 \text{ x } 10^3 \text{ g}}{\text{mg}} \quad \text{x} \quad \frac{1 \text{ mol}}{151.2 \text{ g}} \quad \text{x} \quad \frac{6.02 \text{ x } 10^{23} \text{ molecules}}{1 \text{ mol}} = 1.0 \text{ x } 10^{21} \text{ molecules}$$

3.57   *How are electrons involved in the production of light by luciferin?*

When luciferin is acted on by a particular enzyme in the presence of oxygen gas and ATP, luciferin is pushed into an excited state. When the molecule returns to ground state, light is emitted.

3.61   *K^+ attaches more strongly to valinomycin than Na^+ because of the size of the cavity (binding site) in the center of the compound. Is the cavity too large for Na^+ or is it too small?*

Since K^+ ions are larger than Na^+ ions, the cavity must be too large for the Na^+ ion.

3.63   *In many cities, there is great debate about the issue of fluoridating the water supply. What are the pros and cons of doing so?*

Answers will vary and may include: Fluoride reduces cavities, fluoride may be toxic.

Chapter 4
An Introduction to Organic Compounds

Solutions to Problems

4.1   *Draw the line-bond structure of each molecule.*

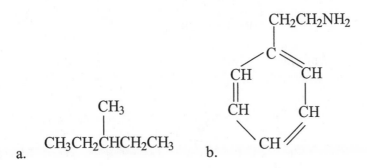

a.

b.

To make a line-bond structure, place a line (bond) between *every* atom.

a.   b.

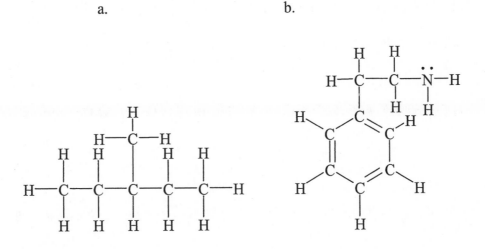

4.3   *What is the shape around the nitrogen and sulfur atoms in the ions shown in Figure 4.2?*

First you must refer to the three drawings in Figure 4.2 . In the first, nitrogen has four attached hydrogens, which makes the predicted shape tetrahedral. In the second drawing, nitrogen has three attached oxygens, which makes the predicted shape trigonal planar.   In the third drawing, the sulfur atom has four attached oxygens which makes the predicted shape tetrahedral.

a. tetrahedral; b. trigonal planar; c. tetrahedral

4.5 *What is the formal charge on each atom in the following molecules or ions?*

a.

$$H \overset{\cdots}{\underset{\cdots}{S}} H$$

b.

$$\left[\; :\overset{\cdots}{O} - C - \overset{\cdots}{O}: \atop :\overset{\cdots}{O}: \; \right]^{2-}$$

c.

$$:\overset{\cdots}{Cl} - N - \overset{\cdots}{Cl}: \atop :\overset{\cdots}{Cl}:$$

a. Formal charge is calculated using the equation:

Formal charge = number of valence electrons for a neutral atom − number of electrons around the atom in a compound

Using this formula you assume that the atom sees its nonbonding electrons and the bonding electrons nearest to it.

  Formal charge on H = 1 − 1 = 0

  Formal charge on S = 6 − 6 = 0

b. Formal charge on C = 4 − 4 = 0

  Formal charge on double bonded O = 6 − 6 = 0

  Formal charge on each single bonded O = 6 − 7 = 1−

c. Formal charge on N = 5 − 5 = 0

  Formal charge on each Cl = 7 − 7 = 0

4.7 *Label any polar covalent bond(s) in the molecules and ions that appear in problem 4.5.*

Since none of the bonds shown in the drawings in problem 4.5 are made between the same atoms, all of the bonds are polar to the same degree.

  a. While H and S have different electronegativities, for our purposes the bonds are not considered to be polar covalent.

338

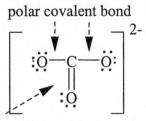

polar covalent bond

b. polar covalent bond

c. While N and Cl have different electronegativities, for our purposes the bonds are not considered to be polar covalent.

4.9   *Specify the shape around each specified atom in Problem 4.5.*

a. The S atom in H₂S
b. The C atom in CO₃²⁻
c. The N atom in NCl₃

a. The S atom has four groups of electrons (two attached hydrogens and two pairs of nonbonding electrons) and is bent.   b. The C atom has three groups of electrons and is trigonal planar. c. The N atom has four groups of electrons and is pyramidal.

a. bent; b. trigonal planar; c. pyramidal

4.11   *Which of the molecules are polar?*

To be polar the molecule must meet two basic criteria; first it must have polar bonds and these bonds cannot be equally distributed in the molecule (molecule must be unsymmetrical). Since all of the molecules have polar bonds, the only criterion to check is bond distribution. The shape around the C atom in each molecule is tetrahedral. Only in c do the polar bonds cancel one another.

a and b.

4.13   *Do hydrogen bonds form between formaldehyde molecules?*

In order to have hydrogen bonding, there must be at least one hydrogen atom attached to a nitrogen, oxygen, or fluorine atom. In examining the formaldehyde structure, $H_2C{=}O$, we see that the hydrogens are not attached to the oxygen, so hydrogen bonds do not form between formaldehyde molecules.

No.

4.15   *Which pairs of molecules can form a hydrogen bond with one another?*

A hydrogen bond is the interaction of a nitrogen, oxygen, or fluorine atom with a hydrogen atom that is covalently bonded to a different nitrogen, oxygen, or fluorine atom. This gives two criteria for hydrogen bonding to occur; first, at least one of the molecules must have a hydrogen atom that is attached to a nitrogen, oxygen, or fluorine atom, second, the other molecule must have a nitrogen, oxygen, or fluorine atom in its structure. Molecule sets c and d are the only ones that meet both criteria and the only sets that can have hydrogen bonding.

c and d.

4.17　*Which of the molecules in Problem 4.15 can form a hydrogen bond with a water molecule?*

A hydrogen bond is the interaction of a nitrogen, oxygen, or fluorine atom with a hydrogen atom that is covalently bonded to different nitrogen, oxygen, or fluorine atom. This gives two criteria for hydrogen bonding to occur; first, at least one of the molecules must have a hydrogen atom that is attached to a nitrogen, oxygen, or fluorine atom, second, the other molecule must have a nitrogen, oxygen, or fluorine atom in its structure. Water meets either criterion, the molecule in part b meets the second criterion, the molecules in parts c and d meet either criterion.

b, c, and d.

4.19　*A protein contains the following groups. Which can form salt bridges with one another?*

A salt bridge is another name for an ionic bond. The term is used to describe ionic bonds that form between charged groups in protein molecules. Therefore, the basis for identifying the parts of the protein that can form a salt bridge is to identify the part of the molecule with a charge. In this case, molecules shown in b are the only ones that both have a charge ($-O^-$ in one and the $-NH_3^+$ in the other).

b.

4.21　*Which share the stronger London force interactions, two $CH_3CH_2CH_2CH_2CH_3$ molecules or two $CH_3CH(CH_3)CH_2CH_3$ molecules?*

London force interactions are related to surface area. As the surface area drops, so do the London force interactions. Each of the molecules in this question have the formula C_5H_{12}, but branching in $CH_3CH(CH_3)CH_2CH_3$ reduces surface area and London force interactions.

Two $CH_3CH_2CH_2CH_2CH_3$ molecules.

4.23   *Which has the higher boiling point, CH₃CH₂CH₂CH₂CH₃ or*
        CH₃CH(CH₃)CH₂CH₃? Explain.

The stronger the interactions between the molecules in a liquid, the higher the
boiling point. As explained by the solution to Problem 4.21, both molecules have
the formula C_5H_{12}, but $CH_3CH_2CH_2CH_2CH_3$ is unbranched and has a greater
surface area. This leads to stronger London force attractions between molecules.
This gives $CH_3CH_2CH_2CH_2CH_3$ a higher boiling point.

4.25   *Arrange the molecules in order, from the highest boiling point to lowest boiling*
        point: decane, propane, butane.

Note that these are all "straight chained" molecules so that branching need not be
considered. In this case you are looking for which one has longest carbon chain.
Decane (10 carbon atoms) is the longest and has the strongest London force
interactions, followed by butane (4 carbon atoms) and propane (3 carbon atoms).
Remember that boiling points increase as the London force attraction increases.

decane; butane; propane

4.27   *Draw a line-bond structure of each alkane.*

        a. CH₃CH₂C(CH₃)₃
        b. CH₃CH₂CH(CH₃)CH(CH₃)CH₂CH₃

Connect every atom in the formula with a single line to represent the
covalent bond between them. Note how the (CH₃) group is written as a side
branch.

        a.   b.

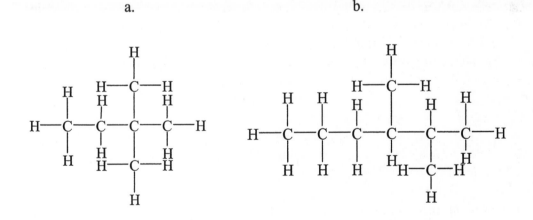

4.29 Find and name the parent chain for each molecule, then give the complete IUPAC name for each.

a.

b.

c.

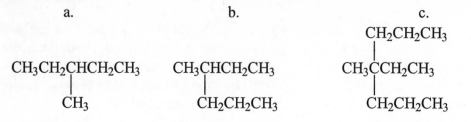

CH₃CH₂CHCH₂CH₃ (with CH₃ below)   CH₃CHCH₂CH₃ (with CH₂CH₂CH₃ below)   CH₂CH₂CH₃ above CH₃CCH₂CH₃ with CH₂CH₂CH₃ below

First, count out the longest chain of carbon atoms. This is the parent chain and you use the appropriate prefix to name it. Remember that the longest is not necessarily the one that goes straight across. Next, identify the substituents. Finally, to complete the IUPAC naming process number the carbons of the parent chain (number from the end nearer the first substituent). These steps are highlighted below.

a. For this molecule, the longest chain has five carbon atoms, making its parent

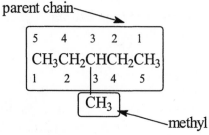

chain pentane. The only is a methyl group and the numbering gives the same number (3) for the methyl position whether you count left to right or right to left. This makes the name of this molecule 3-methylpentane.

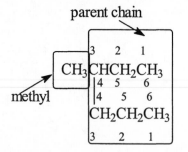

b. Note how the parent chain turns and is not straight across. The longest chain has six carbon atoms making the parent chain hexane. This molecule has only one substituent, the methyl group. In this case numbering top to bottom gives the methyl group a position of 3 but numbering from bottom to top gives a position of 4. The number 3 is the assigned position. This makes the name of the molecule 3-methylhexane.

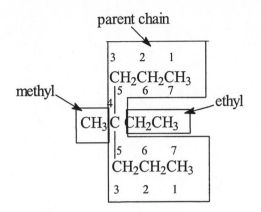

parent chain

methyl

ethyl

c. Note how the parent chain turns and is not straight across. The longest continuous chain is seven carbon atoms long making the parent chain heptane. This molecule has two substituents, one methyl group and one ethyl group (put them in alphabetical order in the name). The carbon they are attached to gets the same assigned number from both directions. This makes the name of the molecule 4-ethyl-4-methylheptane.

a. parent: pentane; name: 3-methylpentane
b. parent: hexane; name: 3-methylhexane
c. parent: heptane; name: 4-ethyl-4-methylheptane

4.31 *Draw and name six of the constitutional isomers with the formula C₇H₁₆.*

Start with the normal straight chain molecule and begin making isomers by shortening the parent chain by one methyl group and placing that methyl group in as many different places as possible. Next, take off two methyl groups and place them on the remaining parent chain to additional structures. Continue the process until you have created six structures. Remember that simply bending or turning the molecule does not make it a different structure. For the molecule C₇H₁₆, there are 9 constitutional isomers.

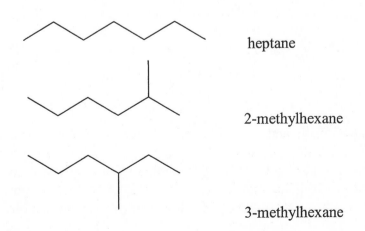

heptane

2-methylhexane

3-methylhexane

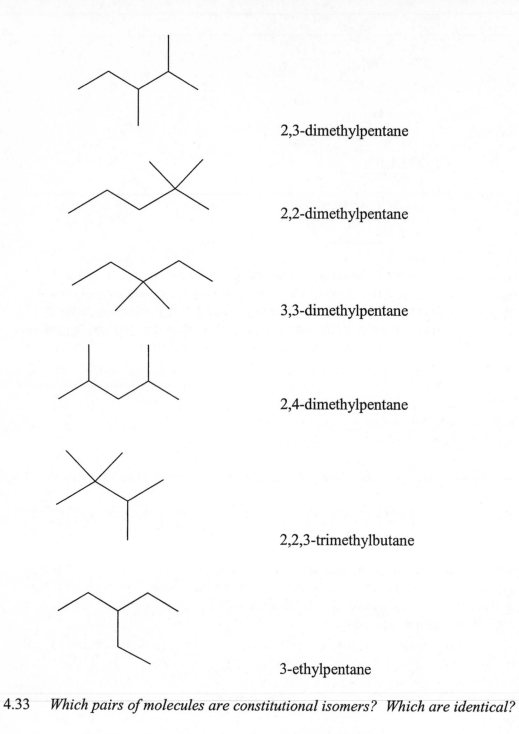

2,3-dimethylpentane

2,2-dimethylpentane

3,3-dimethylpentane

2,4-dimethylpentane

2,2,3-trimethylbutane

3-ethylpentane

4.33   *Which pairs of molecules are constitutional isomers? Which are identical?*

a. $CH_3CH_2CH_2CH_2CH_3$ and $\begin{array}{c} CH_3CHCH_3 \\ | \\ CH_3CH_2CH_3 \end{array}$

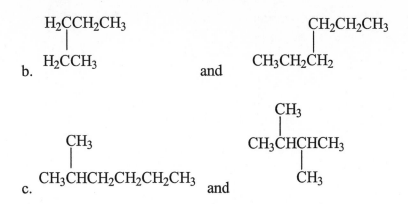

b.
H₂CCH₂CH₃
|
H₂CCH₃

and

CH₂CH₂CH₃
|
CH₃CH₂CH₂

c.
CH₃
|
CH₃CHCH₂CH₂CH₂CH₃

and

CH₃
|
CH₃CHCHCH₃
|
CH₃

One way to check for a constitutional isomer (after you have made sure both molecules have the same molecular formula) is to name the molecules. If they have different names they are constitutional isomers (different structures give different IUPAC names). If the names of two molecules are the same then the molecules are identical and are not constitutional isomers. In a, the first molecule is named hexane and the second molecule is 2-methylpentane, they are constitutional isomers. In b, the first molecule is named hexane (note that turning it does not give it a different name) and the second molecule is hexane (bending it a different way still does not make it something different), which means the molecules are identical. In c, the first molecule is 2-methylpentane and the second is 2,3-dimethylbutane, so they are constitutional isomers.

constitutional isomers, (a and c); identical, (b).

4.35   *Which are constitutional isomers?*

a. *pentane and 2-methylpentane*
b. *2-methylpentane and 3-methylpentane*
c. *2,2-dimethylpropane and pentane*
d. *2,2-dimethylpropane and cyclopentane*

This problem follows the same logic as problem 4.33, except that in this case it is obvious that they all have different names. This means that if they have the same molecular formula they must be constitutional isomers. In part a, pentane has the molecular formula C_5H_{12} and 2-methylpentane has the molecular formula C_6H_{14}. These are not constitutional isomers. In part b, both molecules have the formula C_6H_{14}. They are constitutional isomers. The same is true of the molecules in part c, each has the formula C_5H_{12}. The molecules in part d are not constitutional isomers.

b and c.

4.37   *Which pairs of molecules are constitutional isomers? Which are different
        conformations of the same molecule?*
        a.

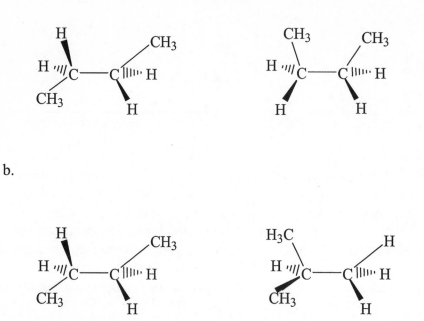

b.

To be constitutional isomers the molecules must have the same molecular formula
(which all of the molecules in both a and b do); and must have different
structures. The molecules in part a fail on the second criterion since both
molecules are butane and have the same structure. In part b, however, the first
molecule is butane and the second is 2-methylpropane. This makes the
molecules constitutional isomers. Since the molecules in part a are the same
molecule with the hydrogens rotated into different special orientations, they
are different conformations of the same molecule.

a. different conformations; b. constitutional isomers.

4.39   a.. *How are constitutional isomers and conformations similar?*

In order to be constitutional isomers or conformations, molecules must have the
same molecular formula.

b. *How are constitutional isomers and conformations different?*

Constitutional isomers have different atomic connections. Confirmations have the
same atomic connections, but different three-dimensional shapes that are
interchanged by bond rotation.

4.41 *Draw and name the three ethylmethylcyclobutane constitutional isomers.*

To be constitutional isomers, the molecules must have different structures and different IUPAC names.

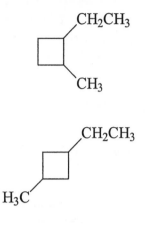

1-ethyl-2-methylcyclobutane

1-ethyl-3-methylcyclobutane

1-ethyl-1-methylcyclobutane

4.43 *Give the correct IUPAC name for each molecule.*

a.

b.

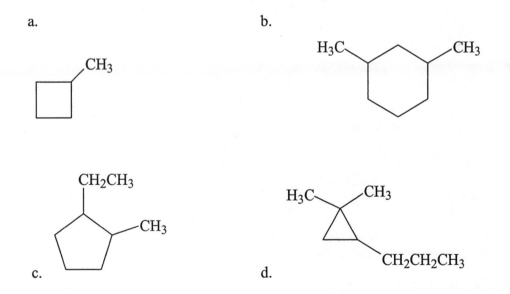

c.

d.

In part a, the parent is cyclobutane. The substituent is a methyl group and is in the 1st position (remember that the position number is left off of the name in cycloalkanes with just one substituent). In part b, the parent is cyclohexane. One methyl group is assigned the carbon 1 position and the second methyl group is assigned position 3. In part c, the parent is cyclopentane. The two substituents, ethyl and methyl are listed in alphabetical order. In this case with the first in the alphabet is assigned the carbon 1 position. This puts the ethyl on carbon 1 and the methyl on carbon 2. In d, the parent is cyclopropane. There are three substituents, two methyl groups and one propyl group. Assigning the smaller number to the methyl groups puts them on carbon 1 and the propyl group is on carbon 2.

a. methylcyclobutane
b. 1,3-dimethylcylohexane
c. 1-ethyl-2-methylcyclopentane
d. 1,1-dimethyl-2-propylcyclopropane

4.45 *Which molecule(s) in Problem 4.43 can exist as cis and trans isomers?*

For a cycloalkane to have *cis* and *trans* isomers, it must have at least two substituents attached to different ring atoms and the substituents must have different spatial orientations (on the same face of the ring – *cis*; on opposite sides – *trans*). In part a, no *cis* or *trans* isomers can exist since can exist since there is only one substituent. In part b, *cis* or *trans* isomers can exist because there are two methyl substituents on different carbons of the ring. In part c, *cis* or *trans* isomers can exist because there are two substituents, ethyl and methyl, on different carbons of the ring. In part d, *cis* or *trans* isomers cannot exist. Whether the propyl group points up or down, there is no "same face" or "opposite face" frame of reference, due to the two identical substituents (methyl groups) on the neighboring ring carbon atom.

b and c.

4.47 *Draw a side view of each cycloalkane.*

a. *trans-1,2-dimethylcyclohexane*
b. *trans-1-ethyl-2-methylcyclohexane*
c. *cis-1,3-diethylcyclopentane*

Trans means that substituents are to be drawn on opposite faces of the ring and *cis* means they are drawn on the same face.

a.

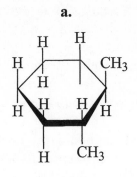

b.

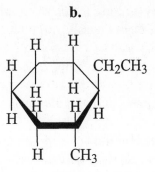

c.

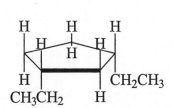

4.49   *Give the complete IUPAC name (including the use of the term cis or trans) for each molecule.*

a.

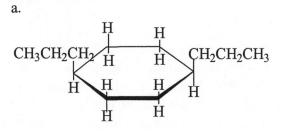

b.

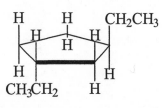

First identify the parent by counting the number of carbon atoms in the ring. Next, identify the substituents and assign numbering positions. Finally, *cis* or *trans* is

349

decided by looking to see if the substituents are on the same face of the ring (*cis*) or opposite faces (*trans*).

a. *cis*-1,4-dipropylcyclohexane ; b. *trans*-1,3-diethylcyclopentane

4.51   *Draw propene, showing the proper three-dimensional shape about each atom.*

Remember that the "ene" ending indicates a hydrocarbon with a double bond. The carbon atoms involved in the double bond are surrounded by a trigonal planar arrangement of atoms. The other carbon atom is surrounded by a tetrahedral arrangement.

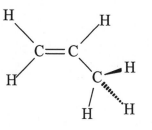

4.53   *Name each molecule.*

a. $CH_2{=}CHCH_2CH_3$

b. $CH_3CHCH{=}CCH_3$
      with CH_3 and CH_3 substituents below

c. $CH_3C{\equiv}CCHCH_3$
      with CH_3 substituent below

When naming hydrocarbons with double (alkene) or triple (alkyne) bonds, you should follow the same basic rules as for the alkanes, except for alkenes use an "ene" ending on the parent chain and for alkynes use an "yne" ending. Also, when naming the parent chain a number is placed in front of the parent chain to indicate the location of the multiple bond. This number takes precedence in the naming so start from the end that makes it smallest. If it is the same from either end, number from the end nearer the first substituent.

a. 1-butene   b. 2,4-dimethyl-2-pentene   c. 4-methyl-2-pentyne

4.55   *Draw each molecule*

a. *3-isopropyl-1-heptene*

b. *2,3-dimethyl-2-butene*

350

c. *5-sec-butyl-3-nonyne*

Draw the parent chain first, placing the multiple bond in the position indicated in front of the parent chain name (remember "ene" means double bond and "yne" means triple bond). Next, starting from the same side used to count off the multiple bond position, count to the position number indicated and draw the substituents in their indicated positions.

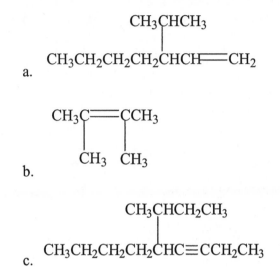

a.

b.

c.

4.57   *The molecule shown is a termite trail marking pheromone. Which double bonds are cis and which are trans?*

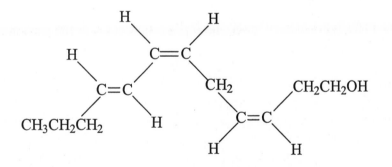

Recall that *cis* means that the atoms or groups of atoms being compared (in this case the attached carbon atoms) are on the same side of a line connecting the two double bonded carbon atoms, and *trans* means they are on opposite sides of that line.

Moving left to right across the molecule; the first double bond is *trans*; the second double bond is *cis*; and the third double bond is *cis*.

4.59   Name each molecule.

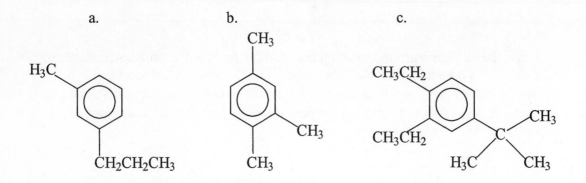

a.

b.

c.

Aromatic rings follow the same naming rules as used earlier in the chapter for cycloalkanes. Assign the number 1 to the substituent that will make the number assignment sequence smallest. When there are only two substituents present, the number 1 is assigned to the one first in the alphabet. Also, in the case of only two substituents, the *ortho* (side by side), *meta* (one carbon between), *para* (on opposite side of the ring) designations may be used.

a. 1-methyl-3-propylbenzene
b. 1,2,4-trimethylbenzene
c. 4-t-butyl-1,2-diethylbenzene

4.61   *Draw each molecule.*

a. *1,2-dipropylbenzene*
b. *m-diisopropylbenzene*
c. *1,2,4-trimethylbenzene*

Draw the benzene ring first. Next, draw the substituents at the carbon number location designated in the name.

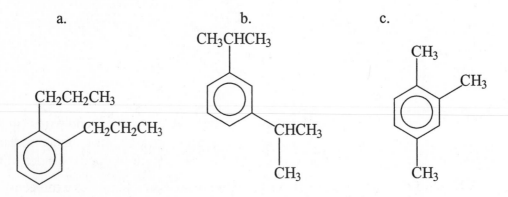

a.

b.

c.

4.63   *List the primary noncovalent attraction between each pair of molecules.*

a. $CH_3CH_2CH_2OH$ and $CH_3CH_2CH_2OH$

352

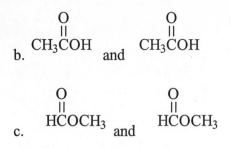

b.

$$\underset{\text{CH}_3\overset{\displaystyle O}{\overset{\|}{\text{C}}}\text{OH}}{}\quad\text{and}\quad\text{CH}_3\overset{\displaystyle O}{\overset{\|}{\text{C}}}\text{OH}$$

c.

$$\text{HC}\overset{\displaystyle O}{\overset{\|}{\text{O}}}\text{CH}_3\quad\text{and}\quad\text{HC}\overset{\displaystyle O}{\overset{\|}{\text{O}}}\text{CH}_3$$

The noncovalent interactions to choose from are hydrogen bonds, ionic bonds, dipole-dipole forces, ion-dipole forces, coordinate-covalent bonds and London forces. Since all of the molecules are covalent in nature, there are no ionic forces in any of them. In a, and b, the molecules have a hydrogen atom attached to an oxygen. This allows them to hydrogen bond. In c., the molecules are polar but cannot form hydrogen bonds. They interact through dipole-dipole forces.

(a) hydrogen bonding;
(b) hydrogen bonding;
(c) dipole-dipole.

4.65 *Dovonex is a prescription drug used to treat psoriasis.*

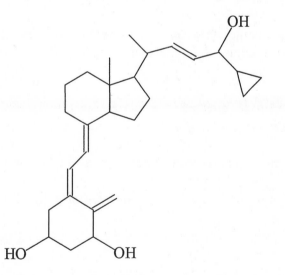

a. *Which functional groups are present in this molecule?*

A functional group is an atom, group of atoms, or bond that gives a molecule a particular set of chemical properties. For this question you are looking for groups other than carbon atoms with single bonds.

Alcohol and alkene

353

b. *Indicate which geometric isomer is present for the alkene group at the top of the structure.*

Geometric isomers come in pairs: *cis* and *trans*. The alkene at the top of the molecule has the attached carbon on opposite sides of the line connecting the two double-bonded carbon atoms. It is a *trans* geometric isomer.

trans

c. *Are any fused rings present in this molecule?*

Fused rings are composed of two or more rings connected to one another by shared carbon-carbon bonds.

Yes.

d. *Three different types of cycloalkane rings are present. Identify them.*

Cyclohexane, cyclopentane and cyclopropane

e. *Is this molecule saturated or unsaturated?*

To be "saturated" the molecule must have only single bonded carbons. This molecule has four double bonds which makes it unsaturated.

unsaturated.

4.67   *Suggest a way to reduce the spread of mad cow disease between cattle.*

Cattle feed should not contain parts from sheep, cattle, or other animals.

4.69   *What properties are important for molecules used as sunscreens?*

The molecules should absorb UV-B radiation and should not be toxic when applied to the skin.

4.71   *Although esters are generally known to have pleasant odors, many large esters have no odor. Explain.*

They are not volatile, so molecules do not reach the nose.

4.73   *The term "organic" can have different meanings. What are two different ways to interpret a sign at the grocery store that reads "organic foods"?*

In the strictest interpretation of "organic", the sign would be interpreted as telling you that there are carbon containing compounds. The second meaning is that the foods were grown utilizing nutrients that came from decaying organic matter.

Chapter 5
Gases, Liquids, and Solids

Solutions to Problems

5.1   a. *What is heat of fusion?*

Heat of fusion is the heat required to melt a solid.

b. *What is heat of vaporization?*

Heat of vaporization is the heat required to evaporate a liquid.

5.3   *If you immerse your arm in a bucket of ice water, your arm gets cold. Where does the heat energy from your arm go and what process is the energy used for?*

The heat energy goes into the ice and the energy is used in the melting process.

5.5   *Calculate the energy required to*

a. *warm 325 g of ice from -10°C to 0°C*

The ice is only heated to the melting point so that the temperature is raised but the ice has not started to change states. To calculate the energy needed for this process, the mass is multiplied by the temperature change and then converted to energy using the factor label method from chapter 1, using the specific heat as the conversion factor. The specific heat of ice is 0.500 cal/g°C.

325 g   x   10°C̶   x $\dfrac{0.500 \text{ cal}}{\text{g·°C̶}}$ = 1625 cal   (Round to 1600)

1.6 x 10³ cal

b. *warm 325 g of water from 10°C to 20°C*

The water is only heated so that the temperature is raised but the water has not changed states. To calculate the energy needed for this process, the mass is multiplied by the temperature change and then converted to energy using the factor label method from chapter 1, using the specific heat as the conversion factor. The specific heat of water is 1.000 cal/g°C.

$$325 \text{ g} \quad \times \quad 10°C \quad \times \quad \frac{1.000 \text{ cal}}{g·°C} \quad = \quad 3250 \text{ cal} \quad \text{(Round to 3300)}$$

3.3×10^3 cal

5.7   *Calculate the heat required to vaporize*

a. *15 g of water*

The water is at its boiling point and is changing states from liquid to gas. To calculate the energy needed for this process, the mass is converted to energy using the factor label method from chapter 1, using the heat of vaporization as the conversion factor. The heat of vaporization for water is 540 cal/g.

$$15 \text{ g} \quad \times \quad \frac{540 \text{ cal}}{g} \quad = \quad 8100 \text{ cal}$$

8.1×10^3 cal

b. *15 g of ethanol*

The ethanol is at its boiling point and is changing states from liquid to gas. To calculate the energy needed for this process, the mass is converted to energy using the factor label method from chapter 1, using the heat of vaporization as the conversion factor. The heat of vaporization for ethanol is 230 cal/g.

$$15 \text{ g} \quad \times \quad \frac{230 \text{ cal}}{g} \quad = \quad 3450 \text{ cal} \quad \text{(Round to 3500)}$$

3.5×10^3 cal

5.9   *Steam can cause severe burns because of the heat released to the skin when the steam condenses to water. How much energy is released when 25 g of steam condense?*

When it condenses, the steam is states from gas to liquid. To calculate the energy needed for this process, the mass is converted to energy using the factor label method from chapter 1, using the heat of vaporization as the conversion factor. The heat of vaporization for water is 540 cal/g. Note: energy released by condensing is the same as the amount required for vaporizing.

$$25 \text{ g} \quad \times \quad \frac{540 \text{ cal}}{g} \quad = \quad 13500 \text{ cal} \quad \text{(Round to 14,000 cal)}$$

1.4×10^4 cal

5.11   *An 11 g piece of copper has a temperature of 23°C. What will the new temperature of the copper be if 45 cal of heat are added?*

The copper is heated so that the temperature is raised but the copper does not change states. To calculate the temperature change for the heating process, the mass is multiplied by the specific heat of copper (0.0924 cal/°C) to cancel units of grams. The answer to this calculation is used as conversion factor to convert 45 cal into units of °C. Finally, the temperature change is added to the original temperature.

$$11 \text{ g} \quad \times \quad \frac{0.0924 \text{ cal}}{\text{g} \cdot °\text{C}} \quad = \quad 1.02 \text{ cal/°C}$$

$$45 \text{ cal} \quad \times \quad \frac{1 \text{ °C}}{1.02 \text{ cal}} \quad = 44°\text{C}$$

$$23°\text{C} \quad + \quad 44°\text{C} \quad = \quad 67°\text{C}$$

5.13   *In chemical terms, what does the term nonspontaneous mean?*

The process will not run by itself unless something keeps it going.

5.15   a. *In terms of enthalpy, is the condensation of steam spontaneous or nonspontaneous?*

The loss of energy is a spontaneous event. Since heat is lost as steam condenses, the process is spontaneous in terms of enthalpy.

Spontaneous

b. *In terms of entropy, is this process spontaneous or nonspontaneous?*

Entropy favors processes that increase in randomness. As steam condenses to water the motion of the molecules is becoming less random and therefore the process is nonspontaneous with respect to entropy.

Nonspontaneous

c. *At 90°C and 1 atm of atmospheric pressure is the condensation of steam spontaneous or nonspontaneous? Is the value of ΔG for this process negative or positive?*

At the pressure given, the boiling point of water is 100°C. This means that at 90°C the water would condense. The process is spontaneous. A spontaneous process has a negative ΔG.

Spontaneous; ΔG is negative

d. *At 110°C and 1 atm of atmospheric pressure is this process spontaneous or nonspontaneous? Is the value of ΔG for this process negative or positive?*

At 1 atm the boiling point of water is 100 °C, so at 110°C water will vaporize. The process is nonspontaneous since energy must be added to make it happen. A nonspontaneous process has a positive ΔG.

Nonspontaneous; ΔG is positive

5.17   a. *Give an example of an element or compound that is a solid at STP.*

Recall that STP means standard temperature (0°C) and pressure (1 atm). Think of something that is a solid under these conditions.

 iron

b. *Give an example of an element or compound that is a liquid at STP.*

 mercury

c. *Give an example of an element or compound that is a gas at STP.*

oxygen

5.19   *A pressure of 13.6 psi is how many*

a. *atmospheres?*

As with any conversion, the first step is to know the equivalency between psi and atmospheres (atm). 1 atm = 14.7 psi. Use this as a conversion factor and convert from psi to atm.

13.6 ~~psi~~   X $\dfrac{1\ atm}{14.7\ psi}$ = 0.925 atm

0.925 atm

b. *torr?*

As with any conversion, the first step is to know the equivalency between psi and torr. 760 torr = 14.7 psi. Use this as a conversion factor and convert from psi to torr.

13.6 ~~psi~~   x $\dfrac{760 \text{ torr}}{14.7 \text{ ~~psi~~}}$ = 703 torr

703 torr

5.21 a. *Estimate the atmospheric pressure, in atmospheres, at an altitude of 5 km.*

Refer to the graph in Figure 5.9. Find 5 km on the left side of the graph and move straight across from left to right until you reach the curve on the graph. Move straight down to the bottom of the graph and approximate that pressure reading.

 0.45 atm

b. *Convert your answer to part (a) into psi and torr.*

As with any conversion, the first step is to know the equivalency between psi, atmospheres (atm), and torr. 1 atm = 760 torr = 14.7 psi Use this as a conversion factor and convert from atm to psi and from atm to torr.

0.45 ~~atm~~   x $\dfrac{14.7 \text{ psi}}{1 \text{ ~~atm~~}}$ = 6.6 psi

0.45 ~~atm~~   x $\dfrac{760 \text{ torr}}{1 \text{ ~~atm~~}}$ = 342 torr   (Rounds to 3.4 x 10^2 torr)

6.6 psi and 3.4 x 10^2 torr

5.23 *At an atmospheric pressure of 760 torr and a temperature of $0°C$, the mercury level in the right arm of the manometer pictured in Figure 5.12 is 5 mm higher than the mercury in the left arm. What is the gas pressure inside the flask?*

The first thing to note is that the manometer pictured in Figure 5.12 is an open-ended manometer. This means that the gas pressure in the gas bulb is pushing against the outside atmospheric pressure of 760 torr. Since the mercury moved up on the right arm this tells you that the pressure of the gas in the bulb is 5 mmHg greater than atmospheric pressure and this much pressure can be added to the atmospheric pressure to get the pressure of the gas. The second thing to note is that the atmospheric pressure is in torr, not mmHg. Since 760 mmHg = 760 torr, the units are interchangeable.

760 torr + 5 torr = 765 torr

765 torr

5.25   a. *Listening to the weather report, you hear that the barometer is falling. What part of the barometer is falling?*

The mercury level column is falling.

b. *What is happening to the air pressure?*

The air pressure is decreasing.

5.27   *At a pressure of 760 torr a balloon has a volume of 1.50 L. If the balloon is put into a container and the pressure is increased to 2500 torr (at constant temperature), what is the new volume of the balloon?*

Identify your variables.

$P_1 = 760$ torr $P_2 = 2500$ torr

$V_1 = 1.50$ L $V_2 = ?$ (this is what you are asked to solve for)

From the gas laws select the one that has these variables ONLY. In this case it will be **Boyle's Law; $P_1V_1 = P_2V_2$** . Rearrange the equation so that the required variable is alone. Replace the variables in the equation with the ones you identified above and algebraically solve for the one missing.

$$V_2 = \frac{P_1V_1}{P_2}$$

$$V_2 = \frac{(760 \text{ torr} \times 1.50 \text{ L})}{2500 \text{ torr}} = 0.456 \text{ L} \quad \text{(Rounds to 0.46 L)}$$

0.46 L

5.29   *At a temperature of 30°C, a balloon has a volume of 1.50 L. If the temperature is increased to 60°C (at constant pressure), what is the new volume of the balloon?*

In gas law problems the temperature must be converted to kelvins.

$K = °C + 273.15 = 30°C + 273 = 303$ K

$K = °C + 273.15 = 60°C + 273 = 333$ K

Identify the variables.

$T_1 = 303$ K $T_2 = 333$ K

$V_1 = 1.50$ L $V_2 = ?$ (this is what you are asked to solve for)

From the gas laws select the one that has these variables ONLY. In this case it will be **Charles' Law; $V_1 / T_1 = V_2 / T_2$**. Rearrange the equation so that the required variable is alone. Replace the variables in the equation with the ones you identified above and algebraically solve for the one missing.

$V_2 = \dfrac{V_1 T_2}{T_1}$

$V_2 = \dfrac{(1.50 \text{ L} \times 333 \text{ K})}{303 \text{ K}} = 1.65 \text{ L}$

1.65 L

5.31   *At a temperature of 30°C, a gas inside a 1.50 L metal canister has a pressure of 760 torr. If the temperature is increased to 60°C (at constant volume), what is the new pressure of the gas?*

In gas law problems the temperature must be converted to kelvins.
$K = °C + 273.15 = 30°C + 273 = 303 \text{ K}$
$K = °C + 273.15 = 60°C + 273 = 333 \text{ K}$

Identify the variables.
$T_1 = 303 \text{ K}$ $T_2 = 333 \text{ K}$
$P_1 = 760 \text{ torr}$ $P_2 = ?$ (this is what you are asked to solve for)

Note: even though the volume was stated it is not included as a variable because the problem states that it remained constant.

From the gas laws select the one that has these variables ONLY. In this case it will be **Gay-Lussac's Law; $P_1/T_1 = P_2/T_2$**. Rearrange the equation so that the required variable is alone. Replace the variables in the equation with the ones you identified above and algebraically solve for the one missing.

$P_2 = \dfrac{P_1 T_2}{T_1}$

$P_2 = \dfrac{(760 \text{ torr} \times 333 \text{ K})}{303 \text{ K}} = 835 \text{ torr}$

835 torr

5.33   *A 2.0 L balloon contains 0.35 mol of $Cl_2(g)$. At constant pressure and temperature, what is the new volume of the balloon if 0.20mol of gas is removed?*

Identify the variables.
Note that 0.20 mol is the amount removed NOT final number of moles.

$n_2 = 0.35 - 0.20 = 0.15 \text{ mol}$

$n_1 = 0.35 \text{ mol}$ $n_2 = 0.15 \text{ mol}$
$V_1 = 2.0 \text{ L}$ $V_2 = ?$ (this is what you are asked to solve for)

From the gas laws select the one that has these variables ONLY. In this case it will be **Avogadro's Law; $V_1 / n_1 = V_2 / n_2$**. Rearrange the equation so that the required variable is alone. Replace the variables in the equation with the ones you identified above and algebraically solve for the one missing.

$$V_2 = \frac{V_1 \, n_2}{n_1}$$

$$V_2 = \frac{(2.0 \text{ L} \times 0.15 \text{ mol})}{0.35 \text{ mol}} = 0.86 \text{ L}$$

0.86 L

5.35   *A balloon with a volume of 1.50 L is at a pressure of 760 torr and a temperature of 30°C. If the balloon is put into a container and the pressure is increased to 2500 torr and the temperature is raised to 60 °C, what is the new volume of the balloon?*

First convert °C to K.
$K = °C + 273.15 = 30°C + 273 = 303 \text{ K}$
$K = °C + 273.15 = 60°C + 273 = 333 \text{ K}$

Identify your variables by the units given.
   $T_1 = 303 \text{ K}$ $T_2 = 333 \text{ K}$
   $P_1 = 760 \text{ torr}$ $P_2 = 2500$
   $V_1 = 1.50 \text{ L}$ $V_2 = ?$ (this is what you are asked to solve for)

From the gas laws select the one that has all these variables. In this case it will be the **Combined Gas Law; $P_1V_1 / n_1T_1 = P_2V_2 / n_2T_2$**. Rearrange the equation so that the required variable is alone. Replace the variables in the equation with the ones you identified above and algebraically solve for the one missing. Note that it is not necessary to include n_1 and n_2 since the number of moles is constant.

$$V_2 = \frac{P_1 V_1 T_2}{P_2 T_1}$$

$$V_2 = \frac{(760 \text{ torr} \times 1.50 \text{ L} \times 333 \text{ K})}{(2500 \text{ torr} \times 303 \text{ K})} = 0.050 \text{ L}$$

0.50 L

5.37   *A 575 mL metal can contains 2.50×10^{2} mol of He at a temperature of 298K. What is the pressure (in atm) inside the can? Is this pressure greater than or less than standard atmospheric pressure?*

Identify the variables.

$$T = 298 \text{ K} \qquad\qquad P = ?$$
$$V = 0.575 \text{ L} \qquad\qquad n = 2.50 \times 10^{-2} \text{ mol}$$

From the gas laws select the one that has just one of each variable. In this case it will be the **Ideal Gas Law; $PV = n\,R\,T$**. R is the gas constant (0.0821 L atm /mol K). Rearrange the equation so that the required variable is alone. Replace the variables in the equation with the ones you identified above and algebraically solve for the one missing.

$$P = \frac{n\,R\,T}{V}$$

Note: since the R constant has the units atm and L in it, before solving for P is it necessary to convert the volume to L.

$$575 \ \cancel{mL} \ \times \ \frac{1 \times 10^{-3} \text{ L}}{\cancel{mL}} = \ 0.575 \text{ L}$$

$$P = \frac{(2.50 \times 10^{-2} \ \cancel{mol} \times 0.0821 \ \cancel{L} \text{ atm} / \cancel{mol} \ \cancel{K} \times 298 \ \cancel{K}}{(0.575 \ \cancel{L})} = 1.06 \text{ atm}$$

1.06 atm (greater than atmospheric pressure)

5.39 *The label on a can of spray paint warns you to keep it away from high temperatures. From the perspective of the ideal gas law, explain why.*

If the number of moles and the volume of a gas remain constant, then an increase in temperature results in an increase in pressure ($PV = nRT$).

5.41 *A 100.0 mL flask contains 400 g of N_2 at $0°C$.*

a. *What is the pressure in atm?*

Identify the variables.
Since grams of N_2 is given instead of moles, first convert 400 g to moles.

$$400 \ \cancel{g} \ \times \ \frac{1 \text{ mol}}{28.0 \ \cancel{g}} \ = \ 14.3 \text{ mol}$$

Also, convert $°C$ to Kelvin.
$$K = °C + 273.15 = 0°C + 273 = 273 \text{ K}$$

And, since the R constant has the units atm and L in it, it is necessary to convert the volume from mL to L.

$$100 \ \cancel{mL} \times \frac{1 \times 10^{-3} \text{ L}}{\cancel{mL}} = 0.100 \text{ L}$$

$$T = 273 \text{ K} \qquad\qquad P = ?$$
$$V = 0.100 \text{ L} \qquad\qquad n = 14.3 \text{ mol}$$

From the gas laws select the one that has just one of each variable. In this case it will be the **Ideal Gas Law; PV = n R T**. R is the gas constant (0.0821 L atm /mol K). Rearrange the equation so that the required variable is alone. Replace the variables in the equation with the ones you identified above and algebraically solve for the one missing.

$$P = \frac{n\,R\,T}{V}$$

$$P = \frac{14.3 \text{ mol} \times 0.0821 \text{ L atm /mol K} \times 273 \text{ K}}{0.100 \text{ L}} = 3210 \text{ atm}$$

$$P = 3.21 \times 10^3 \text{ atm}$$

b. What is the pressure in torr?

Convert the answer from part a to torr. (1 atm = 760 torr)

$$3210 \text{ atm} \times \frac{760 \text{ torr}}{1 \text{ atm}} = 2.44 \times 10^6 \text{ torr}$$

$$P = 2.44 \times 10^6 \text{ torr}$$

c. *What is the pressure in psi?*

Convert the answer from part a to psi. (1 atm = 14.7 psi)

$$3200 \text{ atm} \times \frac{14.7 \text{ psi}}{1 \text{ atm}} = 4.72 \times 10^4 \text{ psi}$$

$$P = 4.72 \times 10^4 \text{ psi}$$

5.43   *A 750 mL flask contains O_2 at a pressure of 0.75 atm and a temperature of 20°C. What mass of O_2 is present?*

Solve for the number of moles and then convert from moles of O_2 to grams of O_2.

Convert °C to Kelvin.
$$K = °C + 273.15 = 20°C + 273 = 293 \text{ K}$$

From the gas laws select the one that has just one of each variable. In this case it will be the **Ideal Gas Law; PV = n R T**. R is the gas constant (0.0821 L atm /mol K).

Since the R constant has the units atm and L in it, before solving for n, it is necessary to convert the volume from mL to L .

$$750 \text{ mL} \times \frac{1 \times 10^{-3} \text{ L}}{\text{mL}} = 0.750 \text{ L}$$

Identify the variables:

T = 293 K	P = 0.75 atm
V = 0.750 L	n = ?

Rearrange the ideal gas equation so that the required variable is alone. Replace the variables in the equation with the ones you identified above and algebraically solve for the one missing.

$$n = \frac{PV}{RT}$$

$$n = \frac{0.75 \text{ atm} \times 0.750 \text{ L}}{0.0821 \text{ L atm /mol K} \times 293 \text{ K}} = 2.34 \times 10^{-2} \text{ mol}$$

Then; $2.34 \times 10^{-2} \text{ mol } O_2 \quad \times \quad \dfrac{32.0 \text{ g } O_2}{1 \text{ mol } O_2} \quad = \quad 0.749 \text{ g of } O_2$

0.749 g

5.45 a. *How many moles of Ne are present in a 1.0 qt flask that has a pressure of 850 torr at a temperature of 35°C?*

This problem gives pressure, volume, and temperature and asks for number of moles. The ideal gas law (PV = nRT) will be used.

First, convert °C to Kelvin.
K = °C + 273.15 = 35°C + 273 = 308 K

And, since the R constant has the units atm and L in it, it is necessary to convert the volume from qt to L.

$$1.0 \text{ qt} \times \frac{1 \text{ L}}{1.06 \text{ qt}} = 0.94 \text{ L}$$

Additionally, torr must be converted to atm.

$$850 \text{ torr} \quad x \quad \frac{1 \text{ atm}}{760 \text{ torr}} \quad = \quad 1.12 \text{ atm}$$

Identify the variables.

        $T = 308 \text{ K}$ $P = 1.12 \text{ atm}$
        $V = 0.94 \text{ L}$ $n = ?$

$$n = \frac{PV}{RT}$$

$$n = \frac{1.12 \text{ atm} \times 0.94 \text{ L}}{0.0821 \text{ L atm /mol K} \times 308 \text{ K}} = 0.042 \text{ mol}$$

0.042 moles of Ne

b. *What is the mass of this Ne?*

Convert the moles calculated in part a. to grams.

$$0.042 \text{ mol Ne} \quad x \quad \frac{20.2 \text{ g Ne}}{1 \text{ mol Ne}} \quad = \quad 0.85 \text{ g of Ne}$$

0.85 g of Ne

5.47   *A mixture of gases contains 0.75 mol of N_2, 0.25 mol of O_2, and 0.25 mol of He.*
       a. What is the partial pressure of each gas (in atm and in torr) in a 25 L cylinder
       at 350 K?

The partial pressure of a gas can be calculated as though it is the only gas in the container.

Identify the variables.
For calculating the partial pressure of N_2:

        $T = 350 \text{ K}$ $P = ?$
        $V = 25 \text{ L}$ $n = 0.75 \text{ mol}$

From the gas laws select the one that has just one of each variable. In this case it will be the **Ideal Gas Law; PV = n R T**. R is the gas constant (0.0821 L atm /mol K). Rearrange the equation so that the required variable is alone. Replace the variables in the equation with the ones you identified above and algebraically solve for the one missing.

$$P = \frac{n R T}{V}$$

$P = \dfrac{0.75 \ \text{mol} \ \text{x} \ 0.0821 \ \text{L atm} \ /\text{mol} \ \text{K} \ \text{x} \ 350 \ \text{K}}{25 \ \text{L}} = 0.86 \ \text{atm}$

$P = 0.86 \ \text{atm for N}_2$

Convert to torr:

$0.86 \ \text{atm} \ \text{x} \ \dfrac{760 \ \text{torr}}{1 \ \text{atm}} = 6.5 \ \text{x} \ 10^2 \ \text{torr}$

$P = 6.5 \ \text{x} \ 10^2 \ \text{torr for N}_2$

Now repeat the calculation for O_2. Since the temperature and volume are the same, you can use the same set-up from above and simply insert the correct number of moles for O_2.

$P = \dfrac{0.25 \ \text{mol} \ \text{x} \ 0.0821 \ \text{L atm} \ /\text{mol} \ \text{K} \ \text{x} \ 350 \ \text{K}}{25 \ \text{L}} = 0.29 \ \text{atm}$

$P = 0.29 \ \text{atm for O}_2$

Convert to torr:

$0.29 \ \text{atm} \ \text{x} \ \dfrac{760 \ \text{torr}}{1 \ \text{atm}} = 2.2 \ \text{x} \ 10^2 \ \text{torr}$

$P = 2.2 \ \text{x} \ 10^2 \ \text{torr for O}_2$

Now repeat the calculation for He. Since the temperature and volume are the same, you can use the same set-up from above and simply insert the correct number of moles for He.

$P = \dfrac{0.25 \ \text{mol} \ \text{x} \ 0.0821 \ \text{L atm} \ /\text{mol} \ \text{K} \ \text{x} \ 350 \ \text{K}}{25 \ \text{L}} = 0.29 \ \text{atm}$

$P = 0.287 \ \text{atm for He}$

Convert to torr:

$0.29 \ \text{atm} \ \text{x} \ \dfrac{760 \ \text{torr}}{1 \ \text{atm}} = 2.2 \ \text{x} \ 10^2 \ \text{torr}$

$P = 2.2 \ \text{x} \ 10^2 \ \text{torr for He}$

b. *What is the total pressure?*

Dalton's Law states that the total pressure for all gases in a container is equal to the sum of their partial pressures. Therefore the total pressure is found by adding up the partial pressures calculated above.

$$P_{total} = P_{N_2} + P_{O_2} + P_{He}$$

$P_{total} = 0.86$ atm $+ 0.29$ atm $+ 0.29$ atm $= 1.44$ atm

1.44 atm; 1.09×10^3 torr

5.49   *For the mixture of gases in Problem 5.47, what are the partial pressures and the total pressure if 0.50 mol of $CO_2(g)$ is added?*

First, since gas pressures are independent of each other, the partial pressures of the first three do not change. Calculate the partial pressure of CO_2. Then add this pressure to the partial pressures of the other gases to find the total pressure.

$$P = \frac{0.50 \text{ mol} \times 0.0821 \text{ L atm /mol K} \times 350 \text{ K}}{25 \text{ L}} = 0.58 \text{ atm}$$

$P = 0.86$ atm N_2, 6.5×10^2 torr
$P = 0.29$ atm O_2, 2.2×10^2 torr
$P = 0.29$ atm He, 2.2×10^2 torr
$P = 0.58$ atm CO_2, 4.4×10^2 torr

Total Pressure $= 2.02$ atm $= 1.54 \times 10^2$ torr

5.51   *In which town does dry air contain the highest partial pressure of O_2, Hot Coffee, Mississippi (alt. 279 ft) or Pie Town, New Mexico (alt. 7778 ft) ?*

Figure 5.9 shows that total atmospheric pressure decreases with altitude. It can be assumed that the partial pressure of O_2 in dry air will also decrease with altitude. The town at the lower altitude will have the higher partial pressure of O_2.

Hot Coffee, Mississippi

5.53   *A patient has 25.0 mL of blood drawn and this volume of blood has a mass of 26.5 g. What is the density of the blood?*

Density is expressed in g/mL. This indicates that the density is found by dividing the mass of the blood by its volume.

Density $= \dfrac{\text{mass}}{\text{volume}} = \dfrac{26.5 \text{g}}{25.0 \text{ mL}} = 1.06$ g/mL

1.06 g/mL

5.55 *At 20°C, how many milliliters of ethanol correspond to 35.2 g? (See Table 5.5.)*

Density can be used as a conversion factor. The density of ethanol is 0.791 g/mL at 20°C.

$$35.2 \text{ g} \quad \text{x} \quad \frac{1\text{mL}}{0.791 \text{ g}} \quad = \quad 44.5 \text{ mL}$$

44.5 mL

5.57 *At 20°C, what volume does 11.2 g of human fat occupy? (see Table 5.5)*

Density can be used as a conversion factor. The density of human fat is 0.94 g/mL at 20°C.

$$11.2 \text{ g} \quad \text{x} \quad \frac{1\text{mL}}{0.94 \text{ g}} \quad = \quad 12 \text{ mL}$$

12 mL

5.59 *What is the density, in g/mL, of $O_2(g)$ at 0°C? (see Table 5.5)*

At 0°C the density of O_2 is 1.43 g/L. Convert this value into g/mL.

$$\frac{1.43 \text{ g}}{1 \text{ L}} \quad \text{x} \quad \frac{1 \text{ x } 10^{-3} \text{ L}}{\text{mL}} \quad = \quad 1.43 \text{ x } 10^{-3} \text{ g /mL}$$

1.43 x 10^{-3} g /mL

5.61 *What is the specific gravity of kerosene at 20°C? (see Table 5.5)*

Specific gravity is the density of a substance divided by the density of water. at the same temperature.

$$\text{Specific gravity of kerosene} = \frac{\text{density of kerosene}}{\text{density of water}}$$

$$\text{Specific gravity of kerosene} = \frac{0.82 \text{ g/mL}}{1.00 \text{ g/mL}} = 0.82$$

0.82

5.63   *If water is placed in a flask and a pump is used to reduce the atmospheric pressure above the water to 17.54 torr, at what temperature does the water boil? (See Table 5.6.)*

The boiling point is the temperature at which the vapor pressure of a liquid is equal to the pressure above it. In this case, the pressure above the liquid is 17.54 torr. Referring to Table 5.6 you can find the temperature that corresponds to a vapor pressure of 17.54 torr. That temperature is 20 °C.

20 °C

5.65   *Estimate the boiling point of water at a pressure of 2.2 atm.*

The boiling point is the temperature at which the vapor pressure of a liquid is equal to the pressure above it. In this case, the pressure above the liquid is 2.2 atm. As the vapor pressure values on Table 5.6 are given in torr, first convert 2.2 atm to torr.

2.2 atm   x $\dfrac{760 \text{ torr}}{1 \text{ atm}}$ =1672 torr   (Rounds to 1700 torr)

Next, referring to Table 5.6 you can estimate the temperature that corresponds to a vapor pressure of 1700 torr. That pressure falls between 1074.56 torr at 110 °C and 1740.93 torr at 125 °C.

120 °C

5.67   *Calculate the energy required for each. (Hint: use information provided in Tables 5.1 and 5.5.)*

a. *warm 35.0 mL of water from 21°C to 29°C*

Use the density of water from Table 5.5 to convert mL of water to grams.

35.00 mL   x $\dfrac{1.00 \text{ g}}{1 \text{ mL}}$ =   35.0 g

Next, use the specific heat of water, 1.000 cal/g°C (Table 5.1) to convert from mass and temperature change to calories. (Temperature change = 29°C - 21°C = 8°C)

35.0 g x 8 °C x $\dfrac{1.000 \text{ cal}}{1 \text{ g}\cdot°C}$ = 300 cal

300 cal

b. *warm 17.5 mL of water from 18°C to 54°C*

Use the density of water from Table 5.5 to convert mL of water to grams.

$$17.5 \text{ mL} \quad \text{x} \quad \frac{1.00 \text{ g}}{1 \text{ mL}} \quad = \quad 17.5 \text{ g}$$

Next, use the specific heat of water, 1.000 cal/g°C (Table 5.1) to convert from mass and temperature change to calories. (Temperature change = 18°C - 54°C = 36°C)

$$17.5 \text{ g} \quad \text{x } 36 \text{ °C} \quad \text{x} \quad \frac{1.000 \text{ cal}}{1 \text{ g °C}} \quad = \quad 630 \text{ cal}$$

630 cal

5.69   *Why do ionic solids typically have a higher melting point than molecular solids?*

Ionic bonds are stronger than intermolecular forces.

5.71   *How is the structure of an amorphous solid different from the structure of a crystalline one?*

Amorphous solids do not have the orderly arrangement of particles found in crystalline structures.

5.73   *During severe bleeding, ADH (a hormone released by the hypothalamus) causes vasoconstriction (shrinking of the blood vessels) to take place. What effect does a decrease in blood vessel volume have on blood pressure?*

A decrease in the volume of a liquid causes an increase in its pressure.

5.75   *If your diet is low in iron-containing foods, your body may not be able to produce normal amounts of hemoglobin and you may develop iron-deficiency anemia. The symptoms of anemia include shortness of breath and fatigue. Explain why insufficient hemoglobin can produce these symptoms.*

Without sufficient hemoglobin, your cells will not be supplied with the O_2 that they require.

5.79   *Why is breathing 100% O_2 in a pressurized hyperbaric chamber more effective at treating gangrene than breathing 100% O_2 at atmospheric pressure?*

At high pressure, more O_2 will dissolve in the blood. The bacteria that cause gangrene cannot survive in an oxygen rich environment.

5.81   *Many sports organizations ban the use and test for the presence of higher than normal levels of erythropoietin (EPO), a naturally occurring compound that stimulates the production of red blood cells. How might EPO enhance athletic performance?*

An increase in the number of red blood cells would allow for more O_2 transport to the cells.

Chapter 6
Reactions

Solutions to Problems

6.1 *Write the following sentence as a balanced chemical equation. Phosphorus reacts with chlorine (Cl_2) to produce phosphorus trichloride.*

Prefixes like "tri" tell you the subscript for the element in the compound's formula. First translate the names to symbols: $P + Cl_2 \rightarrow PCl_3$. Next, check to see if the equation is balanced. It is not because there are 2 Cl atoms on the left and 3 Cl atoms on the right. Place the coefficient 2 in front of the PCl_3 ($2PCl_3$) which makes 6 Cl atoms on the right that can be balanced by placing a 3 in front of the Cl_2 ($3Cl_2$) to make 6 Cl atoms on the left. The $2PCl_3$ also means you have 2 P atoms on the right, so you need to place a 2 in front of the P on the left (2P) in order to have 2 P atoms on that side. This makes the equation balanced.

$$2P + 3Cl_2 \rightarrow 2PCl_3$$

6.3 *What would you observe if you carried out the following reactions?*

a. *$NH_3(g) + HCl(g) \rightarrow NH_4Cl(s)$*

The small letters given with the formulas indicate the physical state of the reactants and products: *(s)* means solid, *(l)* means liquid, *(g)* means gas, and *(aq)* means aqueous (dissolved in water). The *(s)* on the NH_4Cl *(s)* indicates that it is being produced as a solid and, therefore, you would observe a solid being produced during the reaction of the two gases.

The formation of a solid

b. *$HCO_3^-(aq) + H_3O^+(aq) \rightarrow CO_2(g) + 2H_2O(l)$*

In b, the $CO_2(g)$ would be observed as bubbling as the gas is produced.

Bubbling, indicating the formation of a gas.

6.5 *Balance the reaction equations.*

a. *$SO_2 + O_2 \rightarrow SO_3$*

You have 3 O atoms on the right, and 4 on the left. Place a 2 in front of SO_3, ($2SO_3$). This gives 6 O atoms and 2 S atoms on the right, so place a 2 in front of the SO_2 on the left, ($2SO_2$). This balances the S atoms and gives 4 O atoms to go with the O_2, for a total of 6 O atoms on the left.

$$2SO_2 + O_2 \rightarrow 2SO_3$$

b. $NO + O_2 \rightarrow NO_2$

There are 3 O atoms on the left and 2 on the right. Place a 2 in front of the NO (2NO). This means you need 2 in front of the NO_2, ($2NO_2$), so that the N atoms are balanced. This also gives 4 O atoms on both sides and makes the equation balanced.

$$2NO + O_2 \rightarrow 2NO_2$$

6.7   *Balance the reaction equations.*

a. $K + Cl_2 \rightarrow KCl$

There is 1 Cl atom on the right and 2 on the left. Place a 2 in front of the KCl (2KCl). This means that you need 2 in front of the K (2K) so that the K atoms are balanced. This also gives 2 Cl atoms on both sides, so the equation is balanced.

$$2K + Cl_2 \rightarrow 2KCl$$

b. $CH_4 + Cl_2 \rightarrow CH_2Cl_2 + HCl$

There are 3 Cl atoms on the right and 2 on the left. Place a 2 in front of the HCl (2HCl). This means you need 2 in front of the Cl_2 ($2Cl_2$) so that the Cl atoms are balanced. This also gives 4 H atoms on both sides, so the equation is balanced.

$$CH_4 + 2Cl_2 \rightarrow CH_2Cl_2 + 2HCl$$

6.9   *Cave explorers can use lamps that contain calcium carbide (CaC_2) to light their way. When moistened, calcium carbide reacts to form acetylene gas (C_2H_2), which immediately burns to produce light. Balance the equation for the reaction of calcium carbide with water.*

$$CaC_2(s) + H_2O(l) \rightarrow Ca(OH)_2(aq) + C_2H_2(g)$$

There is 1 O atom on the left and 2 on the right. Place a 2 in front of the H_2O ($2H_2O$). This gives two O atoms on both sides. It also gives 4 H atoms on both sides, so the equation balanced.

$$CaC_2(s) + 2\, H_2O(l) \rightarrow Ca(OH)_2(aq) + C_2H_2(g)$$

6.11   *Magnesium reacts with iron(II) chloride according to the equation*

$$Mg(s) + FeCl_2(aq) \rightarrow MgCl_2(aq) + Fe(s)$$

a. *Is magnesium oxidized or is it reduced?*

b. *Is iron(II) ion oxidized or is it reduced?*

c. *What is the oxidizing agent?*

d. *What is the reducing agent?*

Oxidized means to lose electrons and reduced means to gain electrons. To determine which atom is oxidized or reduced, the charge on each atom or ion must be determined. Some of these can be predicted from the periodic table. When that is not possible (transition metals, for example) the charge is calculated using the other ions whose charges are known.

charges: 0   2+ 1-   2+ 1-   0
$$Mg(s) + FeCl_2(aq) \rightarrow MgCl_2(aq) + Fe(s)$$

a. Mg changes from a 0 charge to a 2+ charge, so it is oxidized.

b. Fe^{2+} is changed to Fe*(s)* with 0 charge, so it is reduced.

c. By definition, the species that causes the oxidation to occur is the oxidizing agent. Since Fe^{2+} ion took the electrons from Mg*(s)*, it is the oxidizing agent.

d. By definition the species that causes the reduction to occur is the reducing agent. Since Mg*(s)* gave the electrons to Fe^{2+} ion it is the reducing agent.

6.13   *Aluminum metal reacts with oxygen gas (O$_2$) to form aluminum oxide.*

a. *Write a balanced equation for this oxidation-reduction reaction. (In this problem you need not worry about the physical state of the reactants or product.)*

First write the equation by translating the words to formulas, as discussed in Chapter 3.

$$Al + O_2 \rightarrow Al_2O_3$$

Now balance the equation. There 3 O atoms on the right and 2 on the left. Place a 2 in front of the Al_2O_3 ($2Al_2O_3$). This gives six O atoms on the right. Next, place a 3 in front of O_2 ($3O_2$), which makes six O atoms on the left side. There are also 4 Al atoms on the right side so place a 4 in front of the Al ($4Al$) and the equation balanced.

$$4Al + 3O_2 \rightarrow 2Al_2O_3$$

b. *Which reactant is oxidized?*
c. *Which reactant is reduced?*
d. *What is the oxidizing agent?*
e. *What is the reducing agent?*

Since oxidized means to lose electrons and reduced means to gain electrons, to determine which atom is oxidized or reduced, the charge on each atom or ion must be determined. Some of these can be predicted from the periodic table. When that is not possible (transition metals for example) the charge is calculated using the other ions whose charges are known.

$$\text{charges:} \quad \overset{0}{Al} + \overset{0}{O_2} \rightarrow \overset{3+ \; 2-}{Al_2O_3}$$

b. Al changes from a 0 charge to a 3+ charge so it is oxidized.
c. O_2 with 0 charge on each atom is changed to O^{2-}, it is reduced.
d. By definition, the species that causes the oxidation to occur is the oxidizing agent and since O_2 took the electrons from Al, it is the oxidizing agent.
e. By definition, the species that causes the reduction to occur is the reducing agent and since Al gave the electrons to O_2, it is the reducing agent.

6.15 *Ethanol (CH_3CH_2OH) is mixed with gasoline to produce gasohol, a cleaner burning fuel than gasoline. Write the balanced equation for the complete oxidation of ethanol by O_2 to produce CO_2 and H_2O.*

Write the formulas in equation format by placing the reactants on the left side and the products on the right side of the arrow.

$$CH_3CH_2OH + O_2 \rightarrow CO_2 + H_2O$$

To begin the balancing process, examine the equation carefully to see if it may already be balanced. In this case it is not. Next, note that there are more hydrogens than any other element so this is probably a good element to start with. Since there are 6 H atoms on the left, place a 3 in front of H_2O ($3H_2O$). This makes six hydrogen atoms on both sides. Then, since there are 2 C atoms on the left, place a 2 in front of the CO_2 ($2CO_2$). This leaves only the O atoms to check. On the right there are 4 O atoms from the $2CO_2$ and 3 from the $3H_2O$ for a total of 7 O atoms on the right. On the left there is 1 O atom from the CH_3CH_2OH and 2 O atoms from the O_2. Complete the balancing by placing a 3 in front of the O_2 ($3O_2$). This makes $1 + 6 = 7$ O atoms on the left which balances the equation.

As recommended by the text, this equation can also be balanced by balancing C first, then H, then O.

$$CH_3CH_2OH + 3O_2 \rightarrow 2CO_2 + 3H_2O$$

6.17   a. *"Hydrogen Cars" describes the possibility of using hydrogen as a fuel for cars, buses, and trucks. According to the video clip, one positive aspects of this fuel is that when hydrogen is burned, water and energy are the only products. Write a balanced chemical equation for the reaction of H_2 and O_2 to form H_2O.*

Write the formulas in equation format by placing the reactants on the left side and the products on the right side of the arrow.

$$H_2 + O_2 \rightarrow H_2O$$

To balance the equation note there is 1 O atom on the right and 2 on the left. Place a 2 in front of the H_2O ($2H_2O$), to balance the O atoms with two on each side. This also makes four H atoms on the left, so place a 2 in front of the H_2 ($2H_2$) to give four H atoms on the left and the equation is balanced.

$$2H_2 + O_2 \rightarrow 2H_2O$$

b. *What products are formed when octane (C_8H_{18}), a typical component of gasoline, is burned?*

If a sufficient supply of oxygen is present, the combustion of a hydrocarbon results in the production of carbon dioxide and water.

CO_2 and H_2O

c. *Currently, what is the major disadvantage of using hydrogen as a fuel?*

Visit http://www.wiley.com/college/raymond to view this video.

6.19   *p-Phenylenediamine is a compound that is used in black hair dye. After being applied to and absorbed by hair, this compound is treated with H_2O_2 to produce a black color. Is p-phenylenediamine oxidized or is it reduced by H_2O_2?*

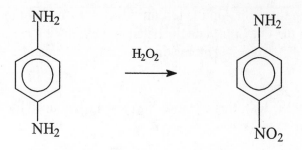

Oxidation can be identified by the loss hydrogen and/or the gain of oxygen. It is easy to see that the NH_2 group on the bottom of the molecule does both of these. p-Phenylenediamine is oxidized.

6.21 *Draw the saturated product expected when 3 mol of H_2, in the presence of Pt, are reacted with 1 mol of the termite trail marking pheromone shown below.*

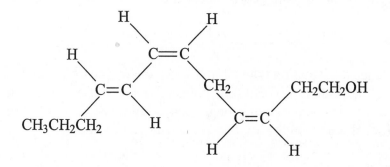

There are 2 H atoms provided by each mole of H_2, therefore 3 mol of H_2 will replace all of the double bonds in 1 mol of the molecule.

$CH_3(CH_2)_{10}CH_2OH$

6.23 *Draw the skeletal structure of the product formed when each alkene is reacted with H_2, in the presence of Pt.*

a. *cyclopentene*

First, draw the structure of cyclopentene using the rules learned in Chapter 4.

Note that the hydrogens added will not show since you are drawing a skeletal structure. Upon reaction with H_2, the double bond is converted to a single bond.

b. *2-ethyl-3-methyl-1-pentene*

First draw the structure of cyclopentene using the rules learned in Chapter 4.

Upon reaction with H_2, the double bond is converted to a single bond. If you draw the answer as a skeletal structure, as is done here, the hydrogens added will not show.

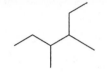

6.25   *Draw the products of each hydrolysis reaction.*

a.

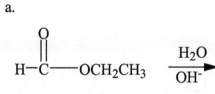

When an ester is hydrolyzed in the presence of OH⁻, the two products are a carboxylate ion and an alcohol.

$$\underset{HC}{\overset{O}{\|}} -O^- \quad + \quad HOCH_2CH_3$$

b.

$$CH_3\overset{O}{\overset{\|}{C}} -OCH_2 \bigcirc \quad \xrightarrow[OH^-]{H_2O}$$

381

When an ester is hydrolyzed in the presence of OH⁻, the two products are a carboxylate ion and an alcohol.

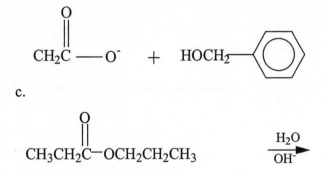

c.

$$CH_3CH_2\overset{\displaystyle O}{\overset{\|}{C}}-OCH_2CH_2CH_3 \qquad \xrightarrow[OH^-]{H_2O}$$

When an ester is hydrolyzed in the presence of OH⁻, the two products are a carboxylate ion and an alcohol.

$$CH_3CH_2\overset{\displaystyle O}{\overset{\|}{C}}-O^- \quad + \quad HOCH_2CH_2CH_3$$

6.27 *Draw the hydration product formed when each alkene is reacted with H_2O in the presence of H^+.*

a. *trans-3-hexene*

Hydration of an alkane is the process of adding a water molecule across a double bond. The end result is that the double bond is replaced by a H atom on one of the originally double bonded carbon atoms and an OH group is placed on the other. To see the product that will form, draw the structure of the molecule named, remove one of the bonds in the double bond and place hydrogen on one side and an OH on the other.

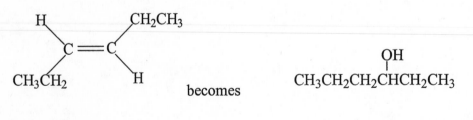

becomes

b. *cis-3-hexene*

See directions in part a.

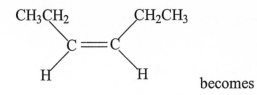

becomes

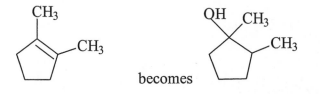

c. *1,2-dimethylcyclopentene*

See directions given in part a.

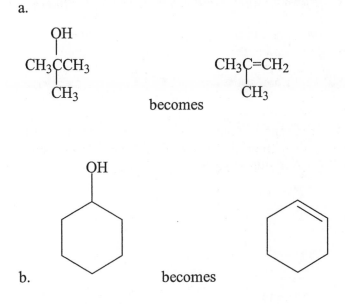

becomes

6.29 *Draw the organic dehydration product formed when each alcohol is heated in the presence of H^+.*

Dehydration is the reverse of hydration. First, identify the location of an H atom and OH group that are on side-by-side carbons. These are removed and replaced with another bond line between the two carbons creating a double bond. The H and OH combine to give water as a second product.

a.

becomes

b.

becomes

c.

$$\underset{\text{OH}}{\underset{|}{\text{CH}_3\text{CH}_2\text{CHCH}_2\text{CH}_3}}$$ becomes $\text{CH}_3\text{CH}_2\text{CH}=\text{CHCH}_3$

6.31 *Aspirin (acetylsalicylic acid) is an ester of acetic acid.*

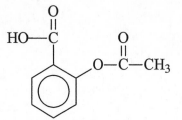

a. *Circle the ester group in aspirin.*

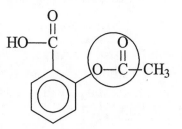

b. *Aspirin is sold with cotton placed in the neck of the bottle to help keep moisture out. When exposed to water, the ester group of aspirin slowly hydrolyzes. Draw the hydrolysis products obtained if aspirin is reacted with H_2O in the presence of OH^-.*

Hydrolysis of an ester in the presence of OH^- results in the formation of the carboxylate ion and an alcohol. In this case, the acetyl group becomes the carboxylate ion and the alcohol is the salicylic acid.

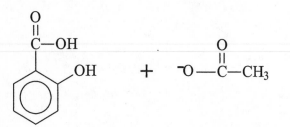

6.33 *For the combustion of methane, beginning with 3.15 mol of CH₄,*

$$CH_4(g) + 2O_2(g) \rightarrow CO_2(g) + 2H_2O(g)$$

a. *How many moles of O_2 are required to completely consume the CH_4?*

Use the mole to mole ratio for CH_4 to O_2 given by the equation as a conversion factor to convert from moles of CH_4 to moles of O_2.

3.15 mol CH₄   X $\dfrac{2 \text{ mol } O_2}{1 \text{ mol } CH_4}$ = 6.30 mol O_2

6.30 mol O_2

b. *How many moles of CO_2 are obtained when the CH_4 is completely combusted?*

Use the mole to mole ratio for CH_4 to CO_2 given by the equation as a conversion factor to convert from moles of CH_4 to moles of CO_2.

3.15 mol CH₄   X $\dfrac{1 \text{ mol } CO_2}{1 \text{ mol } CH_4}$ = 3.15 mol CO_2

3.15 mol CO_2

c. *How many moles of H_2O are obtained when the CH_4 is completely combusted?*

Use the mole to mole ratio for CH_4 to H_2O given by the equation as a conversion factor to convert from moles of CH_4 to moles of H_2O.

3.15 mol CH₄   X $\dfrac{2 \text{ mol } H_2O}{1 \text{ mol } CH_4}$ = 6.30 mol H_2O

6.30 mol H_2O

6.35   *2-Propanol, also known as isopropyl alcohol or rubbing alcohol, can be produced by the reaction:*

$$CH_3-CH{=}CH_2 \quad + \quad H_2O \quad \xrightarrow{\ H^+\ } \quad CH_3-\underset{\overset{|}{\underset{}{}}}{CH}-CH_3$$
$$\overset{OH}{}$$

a. *How many moles of H_2O, are required to completely react with 55.7 mol of propene?*

Use the mole to mole ratio for C_3H_6 to H_2O given by the equation as a conversion factor to convert from moles of C_3H_6 to moles of H_2O.

55.7 mol C₃H₆   X $\dfrac{1 \text{ mol } H_2O}{1 \text{ mol } C_3H_6}$ = 55.7 mol H_2O

55.7 mol H_2O

b. *How many moles of H_2O are required to completely react with 1.66 mol of propene?*

Use the mole to mole ratio for C_3H_6 to H_2O given by the equation as a conversion factor to convert from moles of C_3H_6 to moles of H_2O.

1.66 ~~mol C₃H₆~~   x $\dfrac{1 \text{ mol } H_2O}{1 \text{ mol } C_3H_6}$ =   1.66 mol H_2O

1.66 mol H_2O

c. *How many moles of 2-propanol are expected from the complete reaction of 47.2 g of propene?*

Use the molecular weight of propene to convert from grams to moles. Then, use the mole to mole ratio for C_3H_6 to C_3H_5OH given by the equation as a conversion factor to convert from moles of C_3H_6 to moles of C_3H_5OH.

47.2 ~~g C₃H₆~~   x $\dfrac{\text{mol } C_3H_6}{42.0 \text{ g } C_3H_6}$ =   1.12 mol C_3H_6

1.12 ~~mol C₃H₆~~   x $\dfrac{1 \text{ mol } C_3H_5OH}{1 \text{ mol } C_3H_6}$ =   1.12 mol C_3H_5OH

1.12 mol C_3H_5OH

d. *How many grams of 2-propanol are expected from the complete reaction of 125 g of propene?*

Use the molecular weight of propene to convert from grams to moles. Then, use the mole to mole ratio for C_3H_6 to C_3H_7OH given by the equation as a conversion factor to convert from moles of C_3H_6 to moles of C_3H_7OH. Finally, use the molecular weight of 2-propanol to convert moles of 2-propanol to grams.

125 ~~g C₃H₆~~   x $\dfrac{\text{mol } C_3H_6}{42.0 \text{ g } C_3H_6}$ =   2.98 mol C_3H_6

2.98 ~~mol C₃H₆~~   x $\dfrac{1 \text{ mol } C_3H_7OH}{1 \text{ mol } C_3H_6}$ =   2.98 mol C_3H_7OH

2.98 ~~mol C₃H₇OH~~   x $\dfrac{60.0 \text{ g } C_3H_7OH}{1 \text{ mol } C_3H_7OH}$ =   179 g C_3H_7OH

179 g C_3H_7OH

6.37 *Consider the reaction*

$$KOH(s) + CO_2(g) \rightarrow KHCO_3(s)$$

a. *How many grams of KOH are required to react completely with of 5.00 mol of CO_2?*

Use the mole to mole ratio for CO_2 to KOH given by the equation as a conversion factor to convert from moles CO_2 to moles of KOH.

5.00 ~~mol CO₂~~   x $\dfrac{1 \text{ mol KOH}}{1 \text{ mol } CO_2}$ =   5.00 mol KOH

5.00 ~~mol KOH~~   x $\dfrac{56.1 \text{ g KOH}}{1 \text{ mol KOH}}$ =   281 g KOH

281 g KOH

b. *How many grams of $KHCO_3$ are produced from the complete reaction of 75.9 g of KOH?*

Use the molecular weight of KOH to convert from grams to moles. Then, use the mole to mole ratio for KOH to $KHCO_3$ given by the equation as a conversion factor to convert from moles of KOH to moles of $KHCO_3$. Finally, use the molecular weight of $KHCO_3$ to convert moles of $KHCO_3$ to grams.

75.9 ~~g KOH~~   x $\dfrac{\text{mol KOH}}{56.1 \text{ g KOH}}$ = 1.35 mol KOH

1.35 ~~mol KOH~~   x $\dfrac{1 \text{ mol } KHCO_3}{1 \text{ mol KOH}}$ =   1.35 mol $KHCO_3$

1.35 ~~mol KHCO₃~~   x $\dfrac{100.1 \text{ g } KHCO_3}{1 \text{ mol } KHCO_3}$ =   135 g $KHCO_3$

135 g $KHCO_3$

6.39 *A reaction of iron metal with oxygen gas produces ferrous oxide (FeO).*

$$Fe + O_2 \rightarrow FeO$$

a. *Balance the reaction.*

There is 1 O atom on the right and 2 on the left. Place a 2 in front of the FeO, (2FeO). This gives two O atoms on the right. Since this also makes two Fe atoms on the right, place a 2 in front of the Fe (2Fe) and the equation is balanced.

$$2Fe + O_2 \rightarrow 2FeO$$

b. *When FeO forms, has Fe been oxidized or has it been reduced?*

Since oxidized means loss of electrons and reduced means gain of electrons, to determine which atom is oxidized or reduced, the charge on each atom or ion must be determined. Some of these can be predicted from the periodic table. When that is not possible (transition metals, for example) the charge is calculated using the other ions of known charge.

charges:

$$\overset{0}{2Fe} + \overset{0}{O_2} \rightarrow \overset{2+ \; 2-}{2FeO}$$

b. *Fe is changing from a 0 charge to a 2+ charge as FeO is produced so it oxidized.*

Oxidized

c. *When FeO forms, has O_2 been oxidized or has it been reduced?*

Using the charges calculated in part b, the O_2 with 0 charge on each atom is changed to O^{2-} in FeO. It is reduced.

Reduced

6.41   *One form of phosphorus, called white phosphorus, burns when exposed to air.*

$$P_4(s) + O_2(g) \rightarrow P_4O_{10}(s)$$

a. *Balance the reaction equation.*

There are 2 O atoms on the left and 10 O atoms on the right. Place a 5 in front of the O_2 ($2O_2$). This gives ten O atoms on the left. Since the P atoms are already balanced, the equation is balanced.

$$P_4(s) + 5O_2(g) \rightarrow P_4O_{10}(s)$$

b. *What is the theoretical yield (in grams) of P_4O_{10} if 33.0 g of P_4 are reacted with 40.0 g of O_2?*

The theoretical yield is the amount of product expected when all of the reactant is used up. When it cannot be assumed that the other reactant is in excess, you must first determine which amount given produces the smallest amount of product when used up. Do this by converting each number of grams given to moles of product. Then use the smaller number of moles of product to determine the grams of product that should be produced.

$$33.0 \text{ g P} \times \frac{1 \text{ mol } P_4}{124.0 \text{ g } P_4} \times \frac{1 \text{ mol } P_4O_{10}}{1 \text{ mol } P_4} = 0.266 \text{ mol } P_4O_{10}$$

$$40.0 \text{ g } O_2 \times \frac{1 \text{ mol } O_2}{32.0 \text{ g } O_2} \text{-x} \frac{1 \text{ mol } P_4O_{10}}{5 \text{ mol } O_2} = 0.250 \text{ mol } P_4O_{10}$$

Notice that the O_2 will produce the least amount of product when it is used up and will therefore determine the amount of product that can be produced. Convert the 0.25 mol P_4O_{10} to grams using the molar mass of P_4O_{10}.

$$0.250 \text{ mol } P_4O_{10} \times \frac{284 \text{ g } P_4O_{10}}{1 \text{ mol } P_4O_{10}} = 71.0 \text{ g } P_4O_{10}$$

71.0 g P_4O_{10}

6.43 Many food colors and fabric dyes are manufactured from aniline. Aniline can be produced by the reaction below.

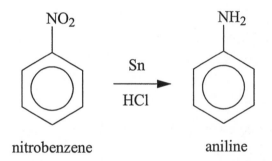

a. *What is the theoretical yield of aniline (in grams), beginning with 150 g of nitrobenzene? (Assume that nitrobenzene is the limiting reactant.)*

Calculate a theoretical yield by using the molecular weight of nitrobenzene to convert grams of nitrobenzene to moles of nitrobenzene. Then convert moles of nitrobenzene to moles of aniline using the mole to mole ratio from the equation. Finally, convert moles of aniline to grams of aniline using the molecular weight of aniline. Recall that the benzene ring has alternating double bonds which allows for only one H atom on each carbon. This makes the molecular formula of nitrobenzene $C_6H_5NO_2$ and the molecular formula of aniline $C_6H_5NH_2$.

$$150 \text{ g } C_6H_5NO_2 \times \frac{1 \text{ mol } C_6H_5NO_2}{123 \text{ g } C_6H_5NO_2} = 1.22 \text{ mol } C_6H_5NO_2$$

$$1.22 \text{ mol } C_6H_5NO_2 \times \frac{1 \text{ mol } C_6H_5NH_2}{1 \text{ mol } C_6H_5NO_2} = 1.22 \text{ mol } C_6H_5NH_2$$

$$1.22 \text{ mol } C_6H_5NH_2 \times \frac{93.0 \text{ g } C_6H_5NH_2}{} = 113 \text{ g } C_6H_5NH_2$$

$$1 \text{ mol } \cancel{C_6H_5NH_2}$$

113 g $C_6H_5NH_2$

b. *Based on the theoretical yield from part a., what is the percent yield if 50.0 g of aniline are obtained?*

Percent yield is what you actually obtained (actual yield) in the experiment divided by what you thought should be produced (theoretical yield) multiplied by 100.

$$\frac{50.0 \text{ g}}{113 \text{ g}} \times 100 = 44.2\%$$

44.2%

6.45 *A person drinks a glass of beer or wine, the first step in metabolizing the ethanol (CH₃CH₂OH) that you have ingested is its enzymatic conversion to acetaldehyde CH₃CHO).*

$$\underset{\begin{array}{c}H\ H\end{array}}{\overset{\begin{array}{c}H\ \ OH\end{array}}{H-C-C-H}} + NAD^+ \xrightarrow{enzyme} CH_3-\overset{O}{\underset{}{C}}-H + NADH^+ + H^+$$

a. *In this reaction is the ethanol oxidized or reduced?*

In the case of an organic molecule, it is not possible to simply look up the charges on the atoms to determine the charges. Oxidation has occurred when H atoms are removed, oxygen is added, or both. Reduction is typically evidenced by the addition of H atoms to, removal of oxygen from, or both. In this case H atoms are removed as the molecule is changed from ethanol to acetaldehyde.

Oxidized

b. *Write this equation by assigning the conversion of NAD⁺ into NADH and H⁺ its own arrow.*

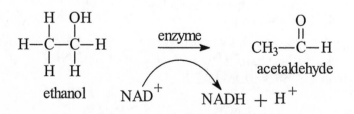

c. *What theoretical yield of acetaldehyde (in grams) is expected from the reaction of 10.0 g of ethanol? (Assume that ethanol is the limiting reactant.)*

Calculate a theoretical yield by using the molecular weight of ethanol to convert grams of ethanol to moles of ethanol. Then convert moles of ethanol to moles of acetaldehyde using the mole to mole ratio from the equation. Finally, convert moles of acetaldehyde to grams of acetaldehyde using the molecular weight of acetaldehyde. The molecular formula of ethanol is C_2H_5OH and the molecular formula of acetaldehyde is C_2H_4O.

$$10.0 \text{ g } C_2H_5OH \times \frac{1 \text{ mol } C_2H_5OH}{46.0 \text{ g } C_2H_5OH} = 0.217 \text{ mol } C_2H_5OH$$

$$0.217 \text{ mol } C_2H_5OH \times \frac{1 \text{ mol } C_2H_4O}{1 \text{ mol } C_2H_5OH} = 0.217 \text{ mol } C_2H_4O$$

$$0.217 \text{ mol } C_2H_4O \times \frac{44.0 \text{ g } C_2H_4O}{1 \text{ mol } C_2H_4O} = 9.55 \text{ g } C_2H_4O$$

9.55 g C_2H_4O

6.47   *Which of the reactions are spontaneous?*

a. *$CO(g) + H_2O(g) \rightarrow CO_2(g) + H_2(g)$* ΔG = -4.1 kcal/mol

Recall from Chapter 5 that reactions with $\Delta G < 0$ are spontaneous, and reactions with $\Delta G > 0$ are nonspontaneous.

Spontaneous

b.   *$2HI(g) \rightarrow H_2(g) + I_2(g)$* ΔG = 0.6 kcal/mol

See part a.

Nonspontaneous

6.49   *Sample Problem 6.11 showed how glucose 6-phosphate is converted into glucose during the manufacture of glucose.*

Glucose 6-phosphate $\rightarrow$ glucose + phosphate + 3.3 kcal

In the breakdown of glucose, glucose is converted into glucose 6-phosphate.

Glucose + ATP $\rightarrow$ glucose 6-phosphate + ADP + 4.0 kcal

a. *Is the conversion of glucose 6-phosphate into glucose + phosphate spontaneous?*

When energy is given off in a reaction, $\Delta G < 0$ and the reaction is spontaneous.

Yes

b. *Is the conversion of glucose + ATP into glucose 6-phosphate + ADP spontaneous?*

When energy is given off in a reaction, $\Delta G < 0$ and the reaction is spontaneous.

Yes

c. *How can ΔG be negative for both the transformation of glucose 6-phosphate into glucose (part a) and the transformation of glucose into glucose 6-phosphate (part b)?*

These are different reactions. The conversion of glucose 6-phosphate into glucose plus phosphate is spontaneous. By bringing ATP into play, the conversion of glucose into glucose 6-phosphate becomes spontaneous.

6.51   *What is the likely effect on the rate of a reaction if*

a. *reactant concentration is decreased?*

Molecules must collide in order to react and the fewer that are present, the less likely it is for collisions to occur and the reaction will be slower. Decreasing the concentration means fewer molecules in a given volume and, therefore, fewer collisions.

Reaction rate decreases.

b. *Temperature is decreased?*

Molecules must collide in order to react. The faster they are moving, the more likely it is for collisions to occur and the reaction will be faster. Also, activation energy is required for a reaction to occur and the higher the temperature the more energy that is available. Decreasing the temperature means fewer molecules colliding at one time and less energy available to ensure reaction.

Reaction rate decreases.

c. *A catalyst is added?*

A catalyst will lower the activation energy and therefore increase the likelihood of a reaction when molecules collide. This means more successful reactions as molecules collide, which speeds up the reaction.

Reaction rate increases.

6.53   *When present in ointments and creams for treating acne, the oxidizing agent benzoyl peroxide ($C_{14}H_{10}O_4$) is safely handled. Pure benzoyl peroxide is quite hazardous to deal with, however. Any heat source can cause it to explode as the compound undergoes rapid oxidation. Complete the balancing of the equation for the combustion of benzoyl peroxide.*

$$2\,C_{14}H_{10}O_4 + O_2 \rightarrow CO_2 + H_2O$$

Place 28 (2 x 14) in front of the carbon on the right to make the number of carbons equal. Next, place 10 in front of the H_2O to make 20 H atoms on the left to equal the 2 x 10 = 20 H atoms on the right. Finally, count the oxygens on the right. 28 x 2 = 56 and 10 x 1 = 10 for a total of 66 O atoms on the right. Since there are only 8 (2 x 4) O atoms in the first compound on the left, the remaining O atoms must be balanced using O_2 molecules. Place 29 in front of the O_2 mole to make 50 O atoms [8 + 58 (on left) = 56 + 10 (on right)].

$$2\,C_{14}H_{10}O_4 + 29\,O_2 \rightarrow 28CO_2 + 10H_2O$$

6.55   *Only 5% of the CO_2 that moves from cells into blood is carried as dissolved CO_2. What happens to the other 95%?*

Approximately 1/3 of the CO_2 attaches to hemoglobin and the remaining part of the 95% is converted into carbonic acid in a hydration reaction catalyzed by the enzyme carbonic anhydrase.

Chapter 7
Solutions, Colloids, and Suspensions

Solutions to Problems

7.1 *A bottle of 190 proof alcohol is 95% ethanol and 5% water. What is the solvent in this solution?*

When both the solute and the solvent are liquids, the one in largest quantity is considered the solvent.

Ethanol is the solvent.

7.3 *A pan full of hot salt water NaCl(aq) is cooled and NaCl(s) precipitates. Explain why this happens.*

The amount of a substance that can dissolve in a solvent is temperature dependent. As the temperature decreases the solubility of the salt goes down and the salt in excess settles out as a precipitate.

7.5 *Give an example of a solution in which*

a. *The solute is a solid and the solvent is a liquid.*

Examples will vary. Some possible answers are: Salt in water, sugar in tea, sugar in coffee.

b. *The solute is a gas and the solvent is a gas.*

Examples will vary. Some possible answers are: CO_2 in the air, O_2 in the air, He in N_2 in a scuba tank.

O_2 in the air

7.7 *When aqueous copper (II) sulfate is mixed with aqueous sodium sulfide, a precipitate forms. Write a balanced equation for this reaction.*

Before attempting to balance the equation make sure that you write the correct formulas. Translate each name given: copper(II) sulfate, $CuSO_4$ (equal numbers of Cu^{2+} and SO_4^{2-} to give a neutral compound). Sodium sulfide is Na_2S(twice as many Na^+ as S^{2-} to give a neutral compound.) Now you have the reactants identified you need to predict the products. In this case Cu^{2+} combines with S^{2-} to form CuS and Na^+ combines with the SO_4^{2-} ion to form Na_2SO_4. Since you are told a precipitate forms and Table 7.1 shows sulfates to be soluble, the CuS must be the solid formed. Now we can write the equation and balance it.

$$CuSO_4(aq) + Na_2S(aq) \rightarrow CuS(s) + Na_2SO_4(aq)$$

Start by counting each atom.

reactant atoms	product atoms
1 Cu atom	1 Cu atom
2 S atoms	2 S atoms
4 O atoms	4 O atoms
2 Na atoms	2 Na atoms

In this case, all of the atoms are balanced on both sides of the equation. The equation is correct as written.

$$CuSO_4(aq) + Na_2S(aq) \rightarrow CuS(s) + Na_2SO_4(aq)$$

7.9   *Predict whether each ionic compound is soluble or insoluble in water.*

The general solubility rules for ions are given in Table 7.1 of the textbook.

a. *$(NH_4)_2SO_4$*   Soluble

b. *K_2SO_4*   Soluble

c. *$CaCO_3$*   Insoluble

d. *$NaNO_3$*   Soluble

7.11   *Complete and balance each precipitation reaction.*

Follow the same basic procedure for each of the equations below. Recombine the reactant cations and anions to form new ionic compounds. Refer to Table 7.1 to see if either compound formed is insoluble. If it is insoluble be sure to write *(s)* with the formula. Use the symbol *(aq)* for those that are water soluble. Then, write the balanced equation.

a. *$CaCl_2(aq) + Li_2CO_3(aq) \rightarrow$*

Ca^{2+} reacts with CO_3^{2-} to form $CaCO_3$ and Li^+ reacts with Cl^- to form LiCl. The solubility Table 7.1 indicates that calcium carbonate is insoluble but lithium chloride is soluble.

$CaCl_2(aq) + Li_2CO_3(aq) \rightarrow CaCO_3(s) + LiCl(aq)$

Now balance the reaction.

> *(NOTE: do not start balancing until you have written the formulas of all reactants and products.)*

$CaCl_2(aq) + Li_2CO_3(aq) \rightarrow CaCO_3(s) + LiCl(aq)$

Start by counting each atom.

reactant atoms	product atoms
1 Ca atom	1 Ca atom
2 Cl atoms	1 Cl atoms
3 O atoms	3 O atoms
2 Li atoms	1 Li atoms

There are 2 Cl atoms on the left and 1 on the right, so multiply LiCl by 2 (2LiCl) and recount.

reactant atoms	product atoms
1 Ca atom	1 Ca atom
2 Cl atoms	**2** Cl atoms
3 O atoms	3 O atoms
2 Li atoms	**2** Li atoms

This makes the atom count the same on both sides, so the equation is balanced.

$CaCl_2(aq) + Li_2CO_3(aq) \rightarrow CaCO_3(s) + 2LiCl(aq)$

b. $Pb(NO_3)_2(aq) + NaCl(aq) \rightarrow$

Pb^{2+} reacts with Cl^- to form $PbCl_2$ and Na^+ reacts with NO_3^- to form $NaNO_3$. The solubility Table 7.1 indicates that $PbCl_2$ is insoluble and that the other compounds are soluble.

$Pb(NO_3)_2(aq) + NaCl(aq) \rightarrow PbCl_2(s) + NaNO_3(aq)$

Now balance the reaction.

Start by counting each atom.

reactant atoms	product atoms
1 Pb atom	1 Pb atom

1 Cl atom	2 Cl atoms
6 O atoms	3 O atoms
1 Na atoms	1 Na atoms
2 N atom	1 N atom

Since there are 2 Cl atoms on the right and only one Cl atom on the left, start by multiplying the NaCl on the left by 2 (2NaCl). Recount the atoms.

reactant atoms	product atoms
1 Pb atom	1 Pb atom
2 Cl atom	2 Cl atoms
6 O atoms	3 O atoms
2 Na atoms	1 Na atoms
2 N atom	1 N atom

The Cl atoms are balanced but there are 2 Na on the left and 1 Na on the right. Multiply $NaNO_3$ by 2 (2NaNO$_3$) and recount.

reactant atoms	product atoms
1 Pb atom	1 Pb atom
2 Cl atom	2 Cl atoms
6 O atoms	**6** O atoms
2 Na atoms	**2** Na atoms
2 N atom	**2** N atom

The atoms are now the same on both sides and you can write the balanced equation.

$$Pb(NO_3)_2(aq) + 2NaCl(aq) \rightarrow PbCl_2(s) + 2NaNO_3(aq)$$

c. $CaBr_2(aq) + K_3PO_4(aq) \rightarrow$

Ca^{2+} reacts with PO_4^{3-} to form $Ca_3(PO_4)_2$ and K^+ reacts with Br^- to form KBr. The solubility Table 7.1 indicates that calcium phosphate is insoluble and that potassium bromide is soluble.

$$CaBr_2(aq) + K_3PO_4(aq) \rightarrow Ca_3(PO_4)_2(s) + KBr(aq)$$

Balance the reaction.

Start by counting each atom.

reactant atoms	product atoms
1 Ca atom	3 Ca atom
2 Br atoms	1 Br atoms
4 O atoms	8 O atoms
3 K atoms	1 K atoms
1 P atom	2 P atoms

When starting with a more complicated equation like this, it can help to start by balancing the atoms with the largest count. Multiply K_3PO_4 by 2 (2 K_3PO_4) and recount the atoms.

reactant atoms	product atoms
1 Ca atom	3 Ca atom
2 Br atoms	1 Br atoms
8 O atoms	8 O atoms
6 K atoms	1 K atoms
2 P atom	2 P atoms

Note that this balances the O atoms and the P atoms but not the K atoms. Next place a 6 in front of the KBr (6KBr) and recount the atoms.

reactant atoms	product atoms
1 Ca atom	3 Ca atom
2 Br atoms	**6** Br atoms
8 O atoms	8 O atoms
6 K atoms	**6** K atoms
2 P atom	2 P atoms

This balances the K atoms but makes 6 Br on the left and 2Br on the right. To balance the Br, place a 3 in front of the $CaBr_2$ (3$CaBr_2$) and recount the atoms.

reactant atoms	product atoms
3 Ca atom	3 Ca atom
6 Br atoms	**6** Br atoms
8 O atoms	8 O atoms
6 K atoms	**6** K atoms
2 P atom	2 P atoms

$$3CaBr_2(aq) + 2K_3PO_4(aq) \rightarrow Ca_3(PO_4)_2(s) + 6KBr(aq)$$

7.13   *In water, ammonium carbonate reacts with hydrogen chloride to produce carbonic acid, which falls apart to form water and carbon dioxide gas. Write balanced equations for these reactions.*

First translate the reactants and products from names to chemical formulas. Ammonium carbonate is $(NH_4)_2CO_3(aq)$, hydrogen chloride is $HCl(aq)$, carbonic acid is $H_2CO_3(aq)$, water is $H_2O(l)$, and carbon dioxide gas is $CO_3(g)$. Notice after you translate the formulas that one product is missing. The ammonium ion, NH_4^+, goes with the Cl^- ion to make $NH_4Cl(aq)$. Write the equation and then balance it.

$$(NH_4)_2CO_3(aq) + HCl(aq) \rightarrow H_2CO_3(aq) + NH_4Cl(aq)$$

Start by counting each atom.

reactant atoms	product atoms
2 N atoms	1 N atom
1 C atom	1 C atoms
3 O atoms	3 O atoms
9 H atoms	6 H atoms
1 Cl atom	1 Cl atom

Remember, a good place to start balancing the equation is with the atom that has the largest count, so place a 2 in front of the NH_4Cl (2 NH_4Cl) and recount the atoms.

reactant atoms	product atoms
2 N atoms	**2** N atom
1 C atom	1 C atoms
3 O atoms	3 O atoms
9 H atoms	**10** H atoms
1 Cl atoms	**2** Cl atoms

This balanced the N atoms but not the H atoms or Cl atoms, so place a 2 in front of the HCl (2HCl) and recount the atoms.

reactant atoms	product atoms
2 N atoms	**2** N atom
1 C atom	1 C atoms

$$3 \text{ O atoms} \qquad\qquad 3 \text{ O atoms}$$

$$\mathbf{10} \text{ H atoms} \qquad\qquad \mathbf{10} \text{ H atoms}$$

$$\mathbf{2} \text{ Cl atoms} \qquad\qquad \mathbf{2} \text{ Cl atoms}$$

This makes the atom count the same on both sides and now the balanced equation can be written.

$$(NH_4)_2CO_3\textit{(aq)} \;+\; 2HCl\textit{(aq)} \;\rightarrow\; H_2CO_3\textit{(aq)} \;+\; 2\,NH_4Cl\textit{(aq)}$$

In the second reaction, carbonic acid falls apart to from water and carbon dioxide gas.

$$H_2CO_3\textit{(aq)} \rightarrow \; H_2O\textit{(l)} \;+\; CO_2\textit{(g)}$$

This reaction equation is balanced.

7.15 *What happens to the solubility of carbon dioxide gas (CO_2) in water in each situation? (Answer as increase, decrease, or no change.)*

a. *The pressure of CO_2 over the solution is increased.*

The solubility of a gas in a liquid increases as the pressure above the liquid increases.

Increases

b. *The temperature is increased.*

The solubility of a gas in a liquid decreases as the temperature of the liquid increases.

Decreases

7.17 *Two unopened bottles of carbonated water are at the same temperature. If one is opened at the top of a mountain and the other at sea level, which will produce more bubbles? Explain.*

The bottle opened at the top of the mountain will produce more bubbles. The solubility of a gas in a liquid decreases and, since the atmospheric pressure is lower at the top of the mountain, more gas will bubble out of solution.

7.19 *When your body metabolizes amino acids, one of the final end products is urea, a water-soluble compound that is removed from the body in urine. Why is urea soluble in water, when hexanamide, a related compound, is not?*

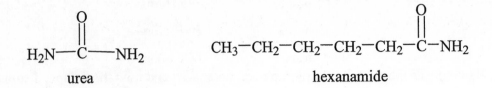

urea   hexanamide

Urea has more atoms capable of forming hydrogen bonds with water and does not contain the nonpolar chain of carbon atoms present in hexanamide.

7.21   *Explain why $CH_3CH_2OCH_2CH_3$ is less soluble in water than its constitutional isomer $CH_3CH_2CH_2CH_2OH$.*

The alcohol $CH_3CH_2CH_2CH_2OH$ is able to form more hydrogen bonds with water.

7.23   *Most general anesthetics have low water solubility, which is sometimes a factor in how long-acting the drugs are. Why do water-insoluble drugs tend to stay in the body for a longer time than water-soluble drugs?*

Water-soluble drugs more readily dissolve into the blood and are carried to the kidneys where they are removed as body waste in the urine.

7.25   *Vitamin D is produced in the skin upon exposure to sunlight. Based on its structure, is vitamin D hydrophilic, hydrophobic, or amphipathic?*

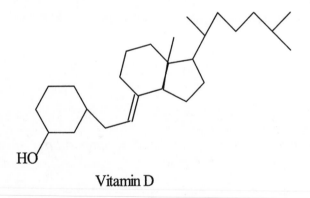

Vitamin D

First recall what each term means; hydrophilic—soluble in water, hydrophobic—insoluble in water, and amphipathic— has both hydrophilic and hydrophobic parts. Note that the vitamin D has an OH group on one end that would make it somewhat like water. However, the large nonpolar part of the molecule is far more dominant. That makes it hydrophobic.

Hydrophobic

7.27   *Some swimmers, football players, sprinters, and other athletes make illegal use of anabolic steroids to increase muscle strength and endurance. One of these drugs, nandrolone decanoate, is slowly released from its site of injection and is metabolized. Six to twelve months after its use, nandrolone decanoate can still be detected in urine. Why does this anabolic steroid remain in the fatty tissues of the body for such a long period of time?*

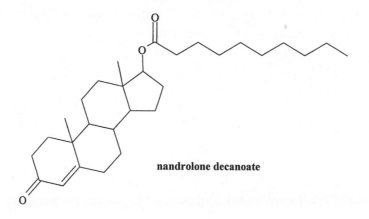

nandrolone decanoate

The drug is hydrophobic and stays dissolved in the fatty (nonpolar) tissues of the body.

7.29   *Explain how you would prepare a saturated aqueous solution of table sugar.*

With stirring, add table sugar to water. Continue until no more will dissolve.

7.31   *You carefully measure 1.0 tsp of milk into some coffee and end up with exactly 1.0 cup of liquid. What is the % (v/v) of milk in this aqueous solution? (See Table 1.1 and 1.2 for conversion factors.)*

In Tables 1.1 and 1.2 you find that 1 tsp = 5 mL and that 1 qt = 0.946 L (946 mL). From your use of English conversions, you should know that 4 cups = 1 qt. To find the volume of the solution, convert from cup to mL. Then you can use the mL equivalent of 1 tsp to calculate the % (v/v) for the solution.

1.0 ~~cup~~   x $\dfrac{1\ \text{qt}}{4\ \text{cups}}$ x $\dfrac{946\ \text{mL}}{\text{qt}}$ =   236.5 mL of solution

To find the percent by volume use the equation

% (v/v)   = $\dfrac{\text{volume of solute}}{\text{volume of solution}}$ x   100

$\dfrac{5.0\ \text{mL solute}}{236.5\ \text{mL of solution}}$ x   100   =   2.1 %(v/v)

$$\frac{160 \text{ ng}}{\text{L}} \quad \times \quad \frac{1 \times 10^{-9} \text{ g}}{\text{ng}} \quad \times \quad \frac{1 \times 10^{-3} \text{ L}}{\text{mL}} \quad = \quad 1.6 \times 10^{-10} \text{ g/mL}$$

2.1 %(v/v)

7.33 *The normal concentration range of vitamin B$_{12}$ in blood serum is 160-170 ng/L. Express this concentration range in parts per million and parts per billion.*

Since parts per million = (g of solute/mL) x 10^6 of solution and parts per billion = (g of solute/mL) x 10^9, convert ng to grams and liters to mL.

$$\frac{160 \text{ ng}}{\text{L}} \quad \times \quad \frac{1 \times 10^{-9} \text{ g}}{\text{ng}} \quad \times \quad \frac{1 \times 10^{-3} \text{ L}}{\text{mL}} \quad = \quad 1.6 \times 10^{-10} \text{ g/mL}$$

$$\frac{170 \text{ ng}}{\text{L}} \quad \times \quad \frac{1 \times 10^{-9} \text{ g}}{\text{ng}} \quad \times \quad \frac{1 \times 10^{-3} \text{ L}}{\text{mL}} \quad = \quad 1.7 \times 10^{-10} \text{ g/mL}$$

Parts per million = (1.6 x10^{-10} g/mL) x 10^6 = 1.6 x10^{-4} ppm and (1.7 x10^{-10} g/mL) x 10^6 = 1.7 x 10^{-4} ppm

Parts per billion = (1.6 x10^{-10} g/mL) x 10^9 = 1.6 x10^{-1} ppb and (1.7 x10^{-10} g/mL) x 10^9 = 1.7 x 10^{-1} ppb

Range in ppm = 1.6 x10^{-4}-1.7 x10^{-4} ppm (0.000016 – 0.00017 ppm)

Range in ppb = 1.6 x10^{-1}-1.7 x10^{-1} ppb (0.16 – 0.17 ppb)

7.35 *If 15.0 g of CaCl$_2$ are present in 250 mL of aqueous solution, what is the concentration in terms of the following?*

a. *molarity*

Molarity is equal to the moles of the solute divided by liters of solution (M = mol/L). Convert grams of CaCl$_2$ to moles and 250 mL to L.

$$\frac{15.0 \text{ g CaCl}_2}{250 \text{ mL}} \quad \times \quad \frac{\text{mol CaCl}_2}{111.1 \text{ g CaCl}_2} \quad \times \quad \frac{1 \text{ mL}}{1 \times 10^{-3} \text{ L}} \quad = \quad 0.54 \text{ mol/L CaCl}_2$$

0.54 M CaCl$_2$

b. *weight/volume percent*

Since percent by weight/volume is the weight of the solute (in grams) divided by the volume of the solution (in mL) x 100, no conversions are necessary.

$$\frac{15.0 \text{ g CaCl}_2}{250 \text{ mL}} \quad \times 100 \quad = 6.0 \text{ %(w/v)}$$

6.0 %(w/v) $CaCl_2$

c. *parts per thousand*

Parts per thousand is the weight of solute (in grams) divided by the volume of the solution (in mLs) x 1000.

$$\frac{15.0 \text{ g } CaCl_2}{250 \text{ mL}} \times 10^3 = 60 \text{ ppt}$$

60 ppt $CaCl_2$

d. *parts per million*

Parts per million is the weight of solute (in grams) divided by the volume of the solution (in mLs) multiplied by 10^6.

$$\frac{15.0 \text{ g } CaCl_2}{250 \text{ mL}} \times 10^6 = 6 \times 10^4 \text{ ppm}$$

6×10^4 ppm $CaCl_2$

e. *parts per billion*

Parts per billion is the weight of solute (in grams) divided by the volume of the solution (in mLs) multiplied by 10^9.

$$\frac{15.0 \text{ g } CaCl_2}{250 \text{ mL}} \times 10^9 = 6.0 \times 10^7 \text{ ppb}$$

6.0×10^7 ppb $CaCl_2$

7.37   *Calculate the molarity of each.*

a. *1.75 mole of NaCl in 15.2 L of solution*

Molarity (M) = moles of solute divided by liters of solution.

$$M = \frac{1.75 \text{mol NaCl}}{15.2 \text{ L}} = 0.115 \text{ mol/L NaCl}$$

0.115 mol/L NaCl or 0.115 M NaCl

b. *270 mg of NaCl in 1.00 mL of solution*

Since molarity (M) = mole of solute divided by liters of solution, convert mg to g, g to mol, and mL to L.

$$\frac{270 \; \text{mg NaCl}}{1.00 \; \text{mL}} \times \frac{1 \times 10^{-3} \; \text{g NaCl}}{\text{mg NaCl}} \times \frac{\text{mol NaCl}}{58.5 \; \text{g NaCl}} \times \frac{1 \; \text{mL}}{1 \times 10^{-3} \; \text{L}} = 4.6 \; \text{mol/L NaCl}$$

4.6 mol/L NaCl or 4.6 M NaCl

7.39 *Lead is a toxic metal that can delay mental development in babies. The Environmental Protection Agency's action level (that level where remedial action must be taken to clean the water) is 15 ppb. Does a 50 mL sample of water that contains 0.35 µg of lead fall above or below this action level?*

First convert the 0.35 mg of lead to grams and then divide that by the 50 mL. Then multiply this by 10^9 to get ppb. Compare your answer to the allowable EPA action level.

$$\frac{0.35 \; \text{µg Pb}}{50 \; \text{mL}} \times \frac{1 \times 10^{-6} \; \text{g}}{\text{µg}} \times 10^9 = 7 \; \text{ppb lead}$$

7 ppb lead; below the action level.

7.41 *Thyroxine, a thyroid hormone, is present in normal blood serum at 58-167 nmol/L. What is the molar concentration of lactate in serum that contains 150 nmol of thyroxine per liter?*

Since molarity is defined as mol per liter, convert nmol to mol.

$$\frac{150 \; \text{nmol}}{\text{L}} \times \frac{1 \times 10^{-9} \; \text{mol}}{\text{nmol}} = 1.5 \times 10^{-7} \; \text{mol/L}$$

1.5×10^{-7} mol/L thyroxine or 1.5×10^{-7} M thyroxine

7.43 *The serum concentration of cortisol, a hormone, is expected to be in the range 8-20 mg/dL. What is the part per million concentration of cortisol in serum that contains 10 mg of cortisol per deciliter?*

Parts per million is the weight of solute (in grams) divided by the volume (in mL) of the solution multiplied by 10^6. Convert mg to g and dL to mL, then multiply by 10^6.

$$\frac{10 \; \text{mg}}{\text{dL}} \times \frac{1 \times 10^{-3} \; \text{g}}{\text{mg}} \times \frac{1 \times 10^{-1} \; \text{dL}}{\text{mL}} \times 10^6 = 100 \; \text{ppm}$$

100 ppm cortisol

7.45 *The normal serum concentration of potassium ion (K⁺) is 3.5-4.9 mEq/L. Convert this concentration range into mmol/L.*

For an ion with a 1+ charge, 1 mol = 1 Eq, so 1 mmol = 1 mEq.

$$\frac{3.5\text{-}4.9 \text{ mEq}}{\text{L}} \quad \times \quad \frac{1 \text{ mmol}}{1 \text{ mEq}} \quad = \quad 3.5\text{-}4.9 \text{ mmol/L}$$

3.5-4.9 mmol/L

7.47 *The normal serum concentration of chloride ion (Cl⁻) is 95-107 mmol/L. Convert this concentration range into mEq/L.*

For an ion with a 1- charge 1 mol = 1 Eq, so 1 mmol = 1 mEq.

$$\frac{95\text{-}107 \text{ mmol}}{\text{L}} \quad \times \quad \frac{1 \text{ mEq}}{1 \text{ mmol}} \quad = \quad 95\text{-}107 \text{ mEq/L}$$

95-107 mEq/L

7.49 *How many milliequivalents of bicarbonate (HCO_3^-) are present in a 75.0 mL blood serum sample with a concentration of 25 mEq/L of HCO_3^-?*

Convert mL to L and multiply by the concentration in mEq/L.

$$75.0 \text{ mL} \quad \times \quad \frac{1 \times 10^{-3} \text{ L}}{\text{mL}} \quad \times \quad \frac{25 \text{ mEq}}{1 \text{ L}} \quad = \quad 1.9 \text{ mEq}$$

1.9 mEq HCO_3^-

7.51 *How many moles of sodium ion (Na⁺) are present in a 10.0 mL blood serum sample with a concentration of 132 mEq/L.*

Convert mL to L and mmol to mEq (for Na⁺, 1 mmol = mEq) and then convert mmol to mol.

$$10.0 \text{ mL} \times \frac{1 \times 10^{-3} \text{ L}}{10^3 \text{ mL}} \times \frac{132 \text{ mEq}}{1 \text{ L}} \times \frac{1 \text{ mmol}}{1 \text{ mEq}} \times \frac{1 \times 10^{-3} \text{ mol}}{\text{mmol}} = 1.32 \times 10^{-3} \text{ mol}$$

1.32×10^{-3} mol Na⁺

7.53 *If 15.0 mL of 3.0 M HCl are diluted to a final volume of 100.0 mL, what is the new concentration?*

Recall the dilution equation is $V_{original} \times C_{original} = V_{final} \times C_{final}$. To use this equation it is not necessary to have a particular volume or concentration unit, provided that both volume units are the same and that both concentration units are the same. The answer will have the same unit as the value that is stated in the problem.

$$V_{original} \times C_{original} = V_{final} \times C_{final}$$

$$C_{final} = \frac{(V_{original} \times C_{original})}{V_{final}}$$

$$C_{final} = \frac{(15.0 \text{ mL} \times 3.0 \text{ M})}{100.0 \text{ mL}} = 0.45 \text{ M}$$

0.45 M

7.55   *A 10.0% (w/v) solution of ethanol is diluted from 50.0 mL to 200.0 mL. What is the new weight/volume percent?*

Recall the dilution equation is $V_{original} \times C_{original} = V_{final} \times C_{final}$.

$$V_{original} \times C_{original} = V_{final} \times C_{final}$$

$$C_{final} = \frac{(V_{original} \times C_{original})}{V_{final}}$$

$$C_{final} = \frac{(50.0 \text{ mL} \times 10\%)}{200.0 \text{ mL}} = 2.50 \text{ \%(w/v)}$$

2.50 %(w/v)

7.57   *How many milliliters of 2.00 M NaOH are needed to prepare 300.0 mL of 1.50 M NaOH?*

Recall the dilution equation is $V_{original} \times C_{original} = V_{final} \times C_{final}$.

$$V_{original} \times C_{original} = V_{final} \times C_{final}$$

$$V_{final} = \frac{(V_{original} \times C_{original})}{C_{final}}$$

$$V_{final} = \frac{(300.0 \text{ mL} \times 1.50 \text{ M})}{2.00 \text{ M}} = 225 \text{ mL}$$

225 mL

7.59   *Calculate the final volume required to prepare each solution.*

a. *Starting with 100.0 mL of 1.00 M KBr, prepare 0.500 M KBr.*

Recall the dilution equation is $V_{original} \times C_{original} = V_{final} \times C_{final}$.

$V_{original} \times C_{original} = V_{final} \times C_{final}$

$$V_{final} = \frac{(V_{original} \times C_{original})}{C_{final}}$$

$$V_{final} = \frac{(100.0 \text{ mL} \times 1.00 \text{ M})}{0.500 \text{ M}} = 200 \text{ mL}$$

200 mL

b. *Starting with 50.0 mL of 0.250 M alanine (an amino acid), prepare 0.110 M alanine.*

Recall the dilution equation is $V_{original} \times C_{original} = V_{final} \times C_{final}$.

$V_{original} \times C_{original} = V_{final} \times C_{final}$

$$V_{final} = \frac{(V_{original} \times C_{original})}{C_{final}}$$

$$V_{final} = \frac{(50 \text{ mL} \times 0.25 \text{ M})}{0.11 \text{ M}} = 114 \text{ mL}$$

114 mL

7.61   *When 1.0 mL of water sample was diluted to 5.0 L, the new solution was found to have a lead concentration of 5.3 ppb. What was the concentration of lead (in ppm) in the original sample?*

Remember that units must be the same in order to use them in the dilution equation. First convert 1.0 mL to L and 5.3 ppb to ppm. Put these values in the dilution equation $V_{original} \times C_{original} = V_{final} \times C_{final}$ and solve for the final concentration.

$$1.0 \text{ mL} \times \frac{1 \times 10^{-3} \text{ L}}{\text{mL}} = 1 \times 10^{-3} \text{ L} = V_{original}$$

$$5.3 \text{ ppb} \times \frac{1 \text{ ppm}}{10^{3} \text{ ppb}} = 5.3 \times 10^{-3} \text{ ppm} = C_{final}$$

$$V_{original} \times C_{original} = V_{final} \times C_{final}$$

$$C_{original} = \frac{V_{final} \times C_{final}}{V_{original}}$$

$$C_{original} = \frac{5.3 \times 10^{-3} \text{ ppm} \times 5.0 \cancel{L}}{1 \times 10^{-3} \cancel{L}} = 27 \text{ ppm}$$

27 ppm

7.63   *How is a colloid different from a solution?*

The particles that make up a colloid are smaller than those present in a suspension. While suspensions settle, colloids do not.

7.65   *Give an example of each.*

a. a solution

Answers may vary but examples are sugar in tea, salt in water, and CO_2 in soda.

b. a suspension

Answers may vary but examples are muddy water and smoke.

c. a colloid

Answers may vary but examples are soapy water, paint, clay, and mustard.

7.67   *A process called active transport moves certain ions and compounds across cell membranes from areas of lower concentration to areas of higher concentration. Does active transport involve diffusion? Explain.*

No. In active transport solutes move in a direction opposite to diffusion.

7.69   *Scale can be removed from a tea kettle by adding some vinegar (a source of H^+) and heating. Write a balanced equation for the reaction of scale (calcium carbonate) and H^+ to produce aqueous calcium ion, carbon dioxide gas, and water.*

First, write the chemical formulas for the reactants and products. Calcium carbonate is $CaCO_3(s)$, hydrogen ion is $H^+(aq)$, calcium ion is Ca^{2+}, water is $H_2O(l)$, and carbon dioxide gas is $CO_2(g)$. Write the equation and then balance it.

$$CaCO_3(s) + H^+(aq) \rightarrow H_2O(l) + CO_2(g) + Ca^{2+}(aq)$$

Start by counting each atom.

reactant atoms	product atoms
1 Ca atom	1 Ca atom
1 C atom	1 C atoms
3 O atoms	3 O atoms
1 H atoms	2 H atoms

The only count that is not the same are the H atoms. Place a 2 in front of the H^+ ($2H^+$) and recount the atoms.

reactant atoms	product atoms
1 Ca atom	1 Ca atom
1 C atom	1 C atoms
3 O atoms	3 O atoms
2 H atoms	2 H atoms

The atom count is the same. The equation is balanced.

$$CaCO_3(s) \ + \ 2H^+(aq) \ \rightarrow \ H_2O(l) \ + \ CO_2(g) \ + \ Ca^{2+}(aq)$$

7.71   *Phenacetin is a prodrug that, in the liver, is converted into acetaminophen, a commonly used pain and fever reducer. While available in many parts of the world, phenacetin is not sold in the U.S. due to concerns about its toxicity. What is the theoretical yield of acetaminophen (in grams) upon reaction of the 325 mg of phenacetin contained in one tablet?*

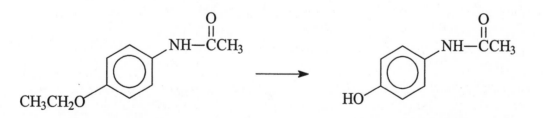

Phenacetin   **Acetaminophen**

According to the equation above, the conversion of phenacetin to acetaminophen has a one mole to one mole relationship. Convert 325 mg to g (grams) and then convert from g of phenacetin to mol of phenacetin using the molecular weight of phenacetin ($C_{10}H_{13}NO_2$). Then use the mole to mole relationship between phenacetin and acetaminophen to convert to mole of acetaminophen. Finally

411

convert moles of acetaminophen to grams using the molecular weight of acetaminophen ($C_8H_9NO_2$).

$$325 \; \cancel{mg} \; C_{10}H_{13}NO_2 \quad \times \quad \frac{1 \times 10^{-3} \; g}{\cancel{mg}} \quad = \quad 0.325 \; g \; C_{10}H_{13}NO_2$$

$$0.325 \; \cancel{g \; C_{10}H_{13}NO_2} \quad \times \quad \frac{1 \; mol \; C_{10}H_{13}NO_2}{179 \; \cancel{g \; C_{10}H_{13}NO_2}} \quad = \quad 1.82 \times 10^{-3} \; mol \; C_{10}H_{13}NO_2$$

$$1.82 \times 10^{-3} \; \cancel{mol \; C_{10}H_{13}NO_2} \quad \times \quad \frac{1 \; mol \; C_7H_6O_3}{1 \; \cancel{mol \; C_{10}H_{13}NO_2}} \quad = \quad 1.82 \times 10^{-3} \; mol \; C_8H_9NO_2$$

$$1.82 \times 10^{-3} \; \cancel{mol \; C_8H_9NO_2} \quad \times \quad \frac{151 \; \cancel{g \; C_8H_9NO_2}}{1 \; \cancel{mol \; C_8H_9NO_2}} \quad = \quad 0.274 \; g \; C_8H_9NO_2$$

Theoretical yield of acetaminophen (in grams) = 0.274 g $C_8H_9NO_2$

7.73 *Kidney disease can lead to nephrotic syndrome, which is characterized, in part, by a low blood serum concentration of albumin (a protein). Explain why low serum albumin levels result in edema, swelling caused by the movement of fluid from the blood into tissue.*

Osmosis causes the movement of water from blood serum (lower solute concentration) to tissues (higher solute concentration).

7.75 *When given to a patient, the drug allopurinol is converted to oxypurinol. Oxypurinol blocks the action of xanthine oxidase, an enzyme that catalyzes one of the steps in the production of uric acid from purines.*

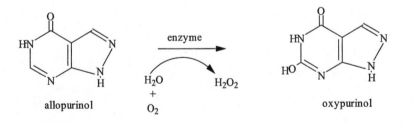

a. *Is allopurinol oxidized or reduced when it is converted to oxypurinol?*

Recall from Chapter 6 that an atom is reduced if it loses oxygen and/or gains hydrogen. The atom is oxidized if it gains oxygen and/or loses hydrogen. In this case, the allopurinol carbon that has double bond with the N atom is losing a hydrogen atom and gaining an OH. This means that it is oxidized.

Oxidized

b. *Might allopurinol be considered a prodrug? (Refer to Health Link: Prodrugs.)*

Allopurinol itself is inactive, but it is converted to an active compound in the body.

Yes.

c. *Suggest an explanation for the fact that oxypurinol is better at blocking xanthine oxidase than is allopurinol. (Hint: look at the structures in the reactions presented just before the chapter summary.)*

The structure of oxypurinol more closely resembles that of xanthine, so it is more likely to interact with xanthine oxidase.

Chapter 8
Lipids and Membranes

Solutions to Problems

8.1   *Why is the melting point of lauric acid (Table 8.1) lower than that of myristic acid?*

Lauric acid has a shorter hydrocarbon tail than mystic acid, so London force interactions between lauric acid molecules are weaker.

8.3   *Draw line-bond structures of the following (see Table 8.1).*

The formula for lauric acid is given on Table 8.1 as $CH_3(CH_2)_{10}CO_2H$. Place a single line to attach each atom except one oxygen in the acid functional group has a double bond.

a.   Lauric acid.

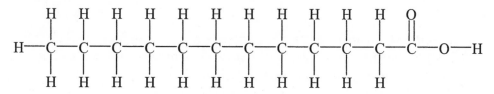

b.   Linolenic acid.

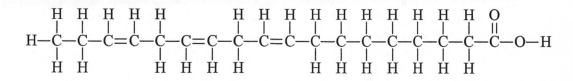

8.5   *Sodium palmitate, $CH_3(CH_2)_{14}CO_2^-Na^+$, has a higher melting point than palmitic acid, $CH_3(CH_2)_{14}CO_2H$. Why?*

The ionic bonds that hold the ions in sodium palmitate to one another are stronger than the noncovalent interactions that hold one palmitic acid molecule to another.

8.7   *In terms of structure, what distinguishes fatty acids from the carboxylic acids that were discussed in Chapter 4?*

Fatty acids are carboxylic acids that typically have between 12 and 20 carbon atoms.

415

8.9 *Draw skeletal structures for the products formed when carnauba wax (Table 8.2) is saponified (hydrolyzed).*

First look up the formula for carnauba wax on Table 8.2 and draw the molecule. Note that the CO_2 is in the middle of the formula, which tells you this is an ester. Hydrolysis breaks the ester bond.

Saponification, the hydrolysis of an ester in the presence of OH^- forms a carboxylate ion and an alcohol.

and

8.11 *The fragrance of spermaceti, a wax produced by whales, once made it important to the perfume industry. One of the main constituents of spermaceti is cetyl palmitate, which is formed from palmitic acid and cetyl alcohol, $CH_3(CH_2)_{14}CH_2OH$. Draw the condensed structure of cetyl palmitate.*

First find the formula for palmitic acid, $CH_3(CH_2)_{14}CO_2H$ in Table 8.1. Next remember that the reaction between a carboxylic acid and an alcohol forms an ester. Remove the –OH from the acid and the –H from the alcohol and join the two ends together.

$$CH_3(CH_2)_{14}\overset{\overset{\textstyle O}{\|}}{C}-OCH_2(CH_2)_{14}CH_3$$

8.13 *Draw the fatty acid and alcohol from which the following kingfisher green wax is made.*

Note that the CO_2 is in the middle of the formula, which tells you this is an ester.

$$CH_3CH_2CH_2\underset{\underset{\textstyle CH_3}{|}}{C}HCH_2\underset{\underset{\textstyle CH_3}{|}}{C}HCH_2\underset{\underset{\textstyle CH_3}{|}}{C}HCH_2\underset{\underset{\textstyle CH_3}{|}}{C}HCH_2\overset{\overset{\textstyle O}{\|}}{C}-OCH_2\underset{\underset{\textstyle CH_3}{|}}{C}HCH_2\underset{\underset{\textstyle CH_3}{|}}{C}HCH_2\underset{\underset{\textstyle CH_3}{|}}{C}HCH_2\underset{\underset{\textstyle CH_3}{|}}{C}HCH_2CH_2CH_3$$

The ester is formed from the following alcohol and carboxylic (fatty) acid:

$$HOCH_2CHCH_2CHCH_2CHCH_2CHCH_2CH_2CH_3$$
$$\quad\quad\ \ |\quad\quad\ |\quad\quad\ |\quad\quad\ |$$
$$\quad\quad CH_3\quad CH_3\quad CH_3\quad CH_3$$

and

$$CH_3CH_2CH_2CHCH_2CHCH_2CHCH_2CHCH_2\overset{\displaystyle O}{\overset{\|}{C}}-OH$$
$$\quad\quad\quad\ |\quad\quad\ |\quad\quad\ |\quad\quad\ |$$
$$\quad\quad\ CH_3\quad CH_3\quad CH_3\quad CH_3$$

8.15 *Write a balanced reaction equation for the complete hydrogenation of palmitoleic acid (Table 8.2).*

Hydrogenation is the process of adding H_2 to carbon atoms of a double bond.

$$CH_3(CH_2)_5CH=CH(CH_2)_7CO_2H \quad \overset{H_2}{\underset{Pt}{\rightarrow}} \quad CH_3(CH_2)_5CH_2CH_2(CH_2)_7CO_2H$$

8.17 *Draw a triglyceride made from glycerol, myristic acid, palmitic acid, and oleic acid. Would you expect this triglyceride to be a liquid or a solid at room temperature? Explain.*

Triglycerides are composed of three fatty acid residues and a glycerol residue. Look up the formulas for each of the fatty acids and attach them to the glycerol molecule in the same way that you drew the esters in the previous problems.

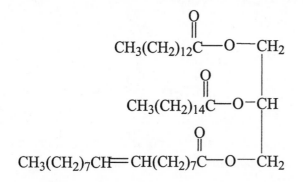

This triglyceride contains more saturated than unsaturated fatty acid residues, so it should be a solid at room temperature.

417

8.19 *Draw the products formed when the triglyceride is saponified.*

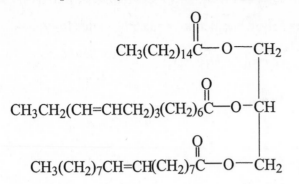

Saponification is another term for ester hydrolysis in the presence of OH^-. To show the products, take the molecule apart the same as any other ester. The products will be the three fatty acid anions and glycerol.

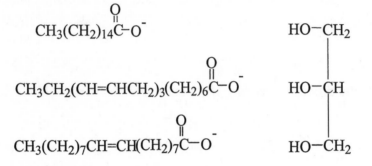

8.21 *Vegetable oils tend to become rancid more rapidly than do animal fats. Why?*

Vegetable oils have more carbon-carbon double bonds than animal fats. It is the oxidation of these bonds that causes oils to become rancid.

8.23 a. *To which class of phospholipids does the compound belong?*

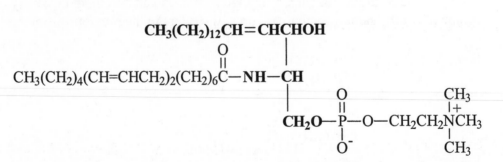

The class sphingolipids is recognized by the incorporation of a sphingosine residue into the structure of the phospholipid, as bolded in the original molecule above.

Sphingolipid

b. *Draw the products obtained when the molecule is saponified into its component parts.*

Saponification results in four products: sphingosine, or fatty acid anion, phosphate, and an alcohol.

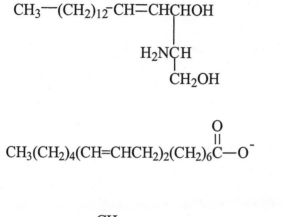

$CH_3(CH_2)_4(CH=CHCH_2)_2(CH_2)_6\overset{\overset{O}{\|}}{C}-O^-$

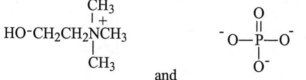

and

8.25   *Lecithin (phosphatidylcholine in Figure 8.15) is an emulsifying agent, but triglycerides are not. Account for the difference.*

Lecithin is amphipathic and triglycerides are hydrophobic. Being amphipathic, lecithins can interact with both nonpolar compounds (oil) and polar ones (water) to keep an emulsion from separating.

8.27   *Draw a sphingomyelin that contains palmitoleic acid.*

Identify the palmitoleic acid formula given in Table 8.1.

$$CH_3(CH_2)_5CH=CH(CH_2)_7\overset{\overset{O}{\|}}{C}OH$$

See Figure 8.1b for general structure of a sphingomyelin and replace the fatty acid residue with one of palmitoleic acid.

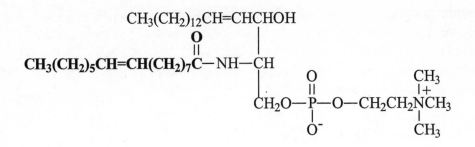

$$CH_3(CH_2)_{12}CH=CHCHOH$$

$$CH_3(CH_2)_5CH=CH(CH_2)_7\overset{\overset{O}{\|}}{C}-NH-CH$$

$$CH_2O-\overset{\overset{O}{\|}}{\underset{\underset{O^-}{|}}{P}}-O-CH_2CH_2\overset{\overset{CH_3}{|+}}{\underset{\underset{CH_3}{|}}{N}CH_3}$$

8.29   *Label the hydrophobic and hydrophilic parts of taurocholate (Figure 8.19).*

Hydrophilic parts will have many –OH, –NH, or –O bonds while hydrophobic will be mostly long chains of hydrocarbons.

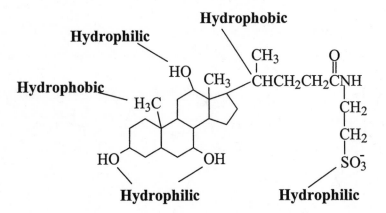

8.31   *Which molecule is the starting point for the synthesis of sex hormones and bile salts?*

Cholesterol

8.33   *Name one male sex hormone and one female sex hormone.*

Male sex hormone – testosterone; female sex hormone – progesterone.

8.35   a. *How do nonsteroidal anti-inflammatory drugs (NSAIDs) act to reduce pain, fever, and swelling?*

They block the action of COX-1 and COX-2, enzymes involved in the conversion of arachidonic acid into prostaglandins and thromboxanes.

b. *How is the action of Celebrex different from that of NSAIDs?*

Celebrex only blocks the action of COX-2.

8.37   *How are facilitated diffusion and active transport different?*

Facilitated diffusion moves substances from areas of higher concentration to areas of lower concentration, and does not require the input of energy. Active transport moves substances in the opposite direction and requires the input of energy.

8.39   *How are facilitated diffusion and diffusion different?*

In facilitated diffusion, diffusion across a membrane takes place with the assistance of proteins.

8.41   *To function properly, membranes must be flexible or fluid. In light of this fact, propose an explanation of why the cell membranes in the feet and legs of a reindeer contain a higher percentage of unsaturated fatty acids than do the cell membranes in the interior of its body.*

Reindeer are typically found in cold, snow covered regions. Since the melting point of fatty acids goes down as the degree of unsaturation goes up, having more unsaturated fatty acids in the feet and legs would reduce the likelihood that the fats would solidify. This would help keep the membranes more flexible.

8.43   *Draw a trans fat that might form if the following triglyceride is subjected to partial hydrogenation*

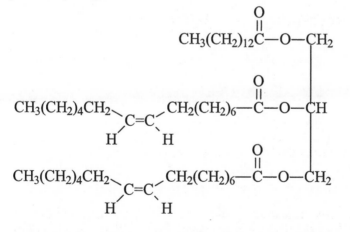

Hydrogenation is the chemical reaction that involves breaking the double bond between two carbons and adding a hydrogen atom to each carbon.

Since the question specifies partial hydrogenation, only one of the double bonds would be hydrogenated. The other bond undergoes *cis* to *trans* conversion For this triglyceride that gives two options as highlighted in the answer.

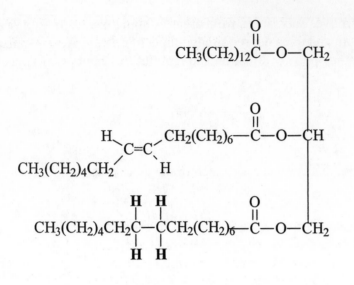

or

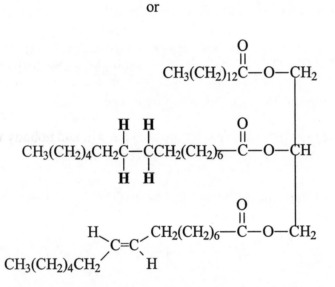

8.45 *Olestra passes through the digestive system untouched, taking some dietary vitamin A, vitamin E, and other fat-soluble (water-insoluble) vitamins with it. Which noncovalent interaction allows these vitamins to be associated with olestra?*

London forces

8.47 *In the body, androstenedione is converted into testosterone. Is androstenedione oxidized or is it reduced in this process?*

Reduction is typically observed in organic compounds as the gain of hydrogen and/or loss of oxygen. Oxidation is the opposite process: a loss of hydrogen and/or a gain of oxygen.

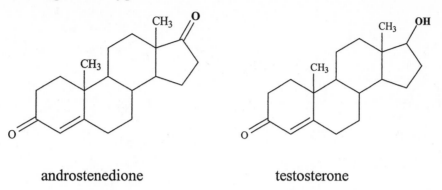

androstenedione   testosterone

As androstenedione is converted to testosterone a hydrogen is added to the structure which means that reduction has occurred.

Reduced

8.51   *Suppose that you are an athlete and that you have just been told that a new performance enhancing synthetic has been developed. If this steroid would never be detectable in drug tests would you take it? Why or why not?*

Answers will vary and may include: Yes, it may help me win and nobody will ever know. No, it would be cheating and may damage my health.

Chapter 9
Acids, Bases, and Equilibrium

Solutions to Problems

9.1   *Based on taste, is it possible to tell whether a food contains acids or bases? Explain.*

Yes. Acids have a sour taste and bases have a bitter taste.

9.3   *Name each of the following as an acid and as a binary compound.*

The names of acids that are binary compounds (they contain hydrogen and only one other nonmetal atom) always start with hydro and end with the nonmetal name, with the ending changed to "ic" instead of the usual "ide" ending. When naming the acid as a binary compound, use the same naming rules learned previously in Chapter 3 but do not put mono, di, tri, etc. in front of the name of the other nonmetal.

a. *HF*

Hydrofluoric acid and hydrogen fluoride

b. *HI*

Hydroiodic acid and hydrogen iodide

9.5   *Give the formula for the conjugate base of each acid.*

Remember that acids release H^+ ions. To obtain the conjugate base, remove one H^+ from the formula of the acid. Note that removing H^+ decreases the charge of the species remaining by one.

a. *HSO_4^-*

SO_4^{2-}

b. *H_3PO_4*

$H_2PO_4^-$

c. *HPO_4^{2-}*

PO_4^{3-}

d. HNO_3

NO_3^-

9.7   *HCO_3^- is amphoteric.*

   a. *Write the equation for the reaction that takes place between HCO_3^- and the acid HCl.*

   The acid HCl donates H^+ to the base HCO_3^-.

   HCO_3^- *(aq)* + HCl*(aq)* → H_2CO_3*(aq)* + Cl^-*(aq)*

   b. *Write the equation for the reaction that takes place between HCO_3^- and the base OH^-.*

   The acid HCO_3^- donates H^+ to the base OH^-.

   HCO_3^- *(aq)* + OH^- *(aq)* → CO_3^{2-} *(aq)* + H_2O*(l)*

9.9   *Identify the Bronsted-Lowry acids and bases for the forward and reverse reactions of each.*

   a. *F^- (aq) + HCl (aq) $\rightleftharpoons$ HF(aq) + Cl^- (aq)*

   Bronsted-Lowry acids donate H^+ and Bronsted-Lowry bases accept H^+. In the forward reaction, HCl gives up H^+ to F^-. In the reverse reaction, HF donates H^+ to Cl^-.

   F^-*(aq)* + HCl *(aq)* $\rightleftharpoons$ HF*(aq)* + Cl^- *(aq)*
   base   acid   acid   base

   b. *CH_3CO_2H (aq) + NO_3^- (aq) $\rightleftharpoons$ $CH_3CO_2^-$ (aq) + HNO_3 (aq)*

   Bronsted-Lowry acids donate H^+ and Bronsted-Lowry bases accept H^+. In the forward reaction, CH_3CO_2H donates H^+ to NO_3^-. In the reverse reaction, HNO_3 donates H^+ to $CH_3CO_2^-$.

   CH_3CO_2H *(aq)* + NO_3^- *(aq)* $\rightleftharpoons$ $CH_3CO_2^-$ *(aq)* + HNO_3 *(aq)*
          acid   base   base   acid

9.11   *Which of the following statements are correct at equilibrium?*

   a. *The concentration of reactants is always equal to the concentration of products.*

b. *No reactants are converted into products.*
c. *The rate of the forward reaction is equal to the rate of the reverse reaction.*

Depending on the reaction, the equilibrium concentration of reactants can be greater than or less than product concentration. Answer (a) is not true. Answer (b) is not true because at equilibrium the reaction continues to take place. The rate of the forward and reverse reactions are equal. The answer to (c) is true. (See the previous sentence.)

c. The rate of the forward reaction is equal to the rate of the reverse reaction.

9.13 *Write the equilibrium constant expression for each reaction. In part b, H_2O is the solvent.*

For the generalized reaction

$$aA + bB \rightleftharpoons cC + dD$$

the equilibrium constant expression is written:

$$K_{eq} = \frac{[C]^c[D]^d}{[A]^a[B]^b}$$

a. $C(s) + CO_2(g) \rightleftharpoons 2CO(g)$

Remember that the concentration of solids and solvents are not included in K_{eq}.

$$K_{eq} = \frac{[CO]^2}{[CO_2]}$$

b. $NH_3(aq) + H_2O(l) \rightleftharpoons NH_4^+(aq) + OH^-(aq)$

Remember that the concentration of solids and solvents are not included in K_{eq}.

$$K_{eq} = \frac{[NH_4^+][OH^-]}{[NH_3]}$$

9.15 *Write the reaction equation from which each equilibrium constant expression is derived (assume that no solids or solvents are present).*

The generalized equilibrium constant expression

$$K_{eq} = \frac{[C]^c[D]^d}{[A]^a[B]^b}$$

is based on the reaction equation

$$aA + bB \rightleftharpoons cC + dD$$

a. $K_{eq} = \dfrac{[NOCl]^2}{[NO]^2[Cl_2]}$

$$2\ NO + Cl_2 \rightleftharpoons 2\ NOCl$$

b. $K_{eq} = \dfrac{[HBr]^2}{[H_2][Br_2]}$

$$H_2 + Br_2 \rightleftharpoons 2\ HBr$$

9.17   *K_{eq} for the reaction below has a value of 4.2×10^{-4}.*

$$CH_3NH_2(aq) + H_2O(l) \rightleftharpoons CH_3NH_3^+(aq) + OH^-(aq)$$

Which are there more of at equilibrium, products or reactants?

Recall that the equilibrium constant is [products]/[reactants]. Since the K_{eq} value is very small this tells you that the reactant concentrations are much larger than the product concentrations.

Reactants.

9.19   *The enzyme carbonic anhydrase catalyzes the rapid conversion of H_2CO_3 into CO_2 and H_2O.*

$$CO_2(g) + H_2O(l) \rightleftharpoons H_2CO_3(aq)$$

a. Write the equilibrium constant expression for this reaction.

See the solution to Problem 9.13 for a description of K_{eq}. Remember that the solvent concentration is not included in K_{eq}.

$K_{eq} = \dfrac{[H_2CO_3]}{[CO_2]}$

b. What effect, if any, does doubling the amount of carbonic anhydrase have on an equilibrium mixture of H_2CO_3, CO_2, and H_2O? Explain.

No effect. The catalyst increases the rate of the forward and reverse reactions to the same extent.

9.21   *When carbon monoxide reacts with hydrogen gas, methanol (CH₃OH) is formed.*

$$CO(g) + 2H_2(g) \rightleftharpoons CH_3OH(g)$$

a. *For the equilibrium above, what is the effect of increasing [CO]? Of decreasing [H₂]? Of increasing [CH₃OH]?*

Increasing [CO] favors the forward reaction (makes more products). **Decreasing [H₂] slows the rate of the forward reaction** (favors the reverse reaction and makes more reactants). **Increasing [CH₃OH] increases the rate of the reverse reaction** (favors the reverse reaction and makes more reactants).

b. *What would be the effect of continually removing CH₃OH from the reaction?*

The reaction would continually move to make CH₃OH and would never reach equilibrium.

9.23   *Calculate the H₃O⁺ concentration present in water when*

a. *[OH] = 4.8 x 10⁻⁸ M*

$K_w = 1.0 \times 10^{-14} = [H_3O^+][OH^-]$

$[H_3O^+] = 1.0 \times 10^{-14} / [OH^-]$

$[H_3O^+] = 1.0 \times 10^{-14} / [4.8 \times 10^{-8}]$

$[H_3O^+] = 2.1 \times 10^{-7}$ M

b. *[OH] = 6.6 x 10⁻² M*

$K_w = 1.0 \times 10^{-14} = [H_3O^+][OH^-]$

$[H_3O^+] = 1.0 \times 10^{-14} / [OH^-]$

$[H_3O^+] = 1.0 \times 10^{-14} / [6.6 \times 10^{-2}]$

$[H_3O^+] = 1.5 \times 10^{-13}$ M

c. *[OH] = 1.5 x 10⁻¹² M*

$K_w = 1.0 \times 10^{-14} = [H_3O^+][OH^-]$

$[H_3O^+] = 1.0 \times 10^{-14} / [OH^-]$

$[H_3O^+] = 1.0 \times 10^{-14} / [1.5 \times 10^{-12}]$

$\mathbf{[H_3O^+] = 6.7 \times 10^{-3} \text{ M}}$

9.25   *In Problem 23, indicate whether each solution is acidic, basic, or neutral.*

Acidic solutions have $[H_3O^+]$ greater than 1×10^{-7} M. Basic solutions have $[H_3O^+]$ less than 1×10^{-7} M. Neutral solutions have $[H_3O^+]$ equal to 1×10^{-7} M.

a. $[H_3O^+] = 2.1 \times 10^{-7}$ M, which is greater than 1×10^{-7} M, so the solution is acidic.

Acidic

b. $[H_3O^+] = 1.5 \times 10^{-13}$ M, which is less than 1×10^{-7} M, so the solution is basic.

Basic

c. $[H_3O^+] = 6.7 \times 10^{-3}$ M, which is greater than 1×10^{-7} M, so the solution is acidic.

Acidic

9.27   *Calculate the OH⁻ concentration present in water when*

a. *$[H_3O^+] = 6.2 \times 10^{-4}$ M*

$K_w = 1.0 \times 10^{-14} = [H_3O^+][OH^-]$

$[OH^-] = 1.0 \times 10^{-14} / [H_3O^+]$

$[OH^-] = 1.0 \times 10^{-14} / [6.2 \times 10^{-4} \text{ M}]$

$\mathbf{[OH^-] = 1.6 \times 10^{-11} \text{ M}}$

b. *$[H_3O^+] = 8.5 \times 10^{-8}$ M*

$K_w = 1.0 \times 10^{-14} = [H_3O^+][OH^-]$

$[OH^-] = 1.0 \times 10^{-14} / [H_3O^+]$

$[OH^-] = 1.0 \times 10^{-14} / [8.5 \times 10^{-8}]$

$\mathbf{OH^-] = 1.2 \times 10^{-7} \text{ M}}$

c. $[H_3O^+] = 1.9 \times 10^{-11} M$

$K_w = 1.0 \times 10^{-14} = [H_3O^+][OH^-]$

$[OH^-] = 1.0 \times 10^{-14} / [H_3O^+]$

$[OH^-] = 1.0 \times 10^{-14} / [1.9 \times 10^{-11}]$

$[OH^-] = 5.3 \times 10^{-4} M$

9.29   *In Problem 9.27, indicate whether each solution is acidic, basic, or neutral.*

Acidic solutions have $[H_3O^+]$ greater than 1×10^{-7} M. Basic solutions have $[H_3O^+]$ less than 1×10^{-7} M. Neutral solutions have $[H_3O^+] = 1 \times 10^{-7}$ M. Refer to Problem 9.27 for $[H_3O^+]$ concentrations.

a. $[H_3O^+] = 6.2 \times 10^{-4}$ M, which is greater than 1×10^{-7} M, so the solution is acidic.

Acidic

b. $[H_3O^+] = 8.5 \times 10^{-8}$ M, which is less than 1×10^{-7} M, so the solution is basic.

Basic

c. $[H_3O^+] = 1.9 \times 10^{-11}$ M, which is less than 1×10^{-7} M, so the solution is basic.

Basic

9.31   *Calculate the pH of a solution in which*

a. $[H_3O^+] = 1 \times 10^{-7} M$

$pH = -\log [H_3O^+]$

$pH = -\log (1 \times 10^{-7})$ On a scientific calculator, enter the number 1×10^{-7}, then press the **log** button. This gives -7.

Then, $- (-7) = 7$.   Report the value with one significant figure (digits to the right of the decimal place).

pH = 7.0

b. $[H_3O^+] = 7.0 \times 10^{-5} M$

$pH = -\log [H_3O^+]$

pH = -log (7.0×10^{-5}) On a scientific calculator, enter the number 7.0×10^{-5} into the calculator, then press the **log** button. This gives -4.15 (reported with two significant figures).

Then, - (-4.15) = 4.15.

pH = 4.15

c. *[OH] = 7.0 x 10^{-5} M*

First calculate the $[H_3O^+]$.

$[H_3O^+]$ = 1.0×10^{-14} / $[OH^-]$

$[H_3O^+]$ = 1.0×10^{-14} / $[7.0 \times 10^{-5}]$ = 1.4×10^{-10} M

Then calculate the pH.

pH = -log $[H_3O^+]$

pH = -log (1.4×10^{-10}) On a scientific calculator, enter the number 1.4×10^{-10}, then press the **log** button. This gives -9.85 (reported with two significant figures).

Then, - (-9.85) = 9.85.

pH = 9.85

d. *[OH] = 1 x 10^{-1} M*

First calculate the $[H_3O^+]$.

$[H_3O^+]$ = 1.0×10^{-14} / $[OH^-]$

$[H_3O^+]$ = 1.0×10^{-14} / $[1 \times 10^{-1}]$ = 1.0×10^{-13} M

Then calculate the pH.

pH = -log $[H_3O^+]$

pH = -log (1.0×10^{-13}) On a scientific calculator, enter the number 1.0×10^{-13}, then press the **log** button. This gives -13.

Then, - (-13) = 13.   Report the value with one significant figure.

pH = 13.0

9.33　*In Problem 9.31, indicate whether each solution in acidic, basic or neutral.*

Acidic solutions have a pH less than 7. Basic solutions have a pH greater than 7. Neutral solutions have a pH = 7.

a. pH = 7.0, which is equal to 7, so the solution is neutral.

Neutral

b. pH = 4.15, which is less than 7, so the solution is acidic.

Acidic

c. pH = 9.85, which is greater than 7, so the solution is basic.

Basic

d. pH = 13.0, which is greater than 7, so the solution is basic.

Basic

9.35　*What is [H_3O^+] in a solution if the pH is*

a. *5.54*

$[H_3O^+] = 10^{-pH}$

On a scientific calculator, enter the number -5.54 then press the **2nd** button followed by the **log** button.
This gives 2.9×10^{-6}.

[H_3O^+] = **2.9 x 10^{-6} M**　　(2 significant figures)

b. *13.8*

$[H_3O^+] = 10^{-pH}$

On a scientific calculator, enter the number -13.8 then press the **2nd** button followed by the **log** button.
This gives 1.6×10^{-14}.

[H_3O^+] = **2 x 10^{-14} M** (1 significant figure)

c. *2.94*

$[H_3O^+] = 10^{-pH}$

On a scientific calculator, enter the number -2.94 then press the 2nd button followed by the **log** button.
This gives 1.1×10^{-3}.

$[H_3O^+] = 1.1 \times 10^{-3} M$ (2 significant figures)

9.37 *Write the chemical equation for the reaction of each weak acid with water. Write the corresponding acidity constant expression.*

Recall that acids donate H^+. Remove one H^+ from the acid given and write the formula remaining, but lower the charge by 1 and make a hydronium ion, H_3O^+, out of the water molecule by adding the H^+ to it. Follow the rules covered previously in problems 9.13 and 9.15 to write the equilibrium expression.

a. NH_4^+

$NH_4^+(aq) + H_2O(l) \rightleftharpoons H_3O^+(aq) + NH_3(aq)$

$K_a = \dfrac{[H_3O^+][NH_3]}{[NH_4^+]}$

b. HPO_4^{-2}

$HPO_4^{2-}(aq) + H_2O(l) \rightleftharpoons H_3O^+(aq) + PO_4^{3-}(aq)$

$K_a = \dfrac{[H_3O^+][PO_4^{3-}]}{[HPO_4^{2-}]}$

9.39 *Calculate the pKa of each acid and indicate which is the stronger acid.*

pK_a is calculated the same way as pH, except that the K_a value is used instead of the $[H_3O^+]$.

a. *HClO, $K_a = 3.0 \times 10^{-8}$*

$pK_a = -\log K_a$

$pK_a = -\log [3.0 \times 10^{-8}]$ On a standard scientific calculator, enter the number 3.0×10^{-8} into the calculator, then press the **log** button. This gives -7.52 (reported with two significant figures).
Then, $- (-7.52) = 7.52$.

$pK_a = 7.52$

b. C_2O_4H, $K_a = 6.4 \times 10^{-5}$

$pK_a = -\log K_a$

$pK_a = -\log [6.4 \times 10^{-5}]$ On a scientific calculator, enter the number 6.4×10^{-5}, then press the **log** button. This gives -4.19 (reported with two significant figures).

Then, - (-4.19) = 4.19.

pK_a = 4.19

The lower the pK_a value, the more hydronium ions produced, meaning that acid strength increases as pK_a decreases.

$C_2O_4H^-$ is the stronger acid

9.41   *0.10 M solutions of each of the following acids are prepared: acetic acid (K_a = 1.8 x 10^{-5}) and hydrofluoric acid (K_a = 6.6 x 10^{-4}). Which acid solution will have the lowest pH? Explain.*

Hydrofluoric acid has a K_a of 6.6×10^{-4} while acetic acid has a K_a of 1.8×10^{-5}. The higher the K_a, the stronger the acid, and the more acidic the solution.

Hydrofluoric acid. Hydrofluoric acid is the stronger acid, so it is better at releasing H^+ (making an acidic solution) than acetic acid.

9.43   *It requires 35.0 mL of 0.250 M KOH to titrate 50.0 mL of an HCl solution of unknown concentration. Calculate the initial HCl concentration.*

Convert the volume of base to liters then use the molarity of the base to convert to moles of base.

$$35.0 \text{ mL KOH} \quad \times \quad \frac{1 \times 10^{-3} \text{ L}}{\text{mL}} \quad \times \quad \frac{0.250 \text{ mol}}{1 \text{ L}} = 8.75 \times 10^{-3} \text{ mol KOH}$$

Next, convert moles of base to moles of acid. Note that KOH has one hydroxide ion and HCl has one hydrogen ion, so they will react 1:1.

$$8.75 \times 10^{-3} \text{ mol KOH} \quad \times \quad \frac{1 \text{ mol HCl}}{1 \text{ mol KOH}} = 8.75 \times 10^{-3} \text{ mol HCl}$$

Finally, convert the volume of the acid to liters and convert moles of HCl to molarity by dividing the moles by the volume.

$$50.0 \text{ mL} \quad \times \quad \frac{1 \times 10^{-3} \text{ L}}{\text{mL}} = 0.0500 \text{ L}$$

$$\frac{8.75 \times 10^{-3} \text{ mol HCl}}{0.0500 \text{ L}} \quad = \quad 0.175 \text{ M}$$

Initial [HCl] = 0.175 M

9.45 *HF is added to water and the solution is allowed to come to equilibrium.*

$$HF + H_2O \ \rightleftharpoons \ F^- + H_3O^+$$

a. *Based on this reaction equation, is HF an acid or is it a base?*

HF is changing to F^-, which means it is donating a hydrogen ion, therefore it is an acid.

Acid

b. *What happens to the pH of the solution if NaF (a water-soluble ionic compound) is added to the solution?*

When NaF is added to water, the equilibrium mixture above, the rate of the reverse reaction will increase. This will lower the concentration of H_3O^+ and raise the pH.

The pH increases.

9.47 *An equilibrium mixture of CH_3CO_2H and $CH_3CO_2^-$ has a pH of 4.8. What happens to the CH_3CO_2H and $CH_3CO_2^-$ concentrations when the pH of the solution is adjusted to 2.8?*

The pH going from 4.8 to 2.8 indicates that H_3O^+ ions are being added. In accordance to Le Châtelier's principle, when H_3O^+ ions are added to the equilibrium, hydrogen ions will form additional CH_3CO_2H by reacting with the $CH_3CO_2^-$. This adds CH_3CO_2H while subtracting $CH_3CO_2^-$.

The concentration of CH_3CO_2H will increase and the concentration of $CH_3CO_2^-$ will decrease.

9.49 *Which weak acid or weak acids from Table 9.4 would be the best choice if you wished to prepare buffers with the following pH values?*

Buffers are prepared by mixing a weak acid with its conjugate base. Buffers are most resistant to pH change when the weak acid and its conjugate base are at equal concentrations when pH = pK_a and are effective when pH = $pKa \pm 1$.

a. *7.0*

On Table 9.4 carbonic acid has a pK_a = 6.36) that is within ± 1 of the pH, and dihydrogen phosphate has a pK_a = 7.21 that is within ± 1 of the pH.

H_2CO_3 (carbonic acid) and $H_2PO_4^-$ (dihydrogenphosphate).

b. *10.0*

On Table 9.4 ammonium ion has a pK_a = 9.25 , hydrocyanic acid has pK_a = 9.31, hydrogen carbonate has a pK_a = 10.25, and methylammonium ion has a pK_a = 10.62, which puts all of their pK_a values within ± 1 of the pH. The two within the range that are closest to the desired pH make the best choices.

HCN (hydrocyanic acid) or HCO_3^- (hydrogencarbonate), or $CH_3NH_3^+$ (methylamine)

c. *4.0*

On Table 9.4 acetic acid has a pK_a = 4.74 , lactic acid has pK_a = 3.85, and hydrofluoric acid has a pK_a = 3.18, which puts all of their pK_a values within ± 1 of the pH. The two within the range that are closest to the desired pH make the best choices.

HF (hydrofluoric acid), $CH_3CH(OH)CO_2H$ (lactic acid), or CH_3CO_2H (acetic acid)

9.51　*A buffer can be prepared using lactic acid [$CH_3CH(OH)CO_2H$] and its conjugate base, lactate [$CH_3CH(OH)CO_2^-$].*

a.　*Write an equation for the reaction that takes place when H_3O^+ is added to this buffer.*

The addition of an acid (increased H_3O^+) will drive the reaction from the weak acid salt toward the weak acid.

$$CH_3CH(OH)CO_2^- + H_3O^+ \rightarrow CH_3CH(OH)CO_2H + H_2O$$

b. *Write equations for the reactions that take place when OH^- is added to this buffer.*

The addition of OH^- will remove H^+ from the H_3O^+ in the solution and the weak acid will generate more H_3O^+.

$$H_3O^+ + OH^- \rightarrow 2H_2O$$

$$CH_3CH(OH)CO_2H + H_2O \rightarrow CH_3CH(OH)CO_2^- + H_3O^+$$

9.53   *In terms of Le Chatelier's principle, describe how the equilibrium*

   $Mb + O_2 \rightleftharpoons MbO_2$ *responds to increases in the concentration of*

   a. *Mb*

   When [Mb] increases, the rate of the forward reaction increases and more MbO_2 is produced.

   b. *O_2*

   When [O_2] increases, the rate of the forward reaction increases and more MbO_2 is produced.

   c. *MbO_2*

   When [MbO_2] increases, the rate of the reverse reaction increases and more Mb and O_2 is produced.

9.55   *Calculate the logarithm of each number. Assume that each is a measured quantity.*

   a. 1×10^{-9}

   On a scientific calculator, enter the number 1×10^{-9}, then press the **log** button. This gives -9. Report the log with one significant figure (digits to the right of the decimal point.)

   -9.0

   b. 1×10^{12}

   On a scientific calculator, enter the number 1×10^{12}, then press the **log** button. This gives 12. Report the log with one significant figure.

   12.0

   c. *3.4×10^{-2}*

   On a scientific calculator, enter the number 3.4×10^{-2}, then press the **log** button. This gives -1.468. Report the log with two significant figures.

   -1.47

   d. *9.7×10^{4}*

On a scientific calculator, enter the number 9.7×10^4, then press the **log** button. This gives 4.98677. Report the log with two significant figures.

4.99

9.57 *Calculate the antilogarithm of each number.*

The numbers in this problem are not measured quantities, so the number of digits reported is arbitrary.

a. *8*

On a scientific calculator, enter the number 8, then press the **2nd** button followed by the **log** button. This gives 1×10^8.

1×10^8

b. *−3*

On a scientific calculator, enter the number -3 (remember to put in 3 then press the sign change, NOT the subtraction button) then press the **2nd** button followed by the **log** button. This gives 1×10^{-3}.

1×10^{-3}

c. *7.9*

On a scientific calculator, enter the number 7.9 into the calculator, then press the **2nd** button followed by the **log** button. This gives 7.9×10^7.

7.9×10^7

d. *15.3*

On a scientific calculator, enter the number 15.3, then press the **2nd** button followed by the **log** button. This gives 2.00×10^{15}.

2.0×10^{15}

9.59 *When you pick a particular flower and immerse it in a pH 5 solution, there is no change in the color. If you place the same flower in a pH 9 solution, the color changes. Why does this happen?*

A variation in pH causes a change in the structure of the compound responsible for the color.

9.61   *Explain how higher than normal levels of HCO_3^- resulting from constipation affect the H_2CO_3 and HCO_3^- buffer and the pH of blood.*

A rise in $[HCO_3^-]$ causes an increase in the rate of the reverse reaction ($H_2CO_3 + H_2O \leftarrow HCO_3^- + H_3O^+$), so $[H_3O^+]$ drops and the pH of blood increases.

9.63   *Suggest an explanation for the fact that the antacids Di-Gel, Mylanta, and Maalox contain both $Al(OH)_3$ and $Mg(OH)_2$.*

The constipating effect of Al^{3+} is offset by the laxative effect of Mg^{2+}

Chapter 10
Carboxylic Acids

Solutions to Problems

10.1 *Match each structure to the correct IUPAC name: 3-methylbutanoic acid, 2-methylpentanoic acid, 2-methylbutanoic acid.*

The carboxylic functional group is dominant in the naming so its carbon is carbon 1. In drawing a, there are four carbons and the methyl group is on the second carbon so that the compound is 2-methylbutanoic acid. In drawing b. also has four carbons but the methyl group is on carbon 3, making its name 3-methylbutanoic acid. In drawing c., the methyl is on the second carbon but the parent chain has five carbons which makes the name 2-methylpentanoic acid.

a.

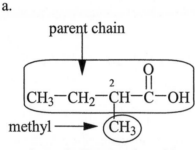

2-methylbutanoic acid

b.

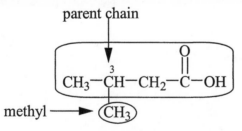

3-methylbutanoic acid

c.

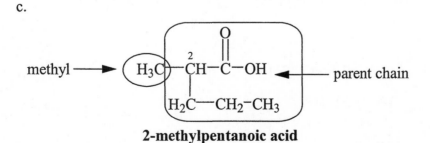

2-methylpentanoic acid

10.3   *Draw each carboxylic acid.*

a. *octanoic acid*

Start by drawing CH_3 then add $(CH_2)_6$ followed by the ending carboxylic functional group.

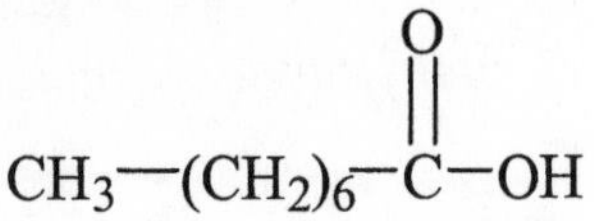

$$CH_3-(CH_2)_6-\overset{\displaystyle O}{\overset{\|}{C}}-OH$$

b. *3,3-dimethylheptanoic acid*

First, draw out the parent chain with seven (heptane) carbons making the last one the carboxylic group indicated by the "oic acid" ending. Add a 2 methyl groups to carbon 3.

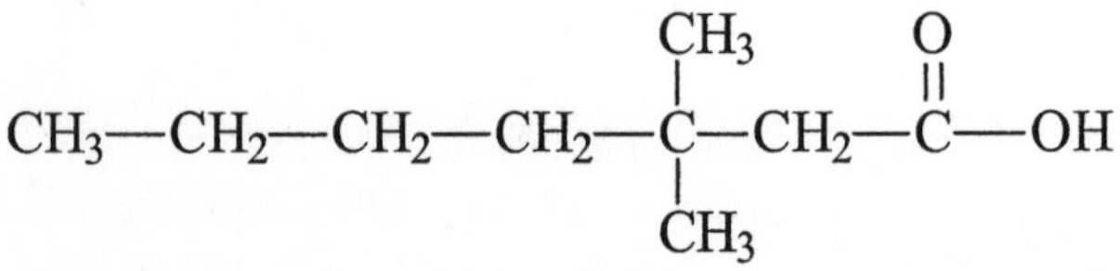

$$CH_3-CH_2-CH_2-CH_2-\overset{\displaystyle CH_3}{\underset{\displaystyle CH_3}{\overset{\|}{C}}}-CH_2-\overset{\displaystyle O}{\overset{\|}{C}}-OH$$

c. *3-isopropylhexanoic acid*

First, draw out the parent chain with six (hexane) carbons making the last one the carboxylic group indicated by the "oic acid" ending. Add an isopropyl group to carbon 3.

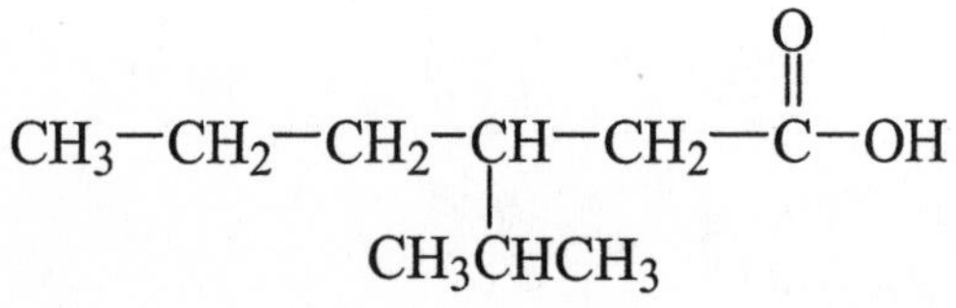

$$CH_3-CH_2-CH_2-\underset{\displaystyle CH_3CHCH_3}{CH}-CH_2-\overset{\displaystyle O}{\overset{\|}{C}}-OH$$

d. *2-bromopentanoic acid*

First, draw out the parent chain with five (pentane) carbons making the last one the carboxylic group indicated by the "oic acid" ending. Add a bromine atom to carbon 3.

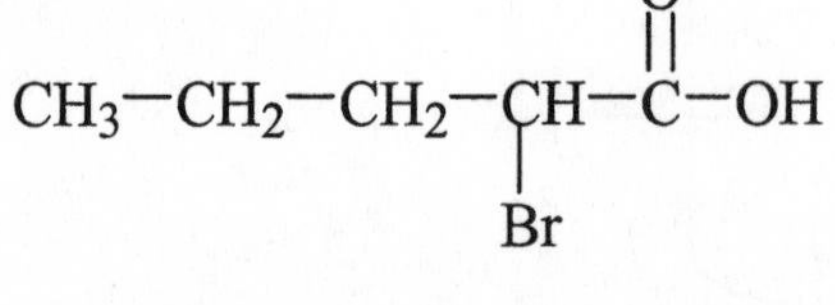

$$CH_3-CH_2-CH_2-\underset{\displaystyle Br}{CH}-\overset{\displaystyle O}{\overset{\|}{C}}-OH$$

10.5   *Draw each phenol.*

a. *2,4-dimethylphenol*

Start by drawing the phenol parent (benzene ring with a –OH in the first position). Counting from the –OH carbon add one methyl group to carbon 2 and one methyl group to carbon 4.

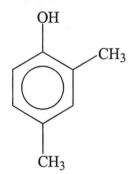

b. *3,5-dimethylphenol*

Start by drawing the phenol parent (benzene ring with a –OH in the first position). Counting from the –OH carbon add one methyl group to carbon 3 and one methyl group to carbon 5.

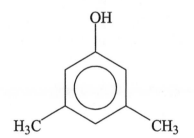

c. *4-propylphenol*

Start by drawing the phenol parent (benzene ring with a –OH in the first position). Counting from the –OH carbon add a propyl group to carbon 4. (a propyl is propane attached by its carbon 1)

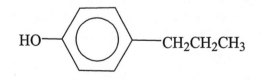

d. *m-propylphenol*

Start by drawing the phenol parent (benzene ring with a –OH in the first position). Counting from the –OH carbon add a propyl group to carbon 3.

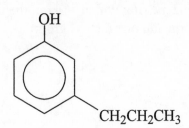

10.7   *If you come into contact with urushiol (Figure 10.3b), the phenol partially responsible for the itching and burning caused by poison ivy, you cannot wash it off with water. Explain.*

The urushiol molecule has a very long hydrocarbon side chain that makes it hydrophobic (water-insoluble).

10.9   *Hexylresorcinol (4-hexyl-3-hydroxyphenol) is an antiseptic used in some mouthwashes and throat lozenges. Draw this compound.*

Start by drawing the phenol parent (benzene ring with a –OH in the first position). Counting from the –OH carbon add another –OH group (called hydroxy when used as a side chain) to carbon 3. Next add a hexyl to carbon 5.

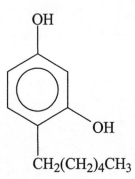

10.11   *Draw m-chlorobenzoic acid and its conjugate base.*

In order to draw *m*-chlorobenzoic acid, draw benzoic acid which is the benzene ring with the carboxyl functional group on carbon 1. Next, add a chloro substituent (chloride) on carbon 3. To make the conjugate base, remove the hydrogen from the carboxylic group and add a negative charge.

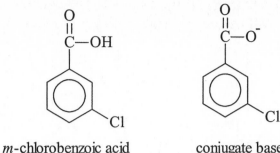

m-chlorobenzoic acid   conjugate base

10.13 *Draw the conjugate base of each carboxylic acid.*

a. *3-methylhexanoic acid*

Draw a six carbon (hexane) with the carbon 1 being the carboxylic functional group. Add a methyl group to carbon 3. Finally, remove hydrogen from the –OH and place a negative charge on the oxygen.

                 Conjugate base

b. *formic acid*

Draw one carbon (methane) as the carboxylic functional group. Remove hydrogen from the –OH and place a negative charge on the oxygen.

$$\overset{\displaystyle O}{\underset{\displaystyle \|}{HC}}\!-\!O^-$$

             Conjugate base

c. *3,5-dibromobenzoic acid*

Start by drawing the carboxylic acid parent chain (benzene ring with a –COOH in the first position). Counting from the carbon containing the –COOH group add a bromo substituent on carbon 3 and another bromo substituent on carbon 5. Remove hydrogen from the –OH and place a negative charge on the oxygen.

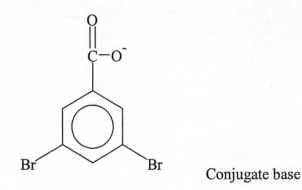

Conjugate base

10.15 *Name each of the conjugate bases in you answer to Problem 10.13.*

To name the conjugate base of a carboxylic acid, the "ic" ending is replaced with "ate" followed by ion instead of acid.

a. 3-methylhexanoate ion

b. formate ion

c. 3,5-dibromobenzoate ion

10.17 *Draw the organic product of each reaction.*

In Chapter 9, you saw that acid and base reactions produced a salt and water. The same is true when an organic acid is reacted with a base. The organic product is the original acid structure with the hydrogen removed from the acid and a negative charge placed on the oxygen, with the cation of the salt drawn beside the O⁻.

a.

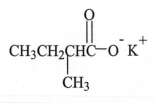

b.

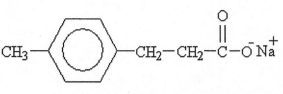

10.19 *Dinoseb, an herbicide and insecticide, is sold as a water-soluble ammonium salt. Draw this ammonium salt, which is produced by reacting Dinoseb with ammonia.*

The organic product has the original acid structure with the hydrogen removed from the acid and a negative charge placed on the oxygen with the cation of the salt drawn beside the O⁻.

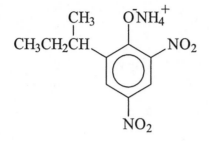

10.21 Gentisic acid (Figure 10.3b) can be used as an anti-rheumatism drug.

a. *Which functional group of gentisic acid is the most acidic, one of the phenol groups or the carboxyl group?*

Carboxylic acids are more acidic than phenols.

The carboxyl group is most acidic.

b. *Draw the ionic compound potassium gentisate (Hint: only one H^+ is removed from gentisic acid.)*

The "ate" ending requires that you draw the organic acid (Figure 10.3b) with the hydrogen removed from the acid group and a negative charge placed on the oxygen, and the cation of the salt drawn beside the O⁻.

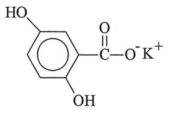

10.23 *2,4-dichlorophenoxide is the conjugate base of 2,4-dichlorophenol. Draw both compounds.*

To draw 2,4-dichlorophenol, start by drawing the phenol parent (benzene ring with a –OH in the first position). Counting from the first carbon add a chloro substituent on carbon 2 and another chloro substituent on carbon 4. To draw 2,4-dichlorophenoxide, remove hydrogen from the –OH and place a negative charge on the oxygen.

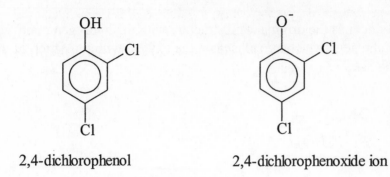

2,4-dichlorophenol 2,4-dichlorophenoxide ion

10.25 *Draw the organic product of each reaction.*

The reaction of a carboxylic acid and alcohol in the presence of an acid catalyst produces an ester. Remove a water molecule and attach the alcohol group.

a.

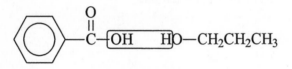

Answer:

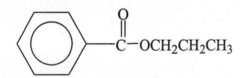

b.

Answer:

CH₃CH₂C-OCH₂—⬡

448

10.27 *Draw the organic product formed when each β-keto acid is decarboxylated.*

When a carboxylic acid is decarboxylated, a carbon dioxide molecule (CO_2) is formed from the carboxyl group.

a.

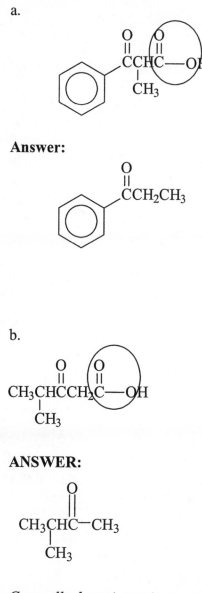

Answer:

10.29 *Geranylhydroquinone is an experimental radioprotective drug, one that protects against the harmful effects of radiation. Draw the organic compound formed when geranylhydroquinone is treated with $K_2Cr_2O_7$.*

$K_2Cr_2O_7$ serves as an oxidizing agent. In this reaction, the alcohol groups are oxidized to ketones as the hydrogens are removed and the oxygens acquire double bonds.

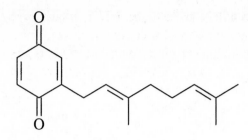

10.31 *Draw the organic product formed when the α-keto acid undergoes an enzyme-catalyzed decarboxylation.*

When a carboxylic acid is decarboxylated, a carbon dioxide molecule (CO_2) is formed from the carboxyl group.

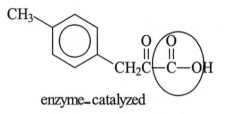

enzyme—catalyzed

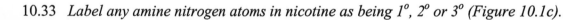

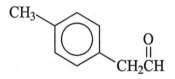

10.33 *Label any amine nitrogen atoms in nicotine as being 1°, 2° or 3° (Figure 10.1c).*

Amines with nitrogen atoms having one bond with a carbon are classified as 1°. Those with two bonds to C are 2° and those with three bonds to C are 3°.

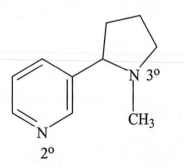

10.35 *Give the IUPAC name of each amine.*

The IUPAC name of amines uses the name of the corresponding alkane parent chain with the "e" replaced with "amine". If a substituent is attached to the nitrogen of the amine instead of the carbons, it is preceded by $N-$. The amine position on the parent chain is given by a number preceding the parent chain.

a. methanamine

b. 2-propanamine

c. N-isopropyl-1-butanamine

d. N-ethyl-N-methyl-1-propanamine

10.37   *Give the common name of each amine in the Problem 10.35.*

The common name of amines puts the name of all substituents attached to the N in front of "amine".

a. methylamine

b. isopropylamine

c. butylisopropylamine

d. ethylmethylpropylamine

10.39   *Identify each compound as a pyridine, a pyrimidine, or a purine (Figure 10.11a).*

A pyridine has only one nitrogen atom replacing a carbon in a benzene ring, a pyrimidine has two, and a purine has two in a benzene ring and a fused ring containing two additional nitrogen atoms.

a. pyrimidine

b. pyridine

c. purine

10.41   *Account for the fact that 1-propanamine is more water soluble that N,N-diethylethanamine.*

Primary amines, like 1-propanamine, have two N–H hydrogen atoms and one nitrogen atom available for forming hydrogen bonds with water. Tertiary amines, like N,N-diethylethanamine, have only one nitrogen atom capable of forming hydrogen bonds. Another factor in the different solubility of these amines is that N,N-diethylethanamine contains more nonpolar ethyl groups.

10.43 *Draw each compound.*

a. *2-hexanamine*

Draw the parent chain hexane, remove one hydrogen from carbon 2 and add the amine group, $-NH_2$.

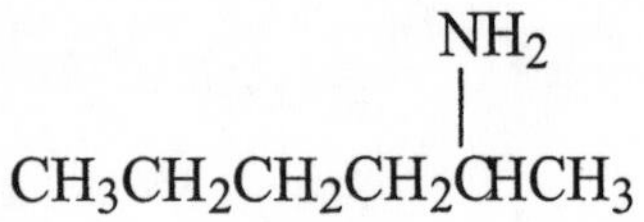

$$\overset{\displaystyle NH_2}{\underset{\displaystyle |}{CH_3CH_2CH_2CH_2CHCH_3}}$$

b. *N-methyl-3-pentanamine*

Draw the parent chain pentane. Remove a hydrogen atom from carbon 3 and add the amine group. Next, remove one hydrogen atom from the amine group and replace it with a methyl group (this is what *N*-methyl indicates).

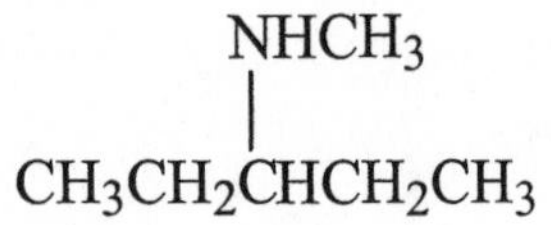

$$\overset{\displaystyle NHCH_3}{\underset{\displaystyle |}{CH_3CH_2CHCH_2CH_3}}$$

c. *N,N-dimethyl-1-butanamine*

Draw the parent chain butane. Remove a hydrogen atom from carbon 1 and add the amine group. Next, remove both hydrogen atoms from the amino group and replace them with two methyl groups (this is what *N,N*-dimethyl indicates).

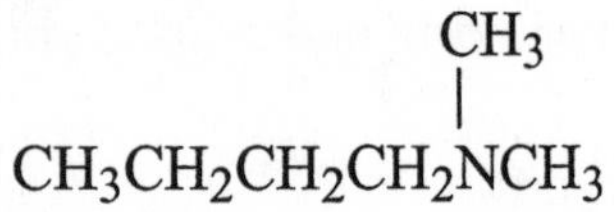

$$\overset{\displaystyle CH_3}{\underset{\displaystyle |}{CH_3CH_2CH_2CH_2NCH_3}}$$

d. *tetramethyammonium ion*

Draw the ammonium ion, NH_4^+. Replace each of the four hydrogen atoms with a methyl group.

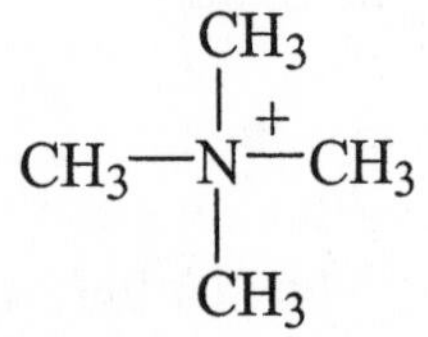

$$CH_3-\overset{\displaystyle CH_3}{\underset{\displaystyle CH_3}{\overset{\displaystyle |}{\underset{\displaystyle |}{N^+}}}}-CH_3$$

10.45 *Draw and name the conjugate acid of*

a. butylamine

Combine NH_2 and a butyl group. Next, add one hydrogen ion to the amino group to make NH_3. The hydrogen ion bound to the nitrogen carried a positive charge so place a "+" to the right of the $-NH_3$. To name the ion, change the amine part of the name to ammonium ion.

$$CH_3CH_2CH_2CH_2NH_3^+ \quad \text{butylammonium ion}$$

b. dipropylamine

Combine the NH and two propyl groups. Next, add one hydrogen ion to the NH to make it NH_2. Next add an isopropyl group to the amine group without removing any of the hydrogen atoms. Place a "+" to the right of the $-NH_2$. To name the ion, change the amine part of the name to ammonium ion.

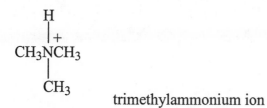

dipropylammonium ion

c. trimethylamine

Combine N and four methyl groups. Place a "+" to the right of the $-N$. To name the ion, change the amine part of the name to ammonium ion.

$$\begin{array}{c} H \\ | \\ | \; + \\ CH_3NCH_3 \\ | \\ CH_3 \end{array}$$

trimethylammonium ion

d. *t*-butylamine

Combine NH_2 and a t-butyl group. Next add one hydrogen ion to the NH_2 to make it NH_3. Place a "+" to the right of the $-NH_2$. To name the ion, change the amine part of the name to ammonium ion.

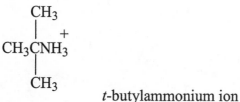

t-butylammonium ion

10.47 *In the early 1980s a street drug sold as "synthetic heroin" appeared in southern California. This drug contained an impurity called MPTP, which caused irreversible symptoms of Parkinson's disease (including immobility, slurred speech, and tremors). The MPTP existed as a hydrochloride salt. Draw this salt.*

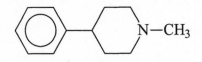

The nitrogen atom of MPTP is basic.

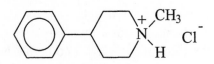

10.49 *Difenpiramide (Problem 10.39) has two nitrogen atoms (one amine and one amide). Label the amide and amine nitrogen atoms in this compound.*

Amines have nitrogen atoms that are bonded to hydrocarbons or hydrogen atoms. Amides have nitrogen atoms bonded to a carboxyl carbon.

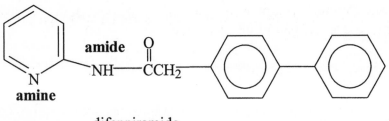

difenpiramide

10.51 *Draw the amide that will be produced by each reaction.*

Amides are produced by heating a carboxylic acid in the presence of an amine or ammonia. Remove a hydrogen from the nitrogen and the –OH group from the carboxylic acid. Attach the nitrogen group to the carboxyl group where the –OH group was originally.

a.

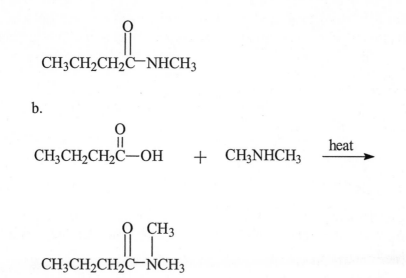

$$\underset{\text{O}}{\overset{\text{||}}{CH_3CH_2CH_2C}}-NHCH_3$$

b.

$$\underset{\text{O}}{\overset{\text{||}}{CH_3CH_2CH_2C}}-OH \quad + \quad CH_3NHCH_3 \quad \xrightarrow{\text{heat}}$$

$$\underset{\text{O} \quad CH_3}{\overset{\text{||} \quad \text{|}}{CH_3CH_2CH_2C}}-NCH_3$$

10.53 *The boiling point of propanamide is 213°C and that of methyl acetate is 57.5°C. Account for this difference in boiling points for these two molecules with very similar molecular weights.*

Propanamide interacts more effectively with water through hydrogen bonding than does methyl acetate.

10.55 *Label the chiral carbon atom in each molecule.*

A chiral carbon must have four different atoms or groups of atoms attached to it.

a.

$$CH_3CH_2CH_2\overset{*}{C}HCH_2CH_3$$
$$\underset{Br}{|}$$

b.

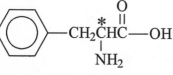

$$-CH_2\overset{*}{C}H\overset{\overset{\text{O}}{\overset{\text{||}}{C}}}{}-OH$$
$$\underset{NH_2}{|}$$

10.57 *Draw both enantiomers of each molecule in Problem 10.55, using wedge bond notation to show the three-dimensional shape about each chiral carbon.*

Enantiomers are molecules that are mirror images of each other and are not superimposable.

a. Note that in each structure the propyl is out of the plane of the paper and the hydrogen points into the plane. The structures are mirror images and they are not superimposable.

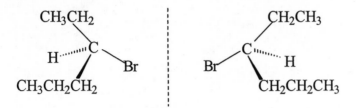

b. Note that in each structure the amino group is out of the plane of the paper and the hydrogen points into the plane. The structures are mirror images and they are not superimposable.

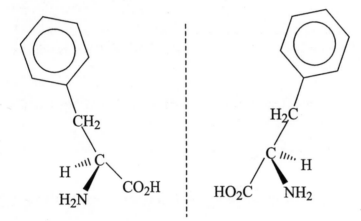

10.59   a. *Draw the enantiomer of each molecule in Figure 10.19b.*

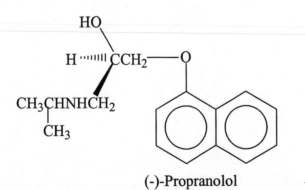

(-)-Propranolol

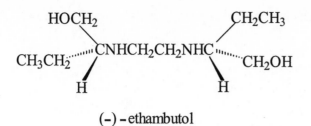

(–) – ethambutol

b. *Draw one diastereomer of (+)-ethambutol (Figure 10.19b)*

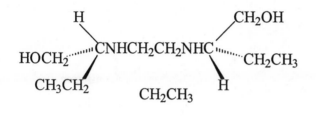

or

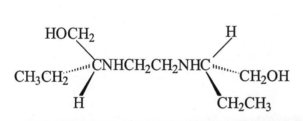

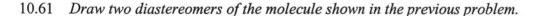

10.61   *Draw two diastereomers of the molecule shown in the previous problem.*

Refer to the drawing in Problem 10.60. A diastereomer is a stereoisomer that is not an enantiomer. Draw one molecule that switches the position of two groups on one chiral carbon atom (here the –CH$_3$ and the –CH$_2$N(CH$_3$)$_2$ on the right side). Then draw its enantiomer.

Answer:

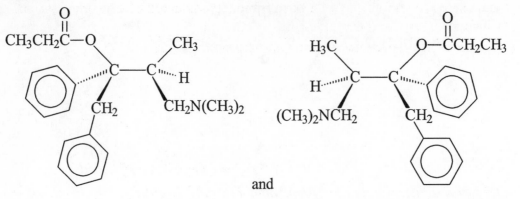

and

10.63 *Olvanil, a compound structurally related to capsaicin (Figure 10.4), has been studied as a potential analgesic. Locate the phenol, amide, and alkene functional groups in the molecule.*

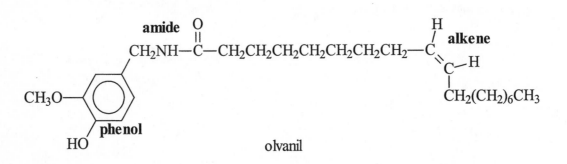

olvanil

10.65 *Precor, used in flea collars and animal sprays, is a hormone that prevents flea pupae (the stage of life between larval and adult forms) from developing.*

a. *Label the chiral carbon atom with an asterisk.*

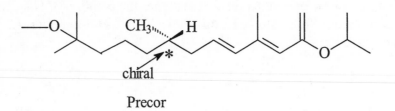

Precor

b. *Precor has low water solubility. Explain why.*

The contribution of the polar ether and ester groups, which can form hydrogen bonds with water, is outweighed by the nonpolar nature of the rest of the molecule.

c. *Draw the products obtained when Precor is hydrolyzed under basic conditions.*

458

Under basic conditions the ester bond is broken producing a carboxylate ion and an alcohol.

In basic conditions the products are

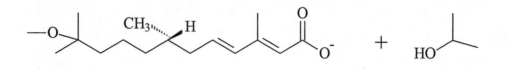

d. *Draw the products obtained when Precor is hydrolyzed under acidic conditions.*

Under acidic conditions the ester bond is broken producing a carboxylic acid and an alcohol.

In acidic conditions the products are

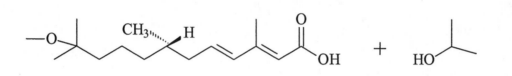

Chapter 11
Alcohols, Ethers, and Related Compounds

Solutions to Problems

11.1 *Identify each alcohol as 1°, 2°, or 3°.*

Alcohols are classified according to the carbon the hydroxyl group is attached to. If the carbon is attached to only one other carbon, the alcohol is 1°, when the carbon is attached to two other carbons it is 2°, and when the carbon is attached to three other carbons the alcohol is 3°.

a. Hydroxyl is attached to a carbon with two other carbons attached to it.
2°

b. Hydroxyl is attached to a carbon with three other carbons attached to it.
3°

c. Hydroxyl is attached to a carbon with one other carbon attached to it.
1°

11.3 *Give the IUPAC name of each alcohol in Problem 11.1.*

a.   The parent chain has five carbon atoms which makes it a pentane.
To name the parent chain drop the "e" and add "ol". The hydroxyl group is on carbon 3 so place a 3- in front of pentanol.

 3-pentanol

b. The parent chain has five carbon atoms which makes it a pentane. To name the parent chain drop the "e" and add "ol". The hydroxyl group is on carbon 3 so place a 3- in front of pentanol. There is also a methyl substituent on carbon three so assign it the number 3 as well.

3-methyl-3-pentanol

c. The parent chain has four carbon atoms which makes it a butane. To name the parent chain drop the "e" and add "ol". The hydroxyl group is on carbon 1 so place a 1- in front of gutanol. There is also a methyl substituent on carbon 2 so assign it the number 2.

2-methyl-1-butanol

11.5   *Draw each alcohol molecule.*

a. *2-hexanol*

Draw a six carbon (hexane) parent chain. Add –OH to carbon 2.

$$\overset{\displaystyle OH}{\underset{\displaystyle |}{CH_3CH_2CH_2CH_2CHCH_3}}$$

b. *3-methyl-1-pentanol*

Draw a five carbon (pentane) parent chain. Add –OH to carbon 1. Add –CH$_3$ to carbon 3.

$$\overset{\displaystyle CH_3}{\underset{\displaystyle |}{CH_3CH_2CHCH_2CH_2OH}}$$

c.   *4-isopropylcyclohexanol*

Draw a six-membered ring (cyclohexane). Add an –OH. The –OH is carbon 1. Count to carbon 4 and add an isopropyl group.

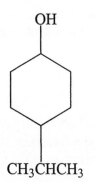

11.7   a. *In some arid parts of the United States a thin layer of 1-octadecdecanol (octadec is the IUPAC prefix for 18) is floated on the top of reservoirs to slow water evaporation. Write the condensed molecular formula of this alcohol.*

To make a condensed molecular formula of a large molecule, the repeating CH$_2$ groups can be accounted for with a subscript. Draw a CH$_3$ followed by (CH$_2$)$_{16}$ and ending with CH$_2$OH. The –OH is on the end because the alcohol was designated 1-.

$CH_3(CH_2)_{16}CH_2OH$

b. *Why is this alcohol insoluble in water?*

The effects of the long nonpolar hydrocarbon chain outweigh those of the polar
–OH group.

11.9   *Give the IUPAC and common name of the molecule.*

In the IUPAC naming system, compounds with a –SH group have the ending
"thiol". After naming the parent chain of the corresponding alkane, the "e" is not
dropped prior to adding thiol to the name. The longest chain is three carbons long
making it a propane and the –SH is on carbon 2 making the parent chain
2- propanethiol. There is also a –CH$_3$, methyl, on carbon 2.

IUPAC name: 2-methyl-2-propanethiol

In the common naming system, compounds with a –SH group are given the
family name mercaptan. All common names with this group end in mercaptan
and the organic group is named as an alkyl group. Since the carbon that is
attached to the –SH has three other carbon groups attached it is a tertiary carbon
and the group has four carbons, it is *t*-butyl.

 Common name: *t*-butyl mercaptan

11.11   *Give the common name of each ether.*

The common naming system gives the names of the alkyl group attached to the O
atom followed by "ether".

a. There is a *t*-butyl group on one side and a methyl on the other.

 t-butyl methyl ether

b. There is a butyl group on one side and an *iso*-propyl group on the other.

butyl isopropyl ether

c. There is a pentyl group on each side. Instead of using pentyl pentyl, the name
dipentyl is used.

 dipentyl ether

11.13   *Account for the fact that dipropyl ether has a higher boiling point than diethyl
ether.*

Dipropyl ether has a longer hydrocarbon chain than diethyl ether and, therefore,
stronger London force interactions.

11.15 *Account for the fact that dipropyl ether has a lower solubility in water than diethyl ether.*

The alcohol ($CH_3CH_2CH_2CH_2OH$) has more opportunities for forming hydrogen bonds than does the ether ($CH_3CH_2OCH_2CH_3$).

11.17 *Thiocitic acid is a growth factor for many bacteria. Which functional groups does this molecule contain?*

Disulfide and carboxylic acid

11.19 *Draw the organic product of each nucleophilic substitution reaction.*

A nucleophilic substitution replaces a leaving group such as chlorine, bromine, or iodine with an electron rich group.

a.

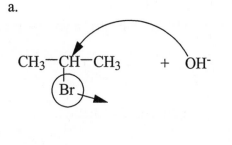

CH_3CHCH_3
$\quad\ \ |$
$\quad\ \ OH$

b.

$CH_3{-}CH_2{-}CH_2{-}CH_2{-}CH_2{-}Br \quad + \quad {^-}OCH_3$

$CH_3CH_2CH_2CH_2CH_2OCH_3$

c.

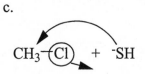

464

CH₃SH

d.

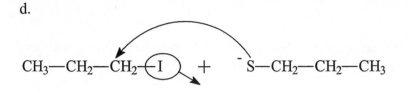

CH₃CH₂CH₂SCH₂CH₂CH₃

11.21 *Name each of the products in Problem 11.19.*

a. The parent chain has three carbons (propane), so drop the "e" and add "ol". The –OH is on the carbon 2 so place 2- in front of propanol.

2-propanol or Isopropyl alcohol

b. The O between two carbon groups indicates this compound is an ether. One group is a methyl and the other is a pentyl.

methyl pentyl ether

c. The SH means this is a thiol. The only group attached is methyl.

methyl mercaptan or methanethiol

d. The S has two organic groups attached to it making the molecule a sulfide. Both groups attached to S are propyl, so the term dipropyl is used with the family name sulfide.

dipropyl sulfide

11.23 *Draw the missing alkyl bromide (RBr) for each reaction.*

In these reactions a given nucleophile reacts with an alkyl bromide to make a product molecule. Examine the product molecule and identify the alkyl group which must be added to the nucleophile to produce it. That alkyl group is supplied by the alkyl bromide.

a. CH₃CH₂Br

b. CH₃Br

c.

CH₃CH₂CH₂CHCH₃
|
Br

11.25 *Write a chemical equation that shows how each compound can be produced from an alkyl bromide (RBr).*

a. $CH_3CH_2Br + {}^-SCH_2CH_2CH_2CH_3 \rightarrow CH_3CH_2SCH_2CH_2CH_2CH_3 + Br^-$

b. $CH_3CH(CH_3)CH_2CH_2Br + {}^-SH \rightarrow CH_3CH(CH_3)CH_2CH_2SH + Br^-$

11.27 *Draw the major product of each reaction.*

The hydration of a double bond results in an –OH group being added to the double bonded carbon with the fewest hydrogens, while an –H is added to the other carbon of the double bond. In part c, the –OH can be added to either carbon atom of the double bond. The same product is obtained either way.

a.

OH
|
CH₃CH₂CCH₂CH₃
|
CH₂CH₃

b.

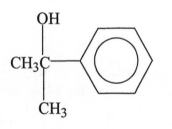

466

c.

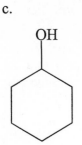

11.29 *Draw the organic product expected from each reaction.*

When an alcohol is oxidized by reaction with $K_2Cr_2O_7$, the CH-OH is converted to C=O. Primary alcohols become carboxylic acids, secondary alcohols become ketones and tertiary alcohols do not react.

a.

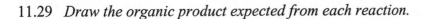

$$CH_3CH_2CH_2\overset{\overset{\displaystyle O}{\|}}{C}-OH$$

b.

$$CH_3CH_2\overset{\overset{\displaystyle O}{\|}}{C}CH_3$$

c.

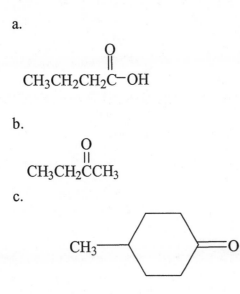

When a thiol is reacted with I_2, the hydrogen will be removed from the –SH on two molecules and the molecules will be joined together to make a disulfide.

d. $CH_3CH_2CH_2SSCH_2CH_2CH_3$

e.

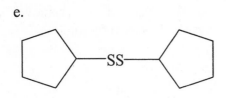

11.31 *Draw each molecule named below and draw the major organic product expected when each is reacted with H⁺ and heat.*

In the presence of heat and H⁺, the –OH group is removed, along with a neighboring hydrogen, to form an alkene. The hydrogen atom removed is from the neighboring carbon atom that carries the fewest H atoms.

a.

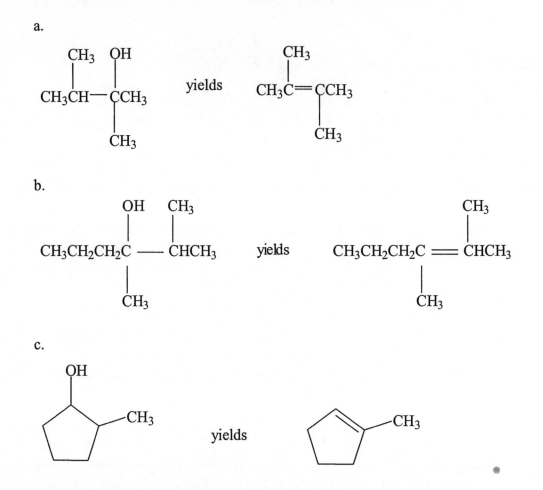

b.

c.

11.33 *Draw the major product of each reaction.*

In the presence of heat and H⁺, the –OH group is removed, along with a neighboring hydrogen, to form a double bond. The hydrogen atom removed is from the neighboring carbon atom that carries the fewest H atoms.

a.

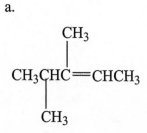

b.

CH₃
|
CH₃C=CHCH₃

11.35   a. *Prednisone is often used to reduce inflammation. Which functional groups does this molecule contain?*

Predisone contains the alcohol, alkene and ketone functional groups.

b. *To which class of lipids does prednisone belong?*

Steroid

11.37   *Name each of the following aldehydes and ketones.*

a. Since the C=O is on the end carbon, this is an aldehyde. In the IUPAC system, the parent chain for the corresponding alkane is named and the "e" is dropped and replaced by "al". Because the C=O is always on carbon 1 for aldehydes, a number is not placed in front of the name of the parent chain. The parent chain is four carbons (butane) and there are two methyl groups located on carbon 3.

  3,3-dimethylbutanal

b. Since the C=O is not on an end carbon, this is a ketone. In the IUPAC system, the parent chain for the corresponding alkane is named and the "e" is dropped and replaced by "one". A number is placed in front of the name of the parent chain to indicate the location of the keto group (Number carbons from the end that gives the C=O the smallest possible number.). The parent chain is four carbons (pentane) and there is one methyl group located on carbon 4.

  4-methyl-2-pentanone

c. Since the C=O is on the carbon 3, this is a ketone. In the IUPAC system, the parent chain for the corresponding alkane is named and the "e" is dropped and replaced by "one". Because the C=O is on carbon 3, a number is placed in front of the name of the parent chain. The parent chain is five carbons (pentane) and there are two methyl groups located on carbon 2 and carbon 4.

2,4,-dimethyl-3-pentanone

d. Since the C=O is on the end carbon this is an aldehyde. In the IUPAC system, the parent chain for the corresponding alkane is named and the "e" is dropped and replaced by "al". Because the C=O is always on carbon 1 for aldehydes, a number is not placed in front of the name of the parent chain. The parent chain is five carbons (pentane) and there is one ethyl group located on carbon 2.

2-ethylpentanal

11.39   *Draw each molecule.*

a. *octanal*

There must be eight carbons (octane). Draw a CH_3 followed by $(CH_2)_6$ and then end the molecule with the aldehyde group (indicated by the "al" ending on the parent chain name).

$$\overset{\displaystyle O}{\overset{\displaystyle \|}{CH_3(CH_2)_6C}}-H$$

b. *2,5-dimethylcyclopentanone*

Draw cyclopentane. Add $= O$ to one carbon atom (this becomes position 1). Add $-CH_3$'s to carbons 2 and 5.

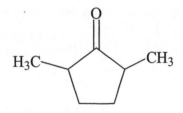

c. *2-octanone*

There must be eight carbons (octane). Draw a CH_3 followed by $(CH_2)_5$ and then add the keto group (indicated by the "one" ending) followed by a CH_3. This puts the C=O on the second carbon, accounting for the 2- in the name.

$$\overset{\displaystyle O}{\overset{\displaystyle \|}{CH_3(CH_2)_5C}}CH_3$$

d. *2-isopropylpentanal*

Draw the parent chain with five carbons (pentane) with C=O on the end carbon (carbon 1). Add an isopropyl group to carbon 2.

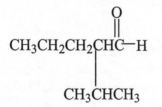

470

11.41 *Acetone has a boiling point of 56°C and isopropyl alcohol has a boiling point of 82°C.*

a. *Draw each molecule*

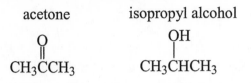

acetone   isopropyl alcohol

b. *Account for the difference in boiling points.*

Alcohol molecules can interact through relatively strong hydrogen bonds, while ketone molecules interact through relatively weak dipole-dipole forces. The stronger the noncovalent interactions, the higher the boiling point.

11.43 *Draw each molecule named below and draw the product (if any) obtained when each is reacted with $K_2Cr_2O_7$. When named as a substituent, -Cl is chloro and -OH is hydroxyl.*

a. When an aldehyde is oxidized by $K_2Cr_2O_7$ the product is a carboxylic acid.

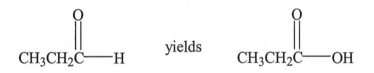

b. Ketones are not oxidized by $K_2Cr_2O_7$.

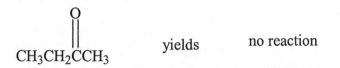

c. First, the alcohol is oxidized by $K_2Cr_2O_7$ to an aldehyde. Then the aldehyde is oxidized to a carboxylic acid.

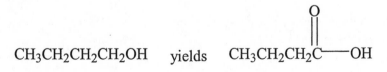

d. When an aldehyde is oxidized by $K_2Cr_2O_7$ the product is a carboxylic acid.

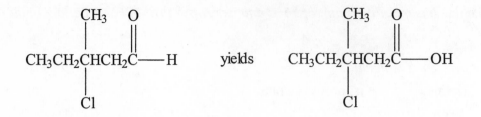

e. When an aldehyde is oxidized by $K_2Cr_2O_7$ the product is a carboxylic acid. This molecule also has a secondary alcohol group which is oxidized to a ketone.

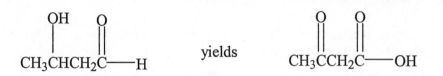

11.45 *Draw the product of each reaction.*

a. This is an oxidation reaction. The oxidizing reagent Cu^{2+} (Benedict's reagent) converts the aldehyde to a carboxylic acid.

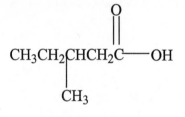

b. The oxidizing reagent Cu^{2+} will not oxidize an alcohol.

No Reaction Product

c. This is an oxidation reaction. The oxidizing reagent Cu^{2+} converts the aldehyde to a carboxylic acid and the alcohol remains unchanged.

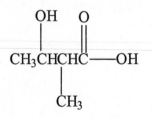

11.47 *Draw the product of each reaction.*

When aldehydes are reacted with H_2 in the presence of a Pt catalyst, the aldehyde is converted to an alcohol.

a.

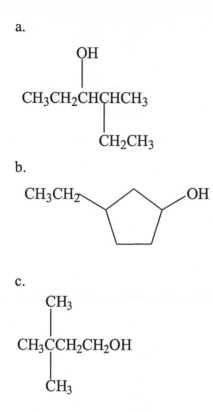

b.

c.

11.49 *The first step in the biochemical breakdown of the steroid progesterone is a reaction involving 20α-hydroxysteroid dehydrogenase (20 α-HSDH), an enzyme which catalyzes the reduction of a ketone group at position 20 in steroids.*

a. Draw the product that forms when progesterone is reduced by the action of 20α-HSDH and NADH.

When a ketone group is reduced, an alcohol is produced. The enzyme 20α-HSDH only reduces the C=O at carbon 20.

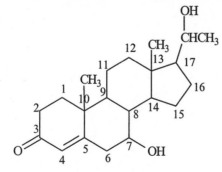

b. Draw the product that forms when 1 mol progesterone is reacted with 3 mol of H_2 in the presence of Pt.

473

In this reaction, a sufficient quantity of H_2 was provided to reduce both C=O groups and to add hydrogen in place of the double bond between carbons 4 and 5.

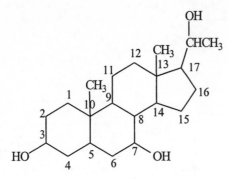

11.51 *Draw the product of each reaction.*

a. When a ketone reacts with an alcohol in the presence of an acid catalyst, the alcohol adds across the C=O bond. In this case, the H of CH_3OH adds to the carbonyl O and CH_3O adds to the carbonyl C.

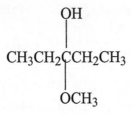

b. When an aldehyde is reacted with two molecules of alcohol in the presence of an acid catalyst, an acetal is formed. In this case the double bonded O is replaced by two CH_3CH_2O- from the alcohol molecules.

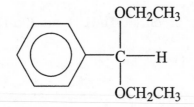

c. Refer to the comments from part a.

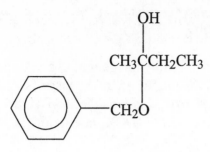

OH
|
CH₃CCH₂CH₃
|
$-$CH₂O

11.53 *Draw the missing reactant for each reaction.*

a. Since the product is a hemiacetal at position 3 of the carbon chain, the original reactant had to be 3-pentanone.

$$\overset{\displaystyle O}{\underset{\displaystyle \|}{}}$$

CH₃CH₂CCH₂CH₃

b. Since the product is an acetal formed by addition of two alcohol molecules, the reactant contained C=O.

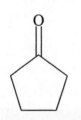

c. Since the product is an acetal formed by addition of just one alcohol molecule, the reactant was a hemiacetal.

O
‖
CH₃C$-$H

11.55 *The toxicity of methanol is mainly due to the aldehyde produced when it is oxidized in the liver. Draw methanol, then draw and name the aldehyde formed on its oxidation.*

Methanol   methanal known also as formaldehyde

CH₃OH   H$-$C$-$H
                        with =O above C

11.59 *In the breakdown of superoxide, which product results from the reduction of O_2^- and which results from the oxidation of O_2^-? Explain.*

Oxidation (the loss of electrons) can be identified by the gain of O and/or the loss of H. Reduction (the gain of electrons) can be identified by the loss of O and/or the gain of H. In going from O_2^- to O_2, superoxide loses an electron (it loses the minus charge) and is oxidized. In going from O_2^- to H_2O_2, superoxide gains H and is reduced.

11.61 *After reviewing the definition of the terms hemiacetal and acetal, identify any hemiacetal or acetal carbons in lactose (also known as milk sugar).*

A hemiacetal carbon atom is attached to a –OH and a –OC. An acetal carbon is attached to two –OC groups.

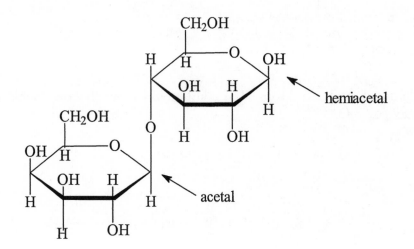

Chapter 12
Carbohydrates
Solutions to Problems

12.1 *How do oligosaccharides differ from polysaccharides?*

Oligosaccharides contain from 2 to10 monosaccharide residues while polysaccharides contain more than 10.

12.3 *Draw an example of each type of monosaccharide*

a. *an aldoheptose*

The name heptose indicates a seven carbon sugar. The "aldo" means it is an aldehyde. Therefore, draw a seven carbon chain with a –OH on every carbon except carbon 1 as this is where the C=O goes.

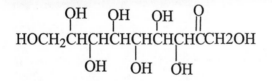

b. *a ketononose*

The name nonose indicates a nine carbon sugar. The "keto" means it is a ketone. Therefore, draw a nine carbon chain with a –OH on every   carbon except carbon 2 as this is where the C=O goes.

$$HOCH_2CHCHCHCHCHCHCCH2OH$$

with OH groups above and below along the chain (OH OH OH on top, O on carbonyl; OH OH OH below)

12.5 a. *How many chiral carbon atoms does D-lyxose contain?*

Examine the molecule for carbons that have four different atoms or group of atoms attached.

3

b. *How many total stereoisomers are possible for this aldopentose?*

Possible number of stereoisomers is found by using the formula 2^n where n = number of chiral carbons. $2^3 = 8$.

8

c. Draw and name the enantiomer of D-lyxose.

The enantiomer of D-lyxose is obtained by drawing the mirror image of D-lyxose.

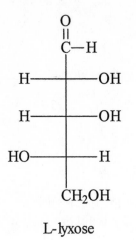

L-lyxose

d. D-ribose is a diastereomer of D-lyxose. Draw two additional D-diastereomers of this monosaccharide.

Diastereomers are stereoisomers that are not mirror images. Move one or more of the –OH groups in directions unlike those found in D-lyxose. Be sure that you haven't drawn L-lyxose (part c) and that have drawn D sugars.

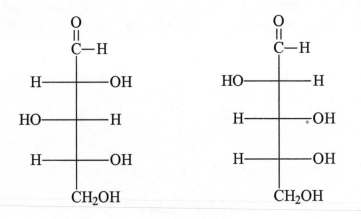

D-diastereomers

12.7 *a. Draw all possible aldotetroses.*

b. Which are D and which are L sugars?

The name tetrose indicates a four carbon sugar. The "aldo" means it is an aldehyde. Therefore, draw a four carbon chain with a –OH on every carbon except carbon 1, as this is where the C=O goes. To make all possible aldotetroses,

478

arrange the –OH groups of carbons 2 and 3 in as many different ways as possible. (Note: Two chiral carbons tell you there are $2^2 = 4$ different structures possible). D structures have the –OH group on the right at carbon 3, while L structures have the –OH group on the left.

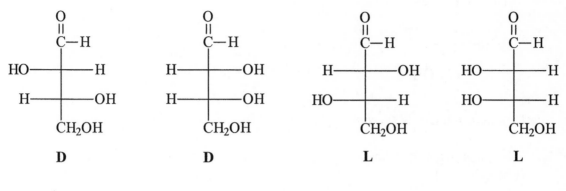

b.

12.9   *Draw each galactose derivative (see Figure 12.4).*

a. *D-2-deoxygalactose*

Draw the galactose as shown in Figure 12.4 and then remove the –OH group from carbon 2 and replace it with hydrogen.

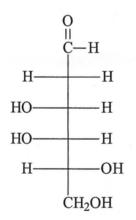

b. *galactitol (the alcohol sugar derived from galactose)*

Draw galactose as shown in Problem 12.4. Convert the aldehyde group into an alcohol.

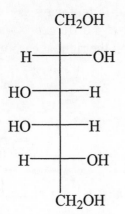

c. *D-galacturonic acid (carbon 6 of galactose is oxidized to a carboxylic acid)*

Draw galactose as shown in Figure 12.4. Convert the alcohol group on carbon 6 to a carboxylic acid.

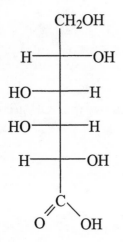

12.11   *Draw the product obtained when*

a. *D-mannose is reduced (see Figure 12.6)*

When aldehydes are reduced, they become alcohols. Draw the structure as shown in Figure 12.6 and convert the aldehyde group to an alcohol.

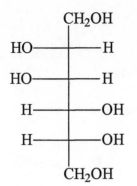

b. *carbon atom 1 of D-mannose is oxidized*

Draw D-mannose as shown in Figure 12.6. Convert the aldehyde group to a carboxylic acid.

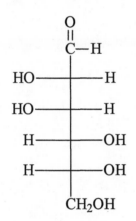

c. *carbon atom 6 of D-mannose is oxidized to a carboxylic acid.*

Draw D-mannose as shown in Figure 12.6. Convert the alcohol on carbon 6 to a carboxylic acid.

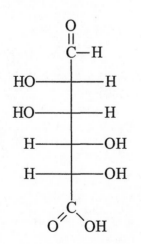

12.13   *Define the term reducing sugar.*

Reducing sugars are sugars that give a positive Benedict's test (in the process of being oxidized, the sugars reduce Cu^{2+} ion present in the reagent).

12.15   *Draw arabinitol, the alcohol sugar formed when D-arabinose is reacted with H_2 and Pt.*

Start with the D-arabinose structure shown in Problem 12.15. Convert the aldehyde group to an alcohol.

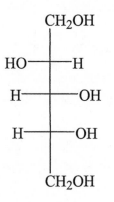

12.17   a. *Draw the D-aldohexose that gives the alcohol sugar below, when treated with H_2 and Pt.*

H_2 and Pt will reduce an aldehyde to an alcohol. Start with the structure shown in the problem. Convert the alcohol into an aldehyde.

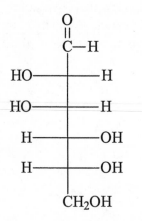

b. *Is the aldohexose a reducing sugar?*

Yes. The aldehyde group can be oxidized by Cu^{2+}.

482

c. *Is the aldohexose a deoxy sugar?*

No. Deoxy means that an –OH group has been replaced by hydrogen. This is not the case for the D-aldohexose.

d. *Is the aldohexose an amino sugar?*

No. Amino sugar means that an –OH has been replaced by –NH$_2$. This is not the case for the D-aldohexose.

12.19 *Draw each molecule.*

a. *α-D-glucopyranose*

A glucopyranose molecule is a glucose that has formed six-membered cyclic hemiacetal. The six-membered ring is drawn with the oxygen at the back and the hemiacetal carbon on the right side. The α-notation indicates that the –OH on the hemiacetal carbon is pointing down.

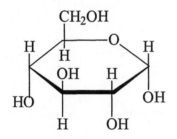

b. *β-D-ribofuranose*

A ribofuranose molecule is a ribose that has formed a five-membered cyclic hemiacetal. The five-membered ring is drawn with the oxygen at the back and the hemiacetal carbon on the right side. The β-notation indicates that the –OH on the hemiacetal carbon is pointing up.

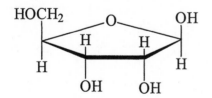

c. *α-D-galactopyranose*

A galactopyranose molecule is a galactose that has formed a six-membered cyclic hemiacetal. The six-membered ring is drawn with the oxygen at the back and the hemiacetal carbon on the right side. The α-notation indicates that the –OH on the hemiacetal carbon is pointing down.

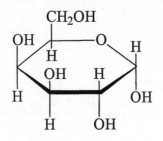

d. *β-D-arabinofuranose*

An arabinofuranose molecule is an arabinose (see Problem 12.15) that has formed a five-membered cyclic hemiacetal. The five-membered ring is drawn with the oxygen at the right and the hemiacetal carbon on the back side. The *β*-notation indicates that the –OH on the hemiacetal carbon is pointing up.

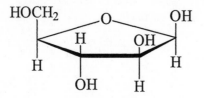

12.21 *D-lyxose (see Problem 12.5) is a diastereomer of D-ribose. Draw α-D-lyxofuranose.*

A lyxofuranose molecule is a lyxose that has formed a five-membered cyclic hemiacetal. The five-membered ring is drawn with the oxygen at the back and the hemiacetal carbon on the right side. The α-notation indicates that the –OH on the hemiacetal carbon is pointing down.

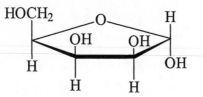

12.23 *Draw β-D-mannopyranose (see Figure 12.6).*

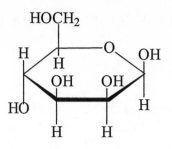

12.25 *Define the term mutarotation.*

Mutarotation is the process of converting back and forth from an α anomer to the open form to the β anomer.

12.27 *Cellobiose is a reducing sugar. Write a reaction equation that shows why.*

When cellobiose mutarotates, the open form contains an aldehyde that can be oxidized by Cu^{2+}.

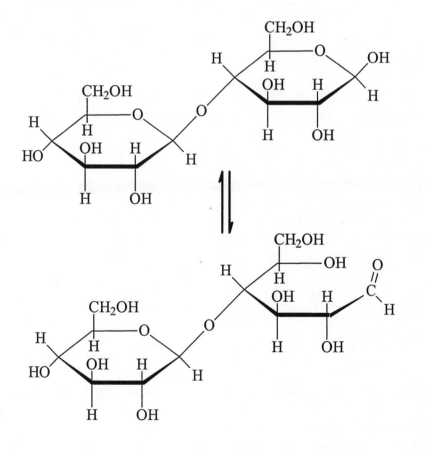

12.29 *Gentiobiose consists of two D-glucose residues joined by a β-(1→6) glycosidic bond.*

a. *Draw this disaccharide.*

Draw two α-D-glucopyranose molecules as described in Problem 12.19a. From one of them, remove the hydrogen from the –OH on the hemiacetal carbon (carbon 1). From the other remove the –OH from carbon 6. Draw a bond connecting the two molecules.

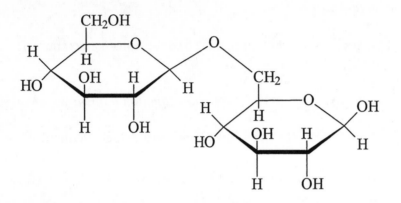

b. *Is this gentiobiose a reducing sugar?*

Yes. When the β hemiacetal (at the right side of the molecule, as drawn) undergoes mutarotation, the resulting aldehyde group can be oxidized by Cu^{2+}.

12.31 *Vanillin β-D-glucoside gives vanilla extract its flavor. Draw the two products obtained when this acetal is hydrolyzed.*

Hydrolysis of an acetal bond produces a hemiacetal and an alcohol.

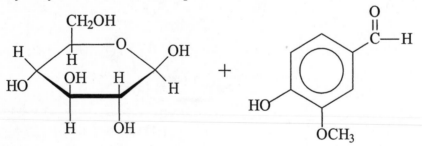

12.33 *Draw a disaccharide consisting of two D-glucose residues that is not a reducing sugar.*

The disaccharide can be made by linking two D-glucose molecules with an α, β-(1↔1) glycosidic bond. The resulting structure contains no hemiacetal group. Therefore, it is unable to mutarotate, and it is not a reducing sugar.

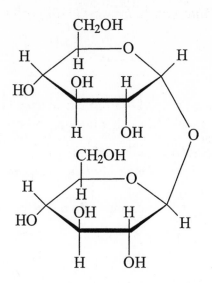

12.35 a. *Is trehalose, a disaccharide found in a wide range of living things, a reducing sugar?*

No. The molecule contains no hemiacetal groups.

b. *Which describes the glycosidic bond in this disaccharide: α,α-(1↔1), α-(1→2), α-(1→3), α-(1→4), α-(1→5), or α-(1→6)?*

The glycosidic bond involves the hemiacetal carbon (carbon 1) of each ring, each of which is α.

α,α-(1↔1)

12.39 *How is the structure of cellulose different from that of amylose?*

Cellulose has β-(1→4) glycosidic bonds and amylose has α-(1→4) glycosidic bonds.

12.41 *How is the structure of glycogen similar to that of amylopectin?*

Glycogen and amylopectin each consist of glucose residues joined by α-(1→4) and α-(1→6) glycosidic bonds.

12.43 *A particular lichen produces the polysaccharide below.*

a. *Is the carbohydrate a homopolysaccharide or a heteropolysaccharide?*

The prefix "homo" means composed of just one type of monosaccharide. The structure shows that polysaccharide is composed of only one type of monosaccharide (glucose).

Homopolysaccharide

b. *Name the glycosidic bond present, as α-(1⟶ 4), β-(1⟶ 2), etc.*

β-(1→ 6)

12.45 *Chondroitin 6-sulfate, identical to chondroitin 4-sulfate (Section 12.5) with the exception of where the sulfate group is attached, is another of the heteropolysaccharides present in connective tissue. Draw chondroitin 6-sulfate.*

To change the molecule from chondroitin 4-sulfate to chondroitin 6- sulfate, move the sulfate ion from carbon 4 to carbon 6.

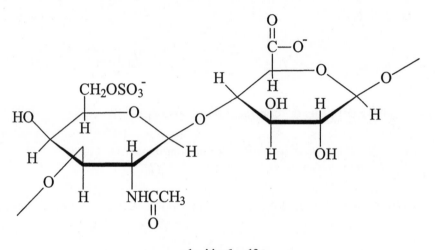

condroitin 6-sulfate

12.47 *Olestra (Figure 8.13) is a derivative of which naturally occurring oligosaccharide?*

Examine Figure 8.13 and match the olestra structure to the oligosacharides shown in Section 12.4

Sucrose

12.49 *What is lactose intolerance and what are some of the options for dealing with it?*

Lactose intolerance is a deficiency in β-galactosidase, the enzyme that catalyzes the hydrolysis of the β-(1→ 4) glycosidic bond in lactose. Persons with this disorder may deal with it by avoiding dairy products, by using dairy products from which lactose has been removed, and by taking tablets containing β-galactosidase.

12.51 *The blood group antigens described in Figure 12.19 also appear in another form, in which the D-galactose drawn in the center is attached to the N-acetyl-D-*

488

glucosamine on the right by a β-(1→ 3) glycosidic bond. Draw the oligosaccharide portion of this alternative structure.

The β-(1→ 4) glycosidic bond shown in Figure 12.19 must be changed into a β-(1→ 3) glycosidic bond.

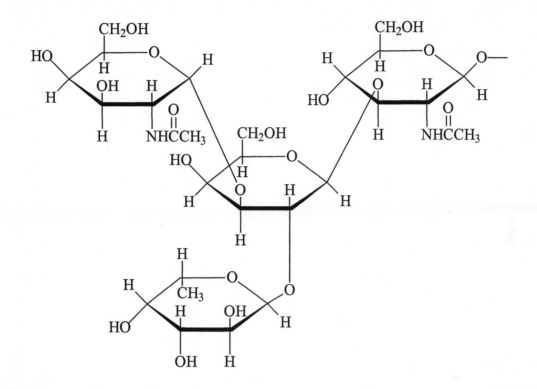

12.53 *Suppose that a new artificial sweetener is approved for use by the Food and Drug Administration, but some public interest groups question its safety (this happened when aspartame was put on the market). Would you use the sweetener?*

Answers will vary and may include: Yes, it has undergone testing and has been approved by the FDA. No sometime in the future it may be found to be unsafe.

Chapter 13
Peptides, Proteins, and Enzymes
Solutions to Problems

13.1 *Draw methionine as it would appear at each of the following pHs.*

a. *pH 1*

At a pH 1 the carboxyl group appears as –CO₂H and the amino group as NH₃⁺.

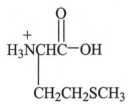

b. *pH 7*

At a pH 7 the carboxyl group appears as –CO₂⁻ and the amino group as NH₃⁺.

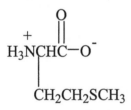

c. *pH 14*

At a pH 14 the carboxyl group appears as –CO₂⁻ and the amino group as NH₂.

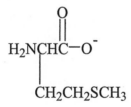

13.3 a. *What is the net charge on arginine at pH 1?*

See Table 13.1 for the structure at pH 7. At pH 1, the carboxyl group appears as –CO₂⁻ and the amino (-NH₂) group as NH₃⁺.

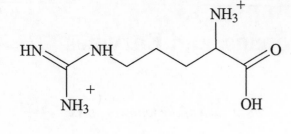

2+

b. *What is net charge on arginine at pH 14?*

See Table 13.1 for the structure at pH 7. At pH 14, the carboxyl group appears as $-CO_2^-$ and the amino group as $^-NH_2$.

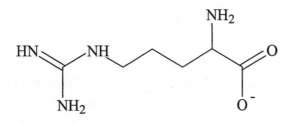

1-

13.5   *Using Fischer projections draw each amino acid as it would appear at pH 7.*

A Fischer projection is drawn with the chiral carbon atom at the intersection of a vertical and a horizontal line. For amino acids, the carboxyl group points up, the R group points down and the amino group points either left (L amino acid) or right (D amino acid).

a. *L-isoleucine*

See Table 13.1 for the structure at pH 7.

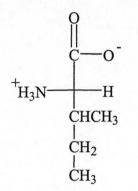

b. *D-aspartic acid*

See Table 13.1 for the structure at pH 7.

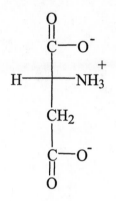

c. *L-tyrosine*

See Table 13.1 for the structure at pH 7.

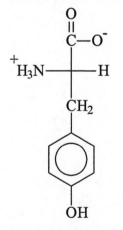

d. *D-phenylalanine*

See Table 13.1 for the structure at pH 7.

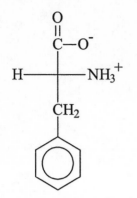

13.7 *Using a Fischer projection, draw each amino acid from Problem 13.5 as it appears at pH 1.*

At a pH of 1, carboxyl groups appear as $-CO_2H$ and amino groups as $-NH_3^+$.

a.

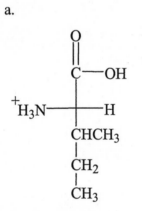

b.

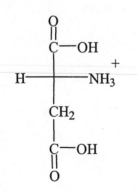

c.

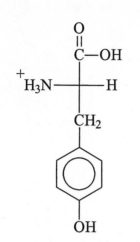

d.

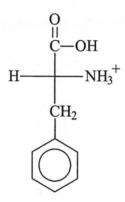

13.9   *Two of the amino acids in Table 13.1 have two chiral carbon atoms. Which are they?*

The two amino acids in Table 13.1 that have two carbons with four different atoms or groups of atoms, are threonine and isoleucine.

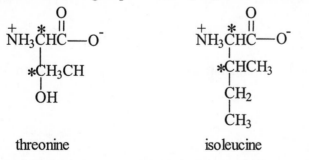

13.11   *Monosodium glutamate (MSG), used to enhance the flavor of certain foods, carries a net charge of zero. It can be formed by reacting glutamic acid   as it appears at low pH (below) with sufficient NaOH. Draw MSG by showing the molecule below as it would appear at pH 7, and attaching Na⁺ to the side chain.*

At a pH of 7, carboxyl groups appear as $-CO_2^-$ and the amino group as $-NH_3^+$.

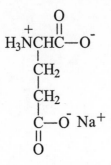

13.13 *Which specific class of bonds holds one amino acid residue to the next in the primary structure of a protein? Are these bonds covalent or noncovalent?*

Peptide bonds, also known as amide bonds. Covalent.

13.15 a. *Is Ala-Phe-Thr-Ser an oligopeptide or a polypeptide?*

The compound has four amino acid residues. Since this is between 2 and 10 it is an oligopeptide.

Oligopeptide

b. *How many peptide bonds does the molecule contain?*

A peptide bond is found between each of the joined residues, therefore there are three bonds.

Three peptide bonds

c. *Which is the N-terminal amino acid?*

The N-terminal residue is written first, therefore alanine is the N-terminal amino acid.

Alanine

13.17 *Name the amino acid residues in the oligopeptide shown in Figure 13.4b.*

Match the side chains of the amino acid residues with drawings in Table 13.1. Start at the N-terminus (one with an unreacted amino group) and list the individual residues.

Cysteine, Tyrosine, Isoleucine, Glutamine, Asparagine, Cysteine, Proline, Leucine, Glycine.

13.19 *Draw Asp-Ser-Lys-Val as it would appear at*

a. *pH 1*

Start with drawings provided with Table 13.1. Asp is at the N-terminus. Draw the individual residues connecting each with a peptide bond. At a pH of 1, carboxyl groups appear as $-CO_2H$ and amino groups appear as $-NH_3^+$.

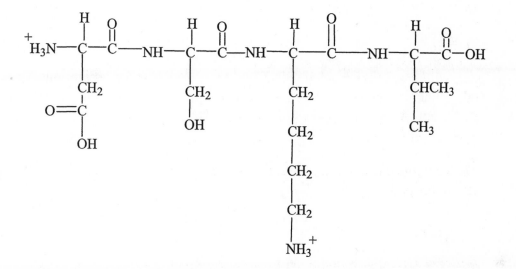

b. *pH 7*

Start with drawings provided with Table 13.1. Asp is at the N-terminus. Draw the individual residues connecting each with a peptide bond. At a pH of 7, carboxyl groups appear as $-CO_2^-$ and amino groups as $-NH_3^+$.

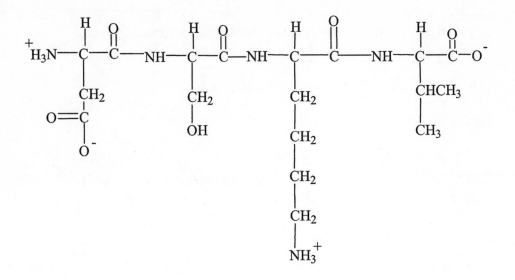

c. *pH 14*

Start with drawings provided with Table 13.1. Asp is at the N-terminus. Draw the individual residues connecting each with a peptide bond. At a pH 14 carboxyl groups appear as $-CO_2^-$ and amino group as NH_2.

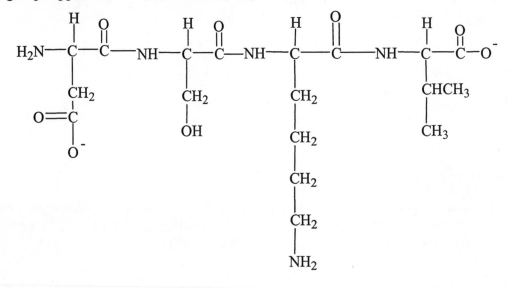

13.21 *What is the net charge on Phe-Asp at each pH?*

Draw the structure for Phe-Asp using the drawings shown in Table 13.1.

a. *pH 1*

At a pH of 1, carboxyl groups appear as $-CO_2H$ and amino groups as $-NH_3^+$. Phe-Asp has net 1+ charge at this pH.

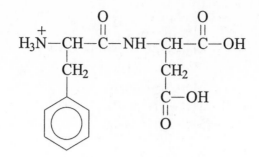

b. *pH 7*

At a pH of 7, carboxyl groups appear as $-CO_2^-$ and amino groups as $-NH_3^+$. Phe-Asp has a net 1- charge at this pH.

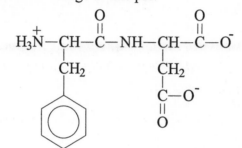

c. *pH 14*

At a pH 14 the carboxyl groups appear as $-CO_2^-$ and amino group as NH_2. Phe-Asp has a net 2- charge at this pH.

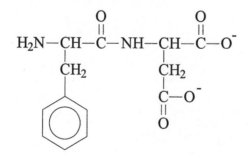

13.23 *How many tetrapeptides can be produced that contain one residue each of serine, methionine, arginine, and tyrosine?*

Starting with a different residue each time, make as many different arrangements as possible. Count the different possible arrangements.

Ser-Met-Arg-Tyr	Met-Ser-Arg-Tyr
Ser-Met-Tyr-Arg	Met-Ser-Tyr-Arg
Ser-Arg-Tyr-Met	Met-Arg-Ser-Tyr
Ser-Arg-Met-Tyr	Met-Arg-Tyr-Ser
Ser-Tyr-Met-Arg	Met-Tyr-Ser-Arg
Ser-Tyr-Arg-Met	Met-Tyr-Arg-Ser
Arg-Tyr-Met-Ser	Tyr-Met-Arg-Ser
Arg-Tyr-Ser Met	Tyr-Met-Ser-Arg
Arg-Met-Tyr-Ser	Tyr-Ser-Arg-Met
Arg-Met-Ser-Tyr	Tyr-Ser-Met-Arg
Arg-Ser-Tyr-Met	Tyr-Arg-Ser-Met
Arg-Ser-Met-Tyr	Tyr-Arg-Met-Ser

Twenty-four tetrapeptides.

13.25　*In an aqueous environment will the following peptide fragment more likely be buried inside a globular protein or located on its surface? Explain.*

Inside. This arrangement allows the nonpolar side chains to interact through London forces and does not disrupt hydrogen bonding between water molecules.

13.27　*List the chemical bonds or forces that are primarily responsible for maintaining*

　　　a. *the primary structure of a protein*

Peptide bonds

　　　b. *the secondary structure of a protein*

Hydrogen bonding

　　　c. *the tertiary structure of a protein*

hydrogen bonding, salt bridges, dipole-dipole interaction, ion-dipole interactions, the hydrophobic effect, and disulfide bonds

　　　d. *the quaternary structure of a protein*

hydrogen bonding, the hydrophobic effect, and salt bridges

13.29　*Which amino acids have side chains that can participate in salt bridge formation (ionic bonds) at pH 7.*

A salt bridge is formed when a positively charged side chain is attracted to a negatively charged side chain. The only amino acids that have side chains that will have a charge at a pH of 7 are: arginine ($=NH_2^+$), aspartic acid ($-CO_2^-$), glutamic acid (CO_2^-), and lysine ($-NH_3^+$).

Aspartic acid, glutamic acid, lysine, and arginine

13.31 *Explain why many proteins have no quaternary structure.*

They consist of a single polypeptide chain.

13.33 *Why do extracellular proteins tend to contain more disulfide bonds than do intracellular ones?*

Disulfide bonds of intracellular proteins tend to be reduced by NADH and other reducing agents present in the cytoplasm of the cell.

13.35 *Draw structures that show*

a. *a salt bridge between Asp and Arg side chains*

Draw the structures of Asp and Arg (See Table 13.1). The salt bridge is the attraction between the $-O^-$ of Asp and the $-NH_2^+$ group of Arg.

$$\text{—CH}_2\overset{\overset{\displaystyle O}{\|}}{C}\text{—O}^- \quad {}^+NH_2$$

$$H_2N\text{—}\overset{\|}{C}\text{—NHCH}_2CH_2CH_2\text{—}$$

b. *hydrophobic interactions between Trp and Val side chains.*

Draw the structures of Trp and Val (See Table 13.1). The hydrophobic interaction is the attraction between the two nonpolar side chains.

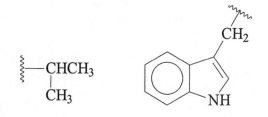

13.37 *What term is used to describe an enzyme that has been unfolded and inactivated without breaking covalent bonds?*

Denaturation

13.39 *Which is/are true for a denatured globular protein?*

  a. *is biologically inactive*
  b. *contains no peptide bonds*
  c. *has an abnormal primary structure*

  a. is biologically inactive

13.41 *List some of the ways to denature a protein.*

  a. change the temperature
  b. change the pH.
  c. add detergent or soap
  d. agitate the protein

13.43 *Is it possible for an enzyme to show both relative specificity and stereospecificity?*

  Relative specificity is the ability to react with a range of substrates, such as many different carbohydrates. Stereospecificity is the ability to react with or form a particular stereoisomer, such as the D isomers of those carbohydrates. Therefore an enzyme can show both relative specificity and sterospecifictiy.

  Yes.

13.45 *The enzyme trypsin, found in the small intestine, operates best at pH 8. Draw a graph for trypsin, similar to that found in Figure 13.21a.*

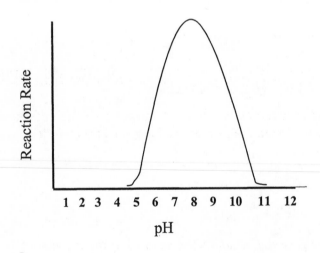

13.47   *A thermophilic bacterium found in a hot spring at Yellowstone Park produces an*
         enzyme with a temperature optimum of 98°C. For this enzyme, draw a graph
         similar to that found in Figure 13.21b.

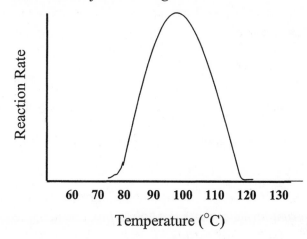

13.49   *An inhibitor is added to a Michaelis-Menten enzyme. How could you distinguish*
         between the inhibitor being competitive or noncompetitive?

         A competitive inhibitor alters K_m but not the V_{max}. A noncompetitive inhibitor
         alters V_{max} but not the K_m.

13.51   *Explain why most competitive inhibitors are structurally similar to a substrate for*
         the enzyme they inhibit, while most noncompetitive inhibitors are not.

         Competitive inhibitors bind at the substrate binding site. Noncompetitive
         inhibitors do not.

13.53   *Describe the effect of positive and negative effectors on allosteric enzymes.*

         A positive effector enhances substrate binding and increases the rate of the
         enzyme catalyzed reaction. A negative effector reduces substrate binding and
         slows the reaction.

13.55   *Explain how addition or removal of a phosphate group can be used to control the*
         activity of an enzyme.

         Phosphorylation or dephosphorylation alters the activity of the enzyme.

13.57   *A person is found to produce trypsinogen molecules that have an abnormal*
         primary structure. The trypsin produced from this zymogen is perfectly normal,
         both structurally and functionally. How is this possible?

The abnormal primary structure is the part of the polypeptide chain removed during activation of trypsinogen.

13.59 *List one unusual amino acid residue found in collagen. Is this residue present in a newly assembled collagen polypeptide? Explain.*

Hydroxyproline (Hyp). No, Hyp is formed by the enzyme-catalyzed oxidation of a hydroxyl group of a proline residue.

13.61 *Describe the quaternary structure of hemoglobin.*

A hemoglobin molecule is a tetramer consisting of four separate polypeptide chains (two identical α chains and two identical β chains). Each α chain consists of 141 amino-acid residues coiled into seven α-helical regions and each β chain consists of 146 amino acid residues coiled to form eight α-helical regions.

13.63 *True or false? If liver cells are the only cells in the human body that produce enzyme x, an increase in serum levels of enzyme x following a virus infection probably indicates that the infection has resulted in some liver damage. Explain.*

True. When cells die their contents are released into the bloodstream. If enzyme *x* is detected in the bloodstream in elevated concentrations it probably indicates liver cells have been damaged.

13.65 *Would orally administered asparaginase be as effective as intravenously injected asparaginase in reducing serum asparaginase levels? Explain.*

No. The asparaginase, like other proteins, would be broken down during digestion.

13.67 *As we will see in the next chapter, some of the information carried by DNA specifies the primary structure of proteins. Can a change in the structure of DNA (a change in the information) result in the altered tertiary structure of a protein that it carries the code for?*

Yes. The tertiary structure of a protein depends on its primary structure.

Chapter 14
Nucleic Acids
Solutions to Problems

14.1   *Which three components go into making a nucleotide?*

Phosphoric acid, a monosaccharide, and an organic base.

14.3   a. *Which monosaccharide is used to make RNA?*

The R in RNA is for "ribo" indicating that the monosaccharide is ribose.

Ribose

b. *Draw this monosaccharide in its β-furanose form.*

Draw the five-membered ribose ring structure with the hemiacetal –OH group pointing upward. See Chapter 12 Carbohydrates if you need review on this structure.

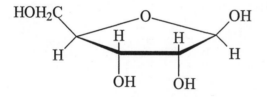

14.5   *Name the four bases that are present in RNA.*

Adenine, guanine, cytosine, and uracil.

14.7   *Which of the bases present in nucleotides are pyrmidines?*

Cytosine, thymine, and uracil.

14.9   *Figure 14.1 shows the phosphorus atom in a phosphate ion as having five covalent bonds. Draw a phosphate ion in which the phosphorus atom has an octet of valence electrons. Specify formal charges.*

Draw the structure shown in Figure 14.1. Replace the double bond with a single bond. This gives two electrons back to oxygen and makes it have a formal charge of 1-. The phosphorous loses that shared pair which makes it have a formal charge of 1+ .

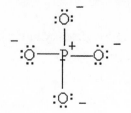

14.11 *Draw the complete structure of guanosine 5'-monophosphate (see Figure 14.6).*

The guanosine molecule is shown in Figure 14.5 and the nucleoside 5' - monophosphate structure is shown in Figure 14.6. Draw the structure for guanosine and then remove the –OH from carbon 5 of the ribose ring and replace it with phosphate.

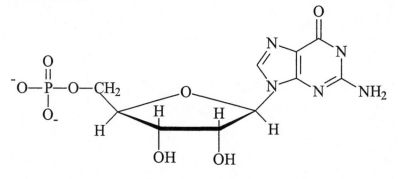

14.13 *How many phosphoester bonds and how many phosphoanhydride bonds are present in the following?*

a. *a nucleotide*

A phosphoester bond connects a phosphate and a nucleoside. There is one phosphate connected to the nucleoside and therefore one phosphoester bond. There are no phosphoanhydride bonds (a chemical bond formed between two phosphate groups) since there is only one phosphate.

One phosphoester bond and no phosphoanhydride bond

b. *nucleoside diphosphate*

There is one phosphate connected to the nucleoside and therefore one phosphoester bond. There are two phosphates and therefore one phosphoanhydride bond.

One phosphoester bond and one phosphoanhydride bond

14.15   a. *Draw 3'-dATP*

3'-dATP is short for 3'-deoxyadenosine triphosphate. Draw dexoxyadenosine.
Remove the –OH on carbon 3 and add three phosphate groups.

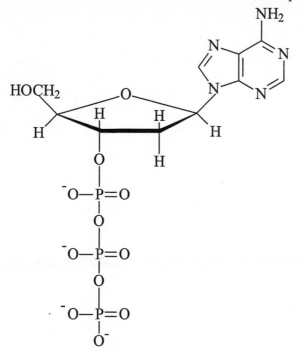

b. Draw 3'ATP

3'ATP is short for adenosine triphosphate. Draw adenosine. Remove the –OH
on carbon 3 and add three phosphate groups.

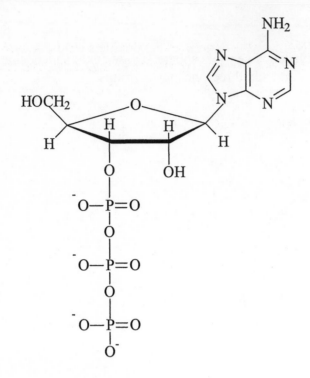

14.17 *In the sugar phosphate backbone of DNA*

a. *which sugar residue is present?*

2-Deoxyribose

b. *which type of bonds connect the sugar and phosphate residues?*

Phosphoester bonds

c. *where are bases attached?*

At C-1'

14.19 *In terms of DNA structure, what do the terms 3'-terminus and 5'-terminus mean?*

At one end of a DNA strand (3'-terminus) the 3' hydroxyl group has no attached nucleotide residue. At the other end (5'-terminus) the 5' hydroxyl group has no attached nucleotide residue.

14.21 *How does a single strand of DNA differ from a single strand of RNA?*

DNA contains 2-deoxyribose residues, while RNA contains ribose residues. The bases in DNA are A, G, C, and T, while those in RNA are A, G, C, and U.

14.23 *Describe the secondary and tertiary structure of DNA.*

The secondary structure of DNA consists of antiparallel double strands twisted into a helix. The tertiary structure of DNA involves supercoiling of the double helix.

14.25 *What force holds double-stranded DNA together?*

Hydrogen bonding between complementary bases.

14.27 a. *Draw the complete structure of the DNA trinucleotide dTAT.*

b. *Label the 3' and 5' ends of the molecule.*

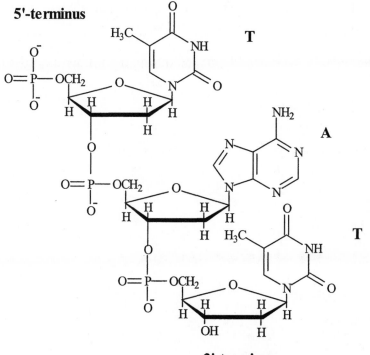

14.29 a. *What are histones?*

Proteins that interact with the phosphate groups of the DNA backbone.

b. *What are nucleosomes?*

Groups of histones wrapped by DNA

c. *What is chromatin?*

A coiled string of nucleosomes.

14.31 *Compare the primary, secondary, and tertiary structure of proteins to that of DNA.*

The primary structure of a protein is sequence of amino acid residues. In DNA, primary structure is the sequence of nucleotide residues. Secondary structure is the arrangement of a protein into an α-helix or β-sheet, and the arrangement of DNA into a double helix. For both proteins and DNA, tertiary structure is the overall three dimensional shape.

14.33 *What does it mean to renature DNA?*

To renature DNA means to allow the DNA to return to its original double stranded formation.

14.35 The primary structure of what molecule is read during the following processes?

a. *DNA replication*

DNA.

b. *transcription*

DNA.

c. *translation*

mRNA.

d. *reverse transcription*

viral RNA.

14.37 a. *What is an origin?*

A site on DNA where the double strand has been pulled apart to expose single strands. DNA replication takes place at an origin.

b. *During DNA replication in plants and animals does just one origin form?*

No, during replication in plants and animals, many origins are opened at one time along a double-stranded DNA helix.

14.39 a. *In which direction along an existing DNA strand does DNA polymerase move?*

510

Polymerases move along an existing DNA strand in a 3' to a 5' direction.

b. *In which direction does DNA polymerase synthesize a new DNA strand?*

Polymerases synthesize new DNA in a 5' to 3' direction.

c. *If an origin opens and a DNA polymerase attaches to each of the exposed single strands of DNA, do the polymerases move in the same direction or in opposite directions as they make new DNA?*

Opposite directions.

14.41　*Name the three types of RNA and describe their function.*

Transfer RNAs carry the correct amino acid to the site of protein synthesis.

Messenger RNAs carry the information that specifies which protein should be made.

Ribosomal RNAs combine with proteins to form ribosomes, the multi-subunit complexes in which protein synthesis takes place.

14.43　*The sequence dGGCAT appears in a template strand of DNA.*

a. *Which base is at the 3'-terminus of this primary structure?*

Unless otherwise noted, the 3'-terminus residue is written on the right of the sequence code. T stands for Thymine.

Thymine.

b. *What is the sequence in the DNA strand that is complementary to this sequence? (Label the 3' and 5' ends.)*

A and T are complementary bases, as are G and C.

3' terminus: dCCGTA : 5' terminus

c. *What is the sequence in the RNA strand that is synthesized from dGGCAT? (Label the 3' and 5' ends.)*

3' terminus: CCGUA : 5' terminus

14.45　a. *What are codons?*

A codon is a series of three bases that specifies a particular amino acid.

b. *Which type of RNA has codons?*

Messenger RNA.

14.47 *Which amino acid is specified by each codon (listed 5' to 3')?*

See Table 14.1. Find the beginning code letter on the right column, locate the last code letter in the far right column. This establishes the row. Find the middle code letter at the top of the table and follow that column down to the row in line with your other two letters and find the listed amino acid.

a. *CCU*

Pro

b. *AGU*

Ser

c. *GUU*

Val

14.49 *Which codon(s) specify each amino acid?*

Find the amino acid on Table 14.1. in the middle of the chart. This establishes the middle code letter. The left and right code letters are found on the same row in the left and right columns. Every row that contains the amino acid is a different codon for it.

a. *Phe*

UUU and UUC

b. *Lys*

AAA and AAG

c. *Asp*

GAU and GAC

14.51 *What types of post-transcriptional modifications do RNAs undergo?*

Transfer RNA is shortened and some of the major bases are converted into minor bases. Messenger RNA may be trimmed at either end, the ends may be altered, and sections of the middle may be removed. Ribosomal RNA is cut into smaller pieces.

14.53 *Which anticodons are complementary to the codons in Problem 14.48? (Label the 3' and 5' ends.)*

a. (listed 3' to 5') AAA

b. (listed 3' to 5') GGG

c. (listed 3' to 5') UUU

d. (listed 3' to 5') CCC

14.55 *Describe the role of mRNA, tRNA, and rRNA in protein synthesis.*

rRNA combines with specific proteins to form a ribosome, the site of protein synthesis. mRNA carries the code that specifies the primary structure of the protein to be made. tRNA carries amino acid residues to where they are needed for addition to the growing protein.

14.57 a. *For the tripeptide in Problem 14.56, again assuming that no post-translational modification took place, what is a different sequence of bases in the mRNA that would code for this tripeptide? (Ignore start and stop codons and label the 3' and 5' ends.)*

The tripeptide can be represented by several different code sequences. Ser has the following codons: UCC, UCG, UCA, UCU, AGU, AGC; Lys has the following codons: AAA, AAG; and Asp has the following codons: GAC, GAU.
Recall that a tripeptide has many different possible combinations of coding anyone of which could be used to answer this question. The possible combinations for Ser – Lys – Asp are shown below. If asked to do both 14.56 and 14.57 for homework make sure the one chosen for this question is different than the one in 14.46.

b. *Assuming that no post-transcriptional modification took place, what is the sequence of bases in the template DNA strand used to make the mRNA? (Label the 3' and 5' ends.)*

The DNA sequences are listed in corresponding order to the answers for part a.

513

a. Possible sequence of mRNA bases **b. Possible sequence of DNA bases**

(5' end)   (3' end)   (3' end)   (5' end)

UCUAAAGAU   dAGATTTCTA
UCCAAAGAU   dAGGTTTCTA
UCAAAAGAU   dAGTTTTCTA
UCGAAAGAU   dAGCTTTCTA
AGUAAAGAU   dUCATTTCTA
AGCAAAGAU   dUCGTTTCTA

UCUAAGGAU   dAGATTCCTA
UCCAAGGAU   dAGGTTCCTA
UCAAAGGAU   dAGTTTCCTA
UCGAAGGAU   dAGCTTCCTA
AGUAAGGAU   dTCATTCCTA
AGCAAGGAU   dTCGTTCCTA

UCUAAAGAC   dAGUTTTCTG
UCCAAAGAC   dAGGTTTCTG
UCAAAAGAC   dAGTTTTCTG
UCGAAAGAC   dAGCTTTCTG
AGUAAAGAC   dTCATTTCTG
AGCAAAGAC   dTCGTTTCTG

UCUAAGGAC   dAGATTCCTG
UCCAAGGAC   dAGGTTCCTG
UCAAAGGAC   dAGTTTCCTG
UCGAAGGAC   dAGCTTCCTG
AGUAAGGAC   dTCATTCCTG
AGCAAGGAC   dTCGTTCCTG

14.59   a. *What are the gene products of the lac operon?*

β-galactose, lactose permease, and transacetylase

b. *How does the repressor protein prevent the lac operon from being expressed?*

The repressor protein binds to operator sites O_1 and O_2, control sites on the DNA near or at the promoter site.

c. *How does the presence of lactose act as an "on" switch for the lac operon?*

Lactose is converted to allolactose by β-galactosidase. Allolactose binds to the repressor protein and changes its shape in such a way that it can no longer attach to the operator sites. This allows the RNA polymerase to bind to the promoter site turning "on" transcription. .

14.61   a. *Draw lactose and allolactose.*

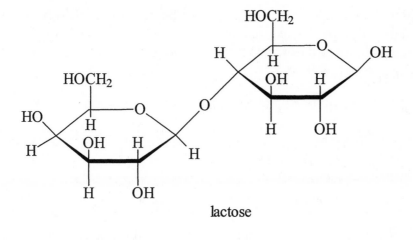

lactose

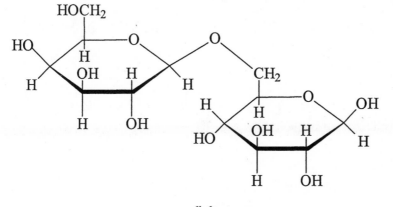

allolactose

b. *What products are obtained when allolactose is hydrolyzed?*

One galactose and one glucose molecule.

14.63   a. *Are all mutations harmful to an individual? Explain.*

b. *Are all mutations passed on to offspring? Explain.*

No. Only mutations that occur in the egg or sperm cells can be passed on to offspring.

14.65   *What is reverse transcriptase? Why do some viruses produce this enzyme?*

Reverse transcriptase is an enzyme that catalyzes the reaction necessary to make DNA copy of a viral RNA. Some viruses produce transcriptase because they contain RNA which, when injected into a host cell must be converted into DNA for viral proteins to be produced.

14.67   *What is the normal function of the BRCA1 and BRCA2 genes?*

The genes BRCA1 and BRCA2 code for proteins indirectly involved in a form DNA repair called recombinational repair.

14.69   *What are short tandem repeats and how are they used in DNA fingerprinting?*

Short tandem repeats are relatively small stretches of DNA that contain short repeating sequences of bases. Restriction enzymes are used to cut the DNA into small segments and the number of base repeats of the STRs are determined.

Chapter 15
Metabolism
Solutions to Problems

15.1 *Define the term metabolic pathway.*

Metabolic pathways are groups of metabolic reactions.

15.3 *In the first step of glycolysis, the following two reactions are coupled*

$$\text{glucose} + P_i \rightarrow \text{glucose 6-phophate} + H_2O \quad \Delta G = +3.3 \text{ kcal/mol}$$
$$\text{ATP} + H_2O \rightarrow \text{ADP} + P_i \quad\quad\quad\quad\quad \Delta G = -7.3 \text{ kcal/mol}$$

a. *Is the reaction glucose + $P_i \rightarrow$ glucose 6-phosphate + H_2O spontaneous?*

A positive ΔG ($\Delta G = +3.3$ kcal/mol) indicates the reaction is nonspontaneous.

No.

b. *Write the overall reaction equation and calculate ΔG for the coupled reaction.*

The overall reaction is the sum of both reactions. Write ALL reacts on the left and ALL products on the right. Cancel out the reactants and products that are the same on both sides. Write the remaining reactants and products as the overall reaction. Add up the ΔG values, +3.3 kcal/mol + (-7.3 kcal/mol) = -4.0 kcal/mol.

glucose + ~~H₂O~~ + ATP + ~~Pᵢ~~ → glucose 6-phosphate + ADP + ~~Pᵢ~~ + ~~H₂O~~

$$\text{glucose} + \text{ATP} \rightarrow \text{glucose 6-phosphate} + \text{ADP} \quad \Delta G = -4.0 \text{ kcal/mol}$$

c. *Is the first step in glycolysis spontaneous?*

A negative ΔG ($\Delta G = -4.0$ kcal/mol) indicates the first step in glycolysis is spontaneous.

Yes.

15.5 *For the reaction described in Problem 15.4, what effect will the enzyme have on the value of ΔG for the reaction?*

No effect.

15. 7   *The abbreviation PP$_i$ is used to represent pyrophosphate ion.*

a. *Draw pyrophosphate.*

Draw two phosphate ions connected by a phosphoanhyrdide bond.

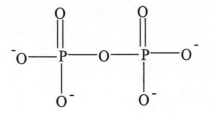

b. *Draw pyrophosphate as it would appear at a very acidic pH.*

A very acidic pH results in H$^+$ attaching to all of the –O$^-$ sites.

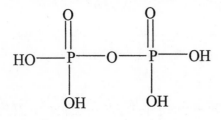

15.9   *What provides the energy used to produce ATP from ADP and P$_i$, anabolism or catabolism?*

Catabolism

15.11   *Define the term reduction and explain how it applies to the difference in the structures of NAD$^+$ and NADH.*

Reduction is the gain of electrons. In organic and biochemical molecules, the gain of hydrogen and /or the loss of oxygen is indication that reduction has taken place. NADH, the reduced form of the coenzyme has one more hydrogen atom than NAD$^+$, the oxidized form of the coenzyme.

15.13   *What type of bond is broken when acetyl-CoA is hydrolyzed?*

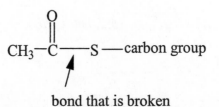

bond that is broken

An ester bond (specifically, a thioester bond).

15.15   *How does anabolism differ from catabolism in terms of the relative size of the reactants that enter a particular pathway and the products that leave it?*

During anabolism, small molecules, such as pyruvate and acetyl-CoA are used to make larger molecules like fatty acids and monosaccharides. In catabolism, larger molecules such as triglycerides, carbohydrates, and proteins, are split into smaller molecules, such as fatty acids, glycerol, monosaccharides, and amino acids.

15.17   *Write a reaction equation for the hydrolysis of UTP to produce PP$_i$.*

$$UTP + H_2O \rightarrow UMP + PP_i$$

15.19   *How do the structures of amylase and amylopectin, the two homopolysaccharides that make up starch, differ?*

Amylopectin contains α-(1 $\rightarrow$ 6) glycosidic bonds. Amylose does not.

15.21   *What two types of glycosidic bonds must be hydrolyzed to break amylopectin down into individual glucose molecules? Does the same enzyme hydrolyze both types?*

The α-(1-6) glycosidic bonds and α-(1-4) glycosidic bonds must be broken. No. The α-(1-4) glycosidic bonds are hydrolyzed by the action of amylase and the α-(1-6) glycosidic bonds are hydrolyzed by a debranching enzyme.

15.23   *What products are obtained when triglycerides are hydrolyzed in the small intestine?*

Fatty acids, glycerol and monoacylglycerides.

15.25   *To which class of lipids do bile acids belong?*

Steroid

15.27   *What is the role of glycolysis?*

Glycolysis converts a glucose molecule into 2 pyruvate ions with accompanying production of 2 ATP and 2 NADH.

15.29   a. *Which step in glycolysis is the major control point?*

Step 3.

b. *What compounds act as positive effectors of the enzyme that catalyzes this reaction?*

Positive effectors are ADP and AMP.

c. *What compounds are negative effectors?*

Negative effectors are ATP and citrate.

15.31 a. *What products are formed when pyruvate undergoes alcoholic fermentation?*

Ethanol and $CO_2(g)$.

b. *What purpose does this reaction have?*

The reaction serves to recycle NADH back into NAD^+, which allows glycolysis to continue.

15.33 *What is the net change in ATP and NADH from the passage of one glucose molecule through glycolysis, followed by the conversion of pyruvates into ethanol and CO_2 molecules?*

One glucose molecule is converted into two pyruvate ions, with the production of 2 ATP and 2 NADH. Conversion of the two pyruvate ions into ethanol and CO_2 requires 2 NADH.

Net change ATP = 2   Net change NADH = 0

15.35 *Draw lactic acid, the acid form of lactate. Which predominates at physiological pH, the acidic form or the basic form of this acid?*

Draw the lactate (see page 472 of the text) and then add H^+ to the $-O^-$ of the carboxylate group.

$$\underset{CH_3-CH-C-OH}{\overset{\overset{\displaystyle OH\ \ \ \ O}{|\ \ \ \ \ \ ||}}{}}$$

The basic form will predominate at physiological pH.

15.37 a. *Is the conversion of pyruvate to lactate an oxidation or a reduction?*

In Figure 15.9, the conversion of pyruvate to lactate shows a ketone being converted to an alcohol. In an organic reaction, the addition of hydrogen indicates reduction has occurred.

Reduction

b. *When this reaction takes place in a cell, what is the oxidizing agent?*

The oxidizing agent in a reaction is the substance that has been reduced. Therefore, pyruvate is the oxidizing agent.

Pyruvate

c. *What is the reducing agent?*

The NADH is losing hydrogen and an electron. In doing so, it reduces the pyruvate and it is therefore the reducing agent.

NADH

15.39 *Is the conversion of pyruvate into acetyl-CoA an oxidation or a reduction?*

Organic oxidation is observed as the loss of hydrogen or the gain of oxygen. In Figure 15.9, the conversion of pyruvate to acetyl-CoA results in the pyruvate losing hydrogen and an electron to NAD^+. Therefore the pyruvate is oxidized.

Oxidation

15.41 *Monosaccharides other than glucose are converted into compounds that are intermediates in glycolysis. Why is this more efficient than having a different catabolic pathway for the conversion of each monosaccharide into pyruvate?*

Each different pathway would require a different set of enzymes.

15.43 *Which three steps in glycolysis cannot be directly reversed during gluconeogenesis?*

Steps 1, 3, and 10

15.45 *From which intermediate shared by glycolysis and gluconeogenesis is glycogen produced?*

Glucose 6-phosphate

15.47 *The alcohol group in isocitrate can be oxidized, but that in citrate cannot. Explain.*

The alcohol group in isocitrate is secondary while, the alcohol group in citrate is tertiary.

15.49 *What is the net change in GTP, NADH, and $FADH_2$ from the passage of two acetyl-CoA through the citric acid cycle?*

The passage of two acetyl-CoA through the citric cycle creates 2 GTP, 6 NADH, and 2 FADH$_2$.

15.51   a. *Which two compounds donate their electrons to electron transport chain?*

NADH and FADH$_2$

b. *Which molecule is the final electron acceptor of the electron transport chain?*

The O$_2$ molecule

15.53   *H$^+$ moves from the intermembrane space, through ATP synthase, and into the mitochondrial matrix. Which term best describes this process: facilitated diffusion or active transport?*

Facilitated diffusion is the process of ions moving through a membrane assisted by a protein.

Facilitated diffusion

15.55   *The electron transport chain and oxidative phosphorylation typically generate how many ATP from 1 NADH?*

2.5 ATP

15.57   *Account for the 30-32 ATPs generated when one glucose molecule is catabolized by glycolysis and the citric acid cycle.*

In glycolysis, one glucose molecule yields 2 pyruvates, 2 ATP and 2 NADH. Depending on the system used to shuttle the electrons from NADH into mitochondria, these 2 NADH become 2 NADH or 2 FADH$_2$. Converting the 2 pyruvates into 2 acetyl-CoAs produces 2 NADH. Two passes through the citric acid cycle starting with acetyl-CoA produces 2 GTP (equivalent to 2 ATP), 2 FADH$_2$, and 6 NADH. In the electron transport chain, each FADH$_2$ generates 1.5 ATP and each NADH generates 2.5 ATP.

15.59   *Describe how fatty acids are activated and then moved from the cytoplasm into the mitochondrion.*

Fatty acids are activated by the attachment of a coenzyme A (CoA) residue. The CoA residue is replaced by a compound called carnitine and the fatty acyl-carnitine is moved into the mitochondria by active transport.

15.61   *In one pass through the β oxidation spiral a fatty acyl-CoA is shortened by two carbon atoms. What other products are formed?*

One acetyl-CoA, one FADH$_2$, and one NADH

15.63 *Calculate the net number of ATPs produced when one 14-carbon fatty acid is activated, enters the mitochondrion, and undergoes complete β oxidation to produce acetyl-CoA and reduced coenzymes.*

Activation in the first step requires one ATP. Each step in the spiral produces one FADH$_2$ = 1.5 ATP and one NADH = 2.5 ATP. This represents ATP production with each pass through the cycle. The number of passes = (number of carbons – 2)/2 so for a 14-carbon fat there are (14-2)/2 = 6 passes with an acetyl-CoA produced on the final pass.

.

FADH$_2$	NADH
6 x 1.5 = 9 ATP	6 x 2.5 = 15 ATP

Total ATP = 24 ATP

Then, as the acetyl-CoA enters the citric cycle, each pass generates 1 GTP (Same as 1 ATP), 3 NADH and 1 FADH$_2$.

For the citric cycle:

Acetyl-CoA	FADH$_2$	NADH
7 x 1 = 7 GTP	7 x 1.5 = 10.5 ATP	(7 x 3) x 2.5 = 52.5 ATP

Total ATP = 70 from citric cycle.

Total ATP for reaction = Total of oxidation spiral + Total of citric cycle - Activation

$$= 24 \text{ ATP } + 70 \text{ ATP - ATP}$$
$$= 93 \text{ ATP}$$

Total ATPs produced = 93

15.65 a. *Name the three ketone bodies.*

Acetoacetate, 3-hydroxybutyrate, and acetone

b. *Are they all ketones?*

No. 3-Hydroxybutyrate contains no ketone group.

15.67 *Why is acidosis (low blood pH) associated with the overproduction of ketone bodies?*

Two of the ketone bodies are carboxylic acids.

15.69 a. *Draw linoleic and linolenic acid.*

$$CH_3(CH_2)_4—CH=CH—CH_2—CH=CH—(CH_2)_7—\overset{\overset{\textstyle O}{\|}}{C}—OH$$

linoleic acid

$$CH_3—CH_2—CH=CH—CH_2—CH=CH—CH_2—CH=CH—(CH_2)_7—\overset{\overset{\textstyle O}{\|}}{C}—OH$$

linolenic acid

b. *Which carbon-carbon bonds in these molecules cannot be formed by human desaturase enzymes?*

The double bonds beyond carbon 10.

15.71 *Where in a cell does fatty acid biosyntheses take place?*

Cytoplasm

15.75 *What is the source of NH_4^+ produced during amino acid catabolism?*

The release of NH_4^+ from amino acids commonly requires two reactions. The first is transamination, wherein an amino group from an amino acid is transferred to an α-keto acid, The second is oxidative deamination, wherein the amino group on the α-keto acid is replaced by a carbonyl group. It is released as NH_4^+.

Oxidative deamination

15.77 *Draw the α-keto acid obtained when cysteine undergoes transamination.*

Draw cysteine and replace the $–NH_3^+$ and hydrogen on the α carbon with a $=O$.

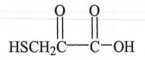

15.79 *Oxidative deamination of glutamate produces which compound besides NH_4^+?*

α-Ketogluterate

15.81 *Why do some scientists believe that mitochondria were once free bacteria?*

Some scientists believe the mitochondria were once free bacteria because they have their own DNA that is different from that of the cell.

15.83 *Hibernating animals produce brown fat which helps them to survive during the winter months. Explain how brown fat makes better use of the animal's triglyceride supply during hibernation than other fat does.*

The presence of the uncoupler thermogenin allows energy released during catabolism to be delivered to the production of heat, rather than production of ATP.

15.85 *Which functional groups are present in rotenone? (See the chapter conclusion.)*

Ether, ketone, aromatic, and alkene

NOTES

NOTES

NOTES

NOTES

NOTES

NOTES

NOTES

NOTES

NOTES

NOTES

NOTES

NOTES

NOTES

NOTES

NOTES

NOTES

NOTES